Advances in Insect Control: The role of transgenic plants

Edited by

NADINE CAROZZI AND MICHAEL KOZIEL

Taylor & Francis
Publishers since 1798

UK Taylor & Francis Ltd, 1 Gunpowder Square, London, EC4A 3DE

USA Taylor & Francis Inc., 1900 Frost Road, Suite 101, Bristol, PA 19007

British Library Cataloguing in Publication Data

A catalogue record for this book is available from the British Library.

ISBN 0-7484-0417-1

Library of Congress Cataloging Publication data are available

Cover design by Youngs Design in Production

Typeset in Times 10/12pt by Santype International Limited, Salisbury, Wilts

Printed in Great Britain by T. J. International Ltd., Cornwall

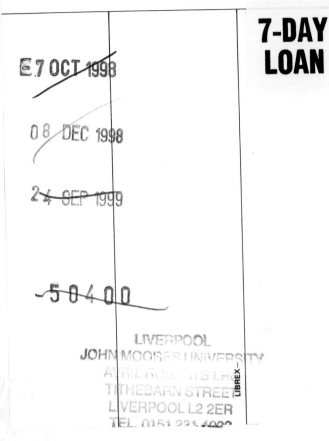

Advances in Insect Control

Contents

Contents

vi

Contents

Contributors

James E. Baker
US Grain Marketing Research
  Laboratory, Grain Marketing and
  Production Research Center,
  USDA-ARS, Manhattan, Kansas,
  66502

Craig A. Behnke
Department of Biochemistry, Kansas
  State University, Manhattan,
  Kansas, 66506

John Bennett
Division of Plant Breeding, Genetics
  and Biochemistry, International Rice
  Research Institute, PO Box 933,
  1099 Manila, Philippines

George Buta
Horticultural Crops Quality
  Laboratory, Agricultural Research
  Service, US Department of
  Agriculture, Beltsville, Maryland,
  20705

Nadine B. Carozzi
Ciba-Geigy Corporation, Agricultural
  Research Unit, Research Triangle
  Park, North Carolina, 27709

Scott Chilton
Department of Botany, North
  Carolina State University, Raleigh,
  North Carolina, 27695

Maarten J. Chrispeels
Department of Biology, University of
  California at San Diego, 9500
  Gilman Drive, La Jolla, California,
  92093-0116

Michael B. Cohen
Entomology and Plant Pathology
  Division, International Rice
  Research Institute, PO Box 933,
  1099 Manila, Philippines

Thomas H. Czapla
Pioneer Hi-Bred International,
  Johnston, Iowa, 50131

Patrick F. Dowd
US Department of Agriculture,
  Agricultural Research Service,
  National Center for Agricultural
  Utilization Research, 1815 N.
  University Street, Peoria, Illinois,
  61604

Nicholas Duck
Ciba-Geigy Corporation, Agricultural
  Research Unit, Research Triangle
  Park, North Carolina, 27709

Stephen Evola
Ciba-Geigy Corporation, Agricultural
    Research Unit, Research Triangle
    Park, North Carolina, 27709

Jeffrey A. Fabrick
Department of Biochemistry, Kansas
    State University, Manhattan,
    Kansas, 66506

Jennifer Feldman
NatureMark Potatoes, 300 East
    Mallard Drive, Boise, Idaho, 83706

Benhzad Ghareyazie
Division of Plant Breeding, Genetics
    and Biochemistry, International Rice
    Research Institute, PO Box 933,
    1099 Manila, Philippines

Sunggi Heu
Plant Molecular Biology Laboratory,
    Agricultural Research Service, US
    Department of Agriculture,
    Beltsville, Maryland, 20705

Lowell Johnson
Department of Plant Pathology,
    Throckmorton Plant Sciences
    Center, Kansas State University,
    Manhattan, Kansas, 66506

Michael R. Kanost
Department of Biochemistry, Kansas
    State University, Manhattan,
    Kansas, 66506

Sanjay K. Katiyar
Division of Plant Breeding, Genetics
    and Biochemistry, International Rice
    Research Institute, PO Box 933,
    1099 Manila, Philippines

Gurdev S. Khush
Division of Plant Breeding, Genetics
    and Biochemistry, International Rice
    Research Institute, PO Box 933,
    1099 Manila, Philippines

Michael G. Koziel
Ciba-Geigy Corporation, Agricultural
    Research Unit, Research Triangle
    Park, North Carolina, 27709

Karl J. Kramer
US Grain Marketing Research
    Laboratory, Grain Marketing and
    Production Research Center, US
    Department of Agriculture-ARS,
    Manhattan, Kansas, 66502

L. Mark Lagrimini
Department of Horticulture and Crop
    Science, The Ohio State University,
    2001 Fyffe Court, Columbus, Ohio,
    43210

Iris McCanna
Plant Molecular Biology Laboratory,
    Agricultural Research Service, US
    Department of Agriculture,
    Beltsville, Maryland, 20705

Subbaratnam Muthukrishnan
Department of Biochemistry, Kansas
    State University, Manhattan,
    Kansas, 66506

Marnix Peferoen
Plant Genetic Systems, N.V., Jozef
    Plateaustraat 22, 9000 Ghent,
    Belgium

John P. Purcell
Monsanto-GG4G, 700 Chesterfield
    Village Parkway North, St Louis,
    Missouri, 63198

Gerald R. Reeck
Department of Biochemistry, Kansas
    State University, Manhattan,
    Kansas, 66506

Rick Roush
Department of Crop Protection, Waite
    Campus, University of Adelaide,
    Glen Osmond, South Australia 5064,
    Australia

Ann Smigocki
Plant Molecular Biology Laboratory,
  Agricultural Research Service, US
  Department of Agriculture,
  Beltsville, Maryland, 20705

Terry Stone
Monsanto Co., 700 Chesterfield
  Parkway North, St Louis, Missouri,
  63198

Gregory W. Warren
Ciba-Geigy Corporation, Agricultural
  Research Unit, Research Triangle
  Park, North Carolina, 27709

Frank White
Department of Plant Pathology,
  Throckmorton Plant Sciences
  Center, Kansas State University,
  Manhattan, Kansas, 66506

Chris Wozniak
Northern Crop Science Laboratory,
  Agricultural Research Service,
  Fargo, North Dakota

Preface

This year will be remembered for the commercial introduction of three crops which are engineered for insect resistance, thereby marking the beginning of a new era in agriculture for the control of insect pests. Plant genetic engineering has moved in the past decade from the laboratory to the market-place as a new and powerful technology to provide both farmers and consumers with improved crops. This technology has the capability to impact agricultural practices more radically than any other single change in the past several thousand years of agriculture. Engineering plants for increased insect and disease resistance will improve both crop quality and yield. Eventually plants will be altered to withstand environmental stress, enhance their nutritional qualities, and be used to produce new and useful proteins.

Engineering plants to resist insects is one of the most urgent demands of plant genetic engineering. Insect pests create very costly losses for farmers and attempts to mitigate insect damage usually require applications of chemical pesticides or spray-on biological pesticides. Insecticidal applications need to be properly timed to coincide with insect infestations and do not always reach burrowing insects. The introduction of plants with built-in resistance to various insect pests will have a tremendous impact on current agricultural practices. Now 1996–97 brings the first market introduction of genetically engineered commercially available insect tolerant corn, potato, and cotton plants expressing a *Bacillus thuringiensis* endotoxin gene. These genetically engineered plants provide resistance against some of the worst crop pests. The first generation of products are still new, but many innovations are visible on the horizon. This book attempts to examine today's technology and successes and to also look at candidates for the next generation of insect control for transgenic crops.

The reality of plant genetic engineering is that although the promise is great, the steps involved in implementing useful technologies are tedious and often fraught with technical difficulties that underscore our ignorance of the workings of various biological processes. Although transformation of tobacco was first reported in 1983, it was still a large hurdle to develop technologies to reproducibly and efficiently introduce new genes into major commercial crops. The use of genetically engineered plants to combat losses caused by insects requires an understanding of not only the crop but an understanding of the equally complex and often less studied insect pest. Plants and insects both hold many mysteries than we have not yet begun to understand. Novel insecticidal proteins described in this book may bring new and different challenges that will no doubt be specific for each crop and insect combination. Novel approaches to introduce multi-gene pathways and/or regulation of existing plant genes will present even more challenges. But while this is a hurdle for the development of commercial varieties, it is the great appeal and challenge for the scientists in this area. It provides an opportunity to contribute to the well being of mankind while unraveling a few of the mysteries of nature.

While agricultural biotechnology is still in its infancy, its promise is bright. It is immensely rewarding to be a contributor and witness to this new day. We hope that this book serves as a rich source of information about both the current status of genetically engineered insect resistant crops that have been recently commercialized to the many new advances and prospects for the future of engineering insect resistant crops.

Dedicated to our children who will enjoy a better world.

Nadine Carozzi and Mike Koziel

Use of Transgenes to Increase Host Plant Resistance to Insects: Opportunities and Challenges

NICHOLAS DUCK and STEPHEN EVOLA

Introduction

Most plant species experience insect predation, although the severity and economic importance of that predation varies significantly among crop species. In the manmade environments of agricultural crop production for food or fiber this predation is almost always to the detriment of the output of the agricultural system yield. In fact, the development of genetic monocultures managed with intensive inputs of nutrients and, in many cases, water, to produce luxuriant growth of a limited number of genotypes per crop, has altered the co-evolutionary balance among crop species and their insect predators and increased the potential for epizootics.

To increase and/or stabilize crop yields two primary approaches have been used to decrease insect predation. The first is the development and application of 'spray-ons', be they synthetic or natural chemicals, microbial fermentation products or living macro-organisms. The second approach is development of crop varieties with increased host plant resistance based on genetic factors found within the crop species or in closely related wild species. Today the development of genetic engineering technology has enabled the introduction of genetic material from any species into major crop plants. Genes encoding insect resistance have been introduced in several major crop species, producing dramatically enhanced insect resistance compared to what had been obtainable by conventional breeding. This book provides many excellent examples of insect controlling principles from a variety of sources and their successful introduction into the genomes of plants, many of which are important crop plants. Some of these principles have reached the marketplace in elite varieties and hybrids with superior agronomic performance. Others are still in the proof-of-concept phase.

The creation of higher levels of insect resistance in transgenic crop plants is not without considerable technical challenges. The first step involves the time consuming process of identification and isolation of insect controlling principles, and the genes that encode them. A critical requirement for these principles is that they possess a sufficiently high specific activity to achieve effective control of insect pests, given the biosynthetic capacities of the host plant and the need to maintain established levels of agronomic performance. Overproduction of certain active principles with low specific activities may disrupt developmental processes and diminish final crop yields. Other insecticidal proteins with high specific activities may not be amenable to production in transgenic plants if they disrupt critical plant physiological processes. Technical solutions involving cellular sequestration of the protein and tissue specific expression may be required. Solutions such as these may add several years to the production of commercially acceptable products. The second but equally important step is the development of transformation systems, particularly with elite genotypes, that permit the stable heritable insertion of foreign DNA. Next the development of breeding and/or evaluation methods must be established to permit the introduction of a desirable transformation event into other lines or varieties and insure that any given transgenic line is unaltered in performance relative to its non-transgenic counterpart. Field trial methods need to be developed to demonstrate convincingly the enhanced resistance of the transgenic plants. In addition, there are challenges that need to be met in the areas of regulatory affairs, production and quality control, resistance management and, for a commercial cultivar, intellectual property. Finally, the transgenic insect resistant crop has to meet with both grower and public approval and acceptance. As the examples in this book demonstrate, the challenges can be met and the benefits of greatly enhanced host plant resistance to insects through genetic engineering are obvious, great and unique.

The Nature and Magnitude of the Insect Pest Problem

Insect Pests of Major Crops

The major crops ('major' defined as harvested mass of at least 20 million metric tons per year) of the world and their most important insect pests are shown in Table 1.1. Although several insect pests can be determined to be significant pests for a given crop, crops vary considerably in the extent to which insecticides are normally used

Table 1.1 Major crops and their chief insect pests

Crop (in order by planted hectares, worldwide)	World production (in million metric tons)[1]	Major insect pests[2,3,4]
Wheat (*Triticum aestivum*)	551	Aphids (*Rhopalosiphum padi, Schizaphis graminum, Sitobion avenae*) Senn pest (*Eurygaster integriceps*) Hessian fly (*Mayetiola destructor*) Wheat bulb fly (*Delia coarctata*)
Rice (*Oryza sativa*)	520	Brown planthopper (*Nilaparvata lugens*) Green planthopper (*Nephotettix virescens*) Yellow rice borer (*Tryporyza incertulas*) Striped rice borer (*Chilo suppressalis*) Gall midge (*Orseolia oryzae*)

Table 1.1—*Continued*

Crop (in order by planted hectares, worldwide)	World production (in million metric tons)[1]	Major insect pests[2,3,4]
Maize (*Zea mays*)	479	European corn borer (*Ostrinia nubilalis*) Asian corn borer (*Ostrinia furnacalis*) Other stem and stalk borers (Chilo and *Sesamia* spp., *Busseola fusca*) Southwestern corn borer (*Diatrea grandiosella*) Corn rootworms (*Diabrotica virgifera virgifera, D. longicornis barberi, D. undecimpunctata*) Black cutworm (*Agrotis ipsilon*) Fall armyworm (*Spodoptera frugiperda*) Corn earworm (*Helicoverpa zea*) Sugar cane borer (*Diatrea saccharalis*) Lesser corn stalk borer (*Elasmopalpus lignosellus*)
Potato (*Solanum tuberosum*)	261	Colorado potato beetle (*Leptinotarsa decemlineata*) Potato leafhopper (*Empoasca fabae*) Potato tuber moth (*Phthorimaea operculella*) Aphids (*Aulacorthum solani, Myzus persicae*) Wireworms (*Agriotes* spp.) Black blister beetles (*Epicauta* spp.)
Cotton (*Gossypium hirsutum*)	21 (lint) 38 (meal)	Bollworm (*Helicoverpa zea*) Tobacco budworm (*Heliothis virescens*) American budworm (*Heliothis armigera*) Cotton boll weevil (*Anthonomus grandis*) Pink bollworm (*Pectinophora gossypiella*) Jassids (*Empoasca* spp.) Cotton leafhopper (*Pseudatomoscelis seriatus*) Plantbugs (*Lygus lineolaris, L. hesperus*)
Soybean (*Glycine max*)	103	Green cloverworm (*Plathypena scabra*) Soybean looper (*Pseudoplusia includens*) Velvetbean caterpillar (*Anticarsia gemmatalis*) Mexican bean beetle (*Epilachna varivestis*) Bean leaf beetle (*Cerotoma trifurcata*) Thrips (*Sericothrips variabilis*) Potato leafhopper (*Empoasca fabae*) Corn earworm (*Helicoverpa zea*) Green stink bug (*Acrosternum hilare*) Threecornered alfalfa hopper (*Spissistilus festinus*) Lesser corn stalk borer (*Elasmopalpus lignosellus*)

Table 1.1—*Continued*

Crop (in order by planted hectares, worldwide)	World production (in million metric tons)[1]	Major insect pests[2,3,4]
Sunflower (*Helianthus annuus*)	23	Blue bugs (*Calidea* spp.) Green stink bug (*Nezara viridula*) American bollworm (*Heliothis armigera*) Chafer grubs (*Schizonycha* spp.)
Rape (*Brassica napus*)	20	Cabbage aphid (*Brevicoryne brassicae*) Brassica pod midge (*Dasyneura brassicae*) Red turnip beetle (*Entomoscelis americana*) Blossom beetles (*Meligethes* spp.) Flea beetles (*Phyllotreta* spp., *Psylliodes chrysocephala*) Cabbage seed weevil (*Ceutorhynchus assimilis*) Cabbage stem weevil (*Ceutorhynchus quadridens*)
Sugar beet (*Beta vulgaris*)	38 (total raw sugar)	Beet leafhopper (*Circulifer tennellus*) Black bean aphid (*Aphis fabae*) Peach-potato aphid (*Myzus persicae*) Mangold fly (*Pegomyia betae*) Silver-Y moth (*Autographa gamma*)
Sugar cane (*Saccharum* interspecific hybrids)	98 (total raw sugar)	Sugar cane borer (*Diatrea saccharalis*)
Grapes (*Vitis vinifera*)	57	Grape leafhoppers (*Erythroneura* spp.) Grape phylloxera (*Viteus vitifolii*) Root mealybugs (*Planococcus* spp.) Grape mealybug (*Pseudococcus maritimus*)
Tomato (*Lycopersicon esculentum*)	69	Tobacco whitefly (*Bemisia tabaci*) Glasshouse whitefly (*Trialeurodes vaporariorum*) Green stink bug (*Nezara viridula*) Tomato leafminer (*Scrobipalpula absoluta*) Tomato fruitworms (*Heliothis* spp.) Onion thrips (*Thrips tabaci*)
Coconut (*Cocos nucifera*)	42	No major insects
Yam (*Dioscorea* spp.)	24	Sweet potato weevil (*Cylas* spp.)
Cassava (*Manihot esculenta*)	154	Thrips (*Frankliniella williamsii, Corynothrips stenopterus, Caliothrips masculinus*) Cassava hornworm (*Erinnyis ello*) Whiteflies (*Bemisia tabaci, Aleurotrachelus* spp.) Mealybugs (*Phenaococcus gossypii, Pseudococcus* spp.) Scales (*Aonidomytilus albus*)

Table 1.1—*Continued*

Crop (in order by planted hectares, worldwide)	World production (in million metric tons)[1]	Major insect pests[2,3,4]
Sweet potato (*Ipomoea batatas*)	126	Sweet potato weevil (*Cylas* spp.)
Sorghum (*Sorghum bicolor*)	58	Sorghum midge (*Contarinia sorghicola*) Shoot fly (*Atherigona soccata*) Stem borer (*Chilo partellus*) Greenbug (*Schizaphis guaminum*)
Millet (several spp.)	29	Millet grain midge (*Geromyia pennisiti*)
Oats (*Avena sativa*)	34	Wheat aphid (*Schizaphis graminum*) Grain aphid (*Sitobium avenae*) Bird-cherry aphid (*Rhopalosiphum padi*) Frit fly (*Oscinella frit*) Cereal leaf beetle (*Oulema melamopa*)
Rye (*Secale cereale*)	27	Bird-cherry aphid (*Rhopalosiphum padi*) Wheat aphid (*Schizaphis graminum*) Grain aphid (*Sitobium avenae*) Frit fly (*Oscinella frit*)

[1] FAO (1991)
[2] Hill (1987)
[3] Dicke and Guthrie (1988)
[4] Niles (1980)

(see Table 1.2). In terms of total insecticides used and proportion of acres treated, cotton, maize and potatoes have the largest commercially important insect problems.

Value of Insect Resistance in Major Crops Estimated Through Crop Losses

Total yield losses from all causes to all crops are roughly estimated to be $500 billion (US) per year (Oerke *et al.*, 1994). Pimentel (1991) estimates the sum of pre-

Table 1.2   Insecticide use on selected crops (Palm, 1991)

Crop	Acres treated %	Amount (% of total used on all crops)
Non-food		
Cotton	61	47
Food		
Maize	35	17
Rice	35	1
Wheat	7	1
Soybeans	8	4
Potatoes	77	2
All crops	6	35 (% of total pesticides)

harvest and postharvest losses to pests (weeds, pathogens and insects) is about 45% (25–35% preharvest, 10–20% postharvest) worldwide. In the US, preharvest losses are estimated to be about 37% and postharvest losses 9%. Of the preharvest losses in the US, approximately 13% result from insect predators. Losses are large and ubiquitous and chemical controls are not fully effective. Yield losses in selected crops, with and without control measures, caused by specific arthropod pests are shown in Table 1.3.

Yield losses vary by crop (see Table 1.4). Worldwide 15.6% of the total production, valued at $90.5 billion (US), of the eight principal food and cash crops (coffee, potato, soybean, maize, barley, cotton, rice, and wheat), was lost to animal predators, primarily insects, in 1988–90. These losses are sustained even though control measures, chiefly insecticide applications, are used on a high proportion of

Table 1.3 Worldwide yield losses resulting from specific anthropod pests

| Crop | Pest | Calculated yield loss (%) | |
		Without control	With control
Barley	Green bug	84	7
Sugar beet	Root maggot	23	8
Maize	Cornfield ant	20	3
	Corn rootworms	16	5
	Garden symphalid	14	7
	Leafhoppers	75	38
	Southwestern corn borer	34	10
	White grubs	43	11
	Wireworms	48	18
Cotton	Aphids	22	<1
	Boll weevil	31	19
	Cabbage looper	35	20
	Cotton aphid	5	0
	Cotton flea hopper	39	13
	Heliothis spp.	91	12
	Pink bollworms	36	10
	Thrips	68	18
Potatoes	Aphids	15	<1
	Armyworms	59	1
	Colorado potato beetle	47	1
	European corn borer	54	2
	Flea beetles	43	<1
	Green peach aphid	4	0
	Leafhoppers	36	0
	Potato aphid	5	0
	Potato leafhopper	43	<1
	Potato psyllid	32	1
	Potato tuberworm	91	1
	Spider mites	5	4
	Wireworms	5	<1
Rapeseed	Cabbage aphids	100	0
Rice	Leafhoppers	42	9
	Rice water weevil	5	2

Table 1.3—*Continued*

Crop	Pest	Calculated yield loss (%) Without control	With control
Soybean	Caterpillar complex	46	0
	Green cloverworm	9	3
	Loopers	26	11
	Mexican bean beetle	26	<1
	Mites	89	0
	Soybean looper	16	9
	Velvetbean caterpillar	17	2
Sugar cane	Sugar cane borer	29	6
Sunflowers	Sunflower moth	40	7
Tomato	Aphids	7	0
	Cabbage looper	15	0
	Colorado potato beetle	93	0
	Cutworms	16	<1
	Leafminers	38	10
	Tomato pinworm	50	5
	White flies	38	10
Wheat (and other small grains)	Banks grass mite	61	18
	Brown wheat mite	100	21
	Cutworms	55	8

the crop acres. Animal losses prevented by various protection methods is estimated to be about 7% of attainable production or about $41 billion (US).

Yield losses also vary considerably by geographic area (Table 1.5). Insects and mites cause significant and rather uniform yield losses over crops (approx. 10–20%) and over geographical areas (approx. 10–20%). It is also clear from these data that the greatest potential for gain in productivity from preventing insect related yield reductions is in rice, most of which is grown in Asia.

Table 1.4 Yield losses (preharvest) by major crop (Oerke *et al.*, 1994)

Crop	Total production $ billion (US) 1988–90	Percentage loss of production due to animal pests
Rice	106.4	20.7
Wheat	74.6	9.3
Barley	13.7	8.8
Maize	44.0	14.5
Potatoes	35.1	16.1
Soybeans	24.2	10.4
Cotton	25.7	15.4
Coffee	11.4	14.9
Total	335.1	15.6

Table 1.5 Yield losses by geographic area (Oerke *et al.*, 1994)

	Production		Loss due to animal
Continent	$ billion (US)	Percentage of total	pests (%)
Africa	13.3	4.0	16.7
North America	50.5	15.1	10.2
Latin America	30.7	9.2	14.4
Asia	162.9	48.6	18.7
Europe	42.6	12.7	10.2
USSR (ex)	31.9	9.5	12.9
Oceania	3.3	1.0	10.7
Total	335.2	100.1	15.6

Yield losses have, in fact, increased since 1965 as a result, in part, of the increased potential of crop productivity in the absence of pests. Better agronomic practices and improved genotypes provide better substrates for insect nutrition. For the eight principal food and cash crops, changes in losses resulting from animal pests (primarily arthropods) between 1965 and 1990 are shown in Table 1.6.

From these data we can conclude that large increases in productivity could result from control of insect pests and that current non-genetic insect control methods are only partially effective in reducing losses. The greatest losses are in the poorest countries, which use the least and the oldest insecticides. Consequently, high levels of sustainable genetic resistance to insect predators would be an excellent preventative solution to this problem of yield losses, particularly for the economically less well developed areas of the world.

Table 1.6 Changes in crop losses resulting from animal pests

	Crop losses (%) due to animal pests[1]	
Crop	1988–90	± vs. 1965[2]
Rice	20.7	−6.8
Wheat	9.3	+4.2
Barley	8.8	+4.9
Maize	14.5	+1.5
Potatoes	16.1	+10.2
Soybeans	10.4	+6.0
Cotton	15.4	−0.6
Coffee	14.9	+2.0

[1] Including losses due to viruses transmitted by insect vectors
[2] Change in percentage losses (1988–90 vs. 1965)

Natural Host Plant Resistance

Mechanisms of genetic resistance to insects have been classified into three categories (Kogan and Ortman, 1978; Painter, 1951):

(1) ANTIXENOSIS in which plant colonization by insects is deterred or reduced, primarily by altering the insects' behavior either through chemical or physical means;

(2) ANTIBIOSIS in which the insects' development is altered after colonization and commencement of feeding; and

(3) TOLERANCE in which the resistant cultivar sustains as much damage as a susceptible cultivar but without loss in quantity or quality of yield.

Antixenosis, also referred to as non-preference, can be based upon physical characteristics such as thickness of cell walls; proliferation of wounded tissues; solidity of stems; presence, density and types of trichomes; surface waxes; incorporation of silica; and anatomical alterations of non-specialized organs. Trichomes and bract structure have proved important for boll weevil resistance in cotton; and trichomes in potatoes and alfalfa for aphids and leafhoppers. Plant surface waxes have been shown to have a deterrent effect on selected insects in sorghum, Brassica spp., apple, rice, and cotton.

Chemicals synthesized by the plant can repel insects or alter their behavior when they occupy the plant such that their reproduction and/or damage to the plant is reduced. Alternatively, the lack of synthesis of chemicals that act as attractants to insects can serve to protect a cultivar. However, a given chemical can have both negative and positive effects on overall insect resistance of a crop species. A good example is that of cucurbitacins, which are feeding attractants for spotted cucumber beetles but repel two-spotted spider mites in Cucurbita species.

Antibiosis mechanisms usually involve toxic chemicals or nutritional deficiencies. The most frequent chemical classes encountered with antifeeding and/or growth reducing effects are terpenoids, alkaloids, flavonoids, cyanogenic glucosides, and hydroxamic acids. Many of these are quite toxic to non-insects, at least at high concentrations. Some can have positive or negative effects on the insects that consume them, depending upon the concentration. Their synthesis can also vary considerably with developmental stage and environment. Clearly, the biosynthetic pathways that produce them are polygenic.

Qualitatively Inherited (Vertical) Insect Resistance

Examples of qualitative, or simply inherited, resistance (Singh, 1986; Smith, 1989) include: greenbug resistance in barley (1 recessive, 1 dominant gene) and wheat (1 recessive); leaf aphid in lettuce (1 dominant); fall armyworm and corn earworm in pearl millet (1 recessive); brown plant hopper in rice (2 dominant, 2 recessive); gall midge in rice (1 dominant); green leafhopper in rice (6 dominant, 1 recessive); white-backed plant hopper in rice (3 dominant, 1 recessive); zigzag leafhopper in rice (3 dominant); greenbug in rye (1 dominant) and sorghum (1 dominant); sweet clover aphid in clover (1 dominant); and Hessian fly in wheat (12 dominant). Worthy of note are the facts that dominant genes are more prevalent than recessive genes, which probably has more to do more with plant species being at least diploid, often polyploid, and many species being outcrossers, than the spectrum of possible muta-tions to insect resistance. Also, the number of resistance genes for a given crop–insect combination is most likely related to a large extent to the effort devoted to finding resistance genes, the typically narrow spectrum of resistance among insect biotypes for a given resistance gene, and the frequently short lifetime of vertical resistance genes.

Challenges in using vertical resistance genes in breeding for insect resistance can be found in the limited variability for the trait in the crop species or its close rela-tives; the narrow spectrum of resistance, even within an insect species; and the usual unadapted nature of resistant germplasm, which means that deleterious linkages of the resistance gene(s) will be likely and the time needed to develop an elite adapted commercially acceptable resistant genotype will be long. When the new cultivar is finished and tested, it could face a limited lifetime because of the empirically based ability of insect biotypes to overcome single native resistance genes.

Quantitatively Inherited (Horizontal) Insect Resistance

Examples of this type of polygenic inheritance (Singh, 1986; Smith, 1989) include: resistance to first brood European corn borer (ECB) in maize, conditioned by 5–6 quantitatively inherited loci and to second brood ECB by another 7 loci; green peach aphid in potato (2 loci); brown planthopper in rice (3–4 loci); and Mexican bean beetle in soybeans (2–3 loci).

Challenges in breeding of horizontal resistances overlap with those in breeding for vertical resistances. Variability in the germplasm is limited. Adaptation in resist-ant genotypes is frequently poor. Deleterious linkages are to be expected. In addi-tion, the ability to screen accurately for resistance in populations segregating for several to many loci makes difficult the transference of the level of resistance in the source to a new cultivar while maintaining the performance expected of elite geno-types. In addition, the insurance of a screen at each generation of breeding requires the provision of insects at the right stages of insect and plant development in the required numbers. This is normally a labor intensive and costly function. Degree of protection against insect damage is not likely to be as high in horizontal resistance as in vertical resistance. However, the benefits can be expected to lie in broadness of resistance (within the insect species and among related insect species) and its persist-ence.

One of the greatest challenges to breeding for insect resistance in crop species, be it vertical or horizontal resistance, is the lack of source germplasm with sufficiently high levels of resistance to make the endeavor worthwhile. For example, in maize there is no source for real resistance to corn rootworms. Finally, the term 'resist-ance' is often relative; it means 'more resistant' than what is currently grown but is frequently a long way from the immunity or near-immunity that is needed to elimi-nate the yield losses that insect predation, as well as associated disease development, cause.

Use of Molecular Markers in Breeding for Host Plant Resistance

With the increasing availability of molecular markers, loci encoding resistance genes are being mapped. Mapping studies have located resistance factors for rice tungro spherical virus and green leafhopper in rice (Sebastian *et al.*, 1996), Russian wheat aphid resistance in barley (Nieto-Lopez and Blake, 1994), bruchid resistance in mungbean (Young *et al.*, 1992), gall midge resistance in rice (Nair *et al.*, 1995), trichome-mediated resistance in potato (Bonierbale *et al.*, 1994), and European corn borer resistance in maize (Christensen *et al.*, 1994).

Molecular markers provide several advantages in breeding for insect resistance. They can be used for identification of the number of genes involved and the magnitude of their contribution to a trait when this is not already known, (e.g., in horizontal resistance). Once the insect resistance genes are identified, markers can be used for screening for the resistance genes in the absence of insects. If sufficient marker loci have been identified that are distributed throughout the genome, breeding insect resistance traits from 'exotic' germplasm is possible with a minimization of maintained linkage with deleterious genes (linkage drag). Finally, marker assisted breeding provides greater efficiency in selecting quantitatively inherited horizontal resistances because partially resistant genotypes will not be mistakenly discarded during the process of assembling fully resistant genotypes.

There are, however, costs associated with molecular markers. The mapping process can be difficult because phenotypes must be as accurate and quantitative as possible to assure that a marker is indeed indicative of the trait. Also molecular marker assays are still expensive and their repeated use during a breeding program adds to the expense. Throughput and turnaround time of molecular marker assays, regardless of cost, are still insufficient for use in large scale breeding programs. There is no guarantee that the resistance loci mapped in one cross will be applicable in the next cross, although that is more likely for vertical resistances than for horizontal resistances. Finally, conversion of elite genotypes with multiple additive genes is still very difficult without sacrificing some yield or other important agronomic trait. Molecular markers can certainly assist in this process, however their widespread use in breeding for insect resistance will require further advances in marker assay technology.

Transgenic Approaches to Host Plant Resistance

Advantages of Transgenic versus Conventional Approaches to Host Plant Resistance

There are several advantages to using transgenic plants for insect control as opposed to breeding for insect resistance from related varieties or ancestors of crop plants. First, transgenic technology allows for the horizontal transfer of genetic information from one species to another, so that a trait that evolved for the benefit of one species can be directly utilized by another species. The two species can be so distantly related that the trait would not normally be transferred by natural mechanisms. The movement of bacterial insecticidal proteins into plants is such an example. The second advantage is that insecticidal genes are often determined by one dominant gene whereas a similar trait in varieties related to the crop of interest are often encoded by multiple genes, possibly on different chromosomes. The direct introduction of a single dominant gene into commercial crop varieties has several additional advantages over breeding for a multigene trait. For instance, if the gene is introduced directly into elite, commercial germplasm, there is no linkage drag that is

often associated with breeding traits from wild relatives. Linkage drag is a consequence of transferring undesirable traits along with the desired traits during breeding. Similarly, linkage drag can be problematic if a transgene is introduced into
non-commercial germplasm, then moved into commercial germplasm by breeding.
Often non-commercial genotypes are easier to work with in tissue culture, consequently transgenes are often introduced into these lines. Even under these conditions, a single dominant gene, transformed into a genotype that is closely related to
the commercial genotype, is likely to carry less linkage drag than a trait that is
determined by multiple genes that are transferred from more distantly related germplasm. Certainly the best case is to introduce the gene of interest (insect resistance)
directly into commercial germplasm.

In the case of transgenic plants expressing *Bacillus thuringiensis* insecticidal proteins, another major advantage over conventional breeding is that the insecticidal
crystal proteins have been used in agriculture for over 50 years, so that their safety
and efficacy have been established. The mode of action has been characterized with
respect to protein structure (Li and Ellar, 1994), specificity of receptor binding
(Hofmann *et al.*, 1988; Van Rie *et al.*, 1989), and solubility and proteolytic activation
in alkaline gut fluids of insects (Hofte and Whiteley, 1989). With conventional breeding, an insect resistance trait is often uncharacterized in terms of its mode of action.
Consequently resistance could be attributed to a wide variety of principles. Some
examples of insect resistance traits in plants include accumulation of alkaloids,
cyanogenic glucosides, glucosinolates and DIMBOA (see Chapter 14). Many of
these compounds occur naturally in wild relatives of domesticated plants and can be
quite toxic to insects and animals alike.

Requirements for Success in Creating and Developing Elite Transgenic Insect Resistant Cultivars

Identification of Active Insecticidal Principle

The process of creating insect resistant cultivars begins with identifying an insecticidal protein that is active against an insect pest. The first such proteins used in
transgenic plants have been the *Bacillus thuringiensis* insecticidal crystal proteins.
However, several other proteins have been identified that may be suitable for
expression in transgenic plants to confer resistance to insects. These proteins can be
grouped into four general classes. The first class is comprised of proteins that have
indirect modes of action. These are proteins that modify dietary constituents depriv-
ing the insect of nutrients or generating toxic compounds, but do not act directly on
the insect gut as a primary target. These include polyphenol oxidase that modifies
dietary proteins (Felton *et al.*, 1992; see also Chapter 6) and invertase and hexosyl-
transferase that modifies dietary sugars (Purcell *et al.*, 1994). The second class has
stoichiometric properties and includes binding proteins such as lectins (see Chapter
8), and enzyme inhibitors such as proteinase inhibitors (see Chapter 10) and α-
amylase inhibitors (see Chapter 9). These proteins deprive insects of necessary nutrients either by directly binding the nutrients and sequestering them, or by interfering
with the digestive process. Lectins can bind carbohydrates in the diet and on surfaces in the insect gut such as the peritrophic matrix and the apical membrane of

gut epithelial cells. Proteinase inhibitors and amylase inhibitors interfere with digestive enzymes of the insect gut. These processes make digestion less efficient. A third class is comprised of proteins that attack membranes and includes the *Bacillus thuringiensis* insecticidal crystal proteins, cholesterol oxidase (see Chapter 6), and lipid acyl hydrolases (Strickland *et al.*, 1995). The general mode of action for these proteins is that they attack the integrity of the apical gut epithelial brush border, leading to cell lysis and insect death. A fourth class of insecticidal proteins is represented by binary toxins. These proteins have two components, both of which are commonly required for toxicity. In most binary systems, one component has receptor binding properties, which determines specificity, and interacts with a receptor that is internalized. Once inside the cell the second component manifests a cytotoxic activity leading to cell death. The mosquitocidal toxins from *Bacillus sphaericus* (Porter *et al.*, 1993) and the vegetative insecticidal proteins, Vip1 and Vip2, from *Bacillus cereus* (see Chapter 7) represent two completely different classes of insecticidal binary toxins.

The importance of understanding the mode of action of insecticidal proteins is critical for determining whether the protein will be effective in a transgenic plant strategy for insect control. For instance, proteins that are active only at high concentrations may not be expressed in transgenic plants at high enough levels to render any insect resistance. This is a key limitation for insecticidal proteins that have stoichiometric activities, such as lectins and enzyme inhibitors. Furthermore, if the activity of the insecticidal protein causes phytotoxicity this too can be problematic. Finally, if the activity has an indirect mode of action, by modifying a constituent of the diet, that constituent must be present in plants for the insecticidal protein to be effective.

Once an insecticidal protein is discovered, the process of moving the activity into transgenic plants begins. This starts with purifying the protein responsible for the activity. Purified proteins have several important uses. First, pure proteins are required to determine accurately the LC_{50} for an insecticidal protein. The LC_{50} can be used to determine whether the protein can be expressed at high enough levels in a transgenic plant to be active. Generally proteins which are active at less than 10 ppm are required for transgenic plant strategies. Also, N-terminal sequence determined from the purified protein can be used to search databases such as GENBANK to determine whether the protein activity has already been identified. Finally, N-terminal sequence can be used for designing degenerate DNA oligos to aid in cloning the gene encoding the protein. Libraries made from DNA of the organism that produces the insecticidal protein can then be screened to identify and clone the gene. Often a synthetic gene is then constructed for plant expression. As in the case of *Bacillus*, genes encoding bacterial proteins have a high A + T content and do not express well in plants. Several explanations have been suggested for why these genes do not express well. These include unfavorable codon usage for plants and the presence of improper plant processing signals in the native gene such as splice sites or polyadenylation signals. Generally these can be avoided by increasing the G + C content of the gene and using plant preferred codons. Once the coding region is optimized for plant expression the gene must be placed downstream of a plant promoter and followed by a plant termination signal.

The final requirement for producing insect resistant cultivars is a plant transformation system. Two transformations systems are commonly used at the present time. *Agrobacterium* mediated transformation (Walkerpeach and Velten, 1994) can be used primarily with dicotyledonous crops, and microprojectile bombardment (Christou, 1994) can be used with almost any crop species. With *Agrobacterium* the insecticidal gene and a selectable marker gene are incorporated into a plasmid flanked by T-DNA border sequences. The T-DNA border sequences have *cis*-acting functions, which direct excision of the T-DNA, mobilization into plant cells and integration into the plant genome. These activities require *trans*-acting factors that are products of the *vir* genes of *Agrobacterium*. A selectable marker, usually anti-biotic resistance, is also transferred allowing for selection of transformed plant cells. With microprojectile bombardment, DNA encoding the insecticidal gene and a selectable marker, such as herbicide resistance, is coated onto gold particles that are accelerated toward embryogenic plant tissue. Once a particle lodges in the nucleus of a plant cell, the DNA can recombine into the plant genome. One key technology needed is the ability to produce embryogenic cultures that can regenerate into a mature fertile plant for any given crop species. Transformation of elite commercial cultivars is of greatest value because no further breeding is necessary.

Procedures for developing transformation events into elite cultivars accepted and used by growers depend on the method of propagation and basic genetics of the crop species. For a crop species such as potato, in which new commercial cultivars are rare due to sterility problems or very complex inheritance, and existing cultivars are long-lived, foreign insect resistance genes must be transformed into the com-mercial cultivars directly. Transformation events recovered in whole plants with the desired new phenotype must be propagated vegetatively and thoroughly analyzed phenotypically in multiple environments over multiple years to insure that the transgenic line is true to the phenotype of the isogenic untransformed cultivar.

For sexually propagated crops the foreign gene can, in principle, be introduced into any genotype of the species and the chromosomally integrated transgene then transferred to other lines by backcrossing. The only difference between inbred and hybrid crops, then, is the need to make hybrids with converted transgenic inbreds to assess the performance of the new transgenic product. In practice, the genotype receiving the foreign gene is important for early and accurate evaluation of the transformation event and should be ideally an elite inbred.

All transformation events must meet several criteria for development into com-mercial products:

— The event must confer upon the recipient genotype(s) the desired trait, meeting both qualitative and quantitative goals, for example, season-long protection from feeding by European corn borer larvae in maize.

— The event must not create a mutation with a deleterious effect on plant per-formance.

— The insertion site should not be so close to deleterious genes that the linkage cannot be broken with population sizes consistent with breeding programs and detected with current molecular marker technology during one or two gener-ations.

— The insertion site should also not be too close to major genes determining general combining ability for yield or other critical agronomic traits for a given heterotic group.

The determination that these criteria are met is most quickly and easily made if genes are inserted into elite inbred lines, whether the end product is a self-pollinated inbred line or a cross-pollinated hybrid. Isogenic lines can be created and field tested readily for this purpose. During evaluation of several transformation events in one cultivar of cotton for primary and secondary yield components, crop-maturity characteristics and fiber properties showed variation for these properties among events but the variation, relative to the untransformed cultivar, was more positive than negative (Wilson *et al.*, 1994).

Conversions of Lines and Demonstration of Maintained Performance

Once a transformation event has been regenerated into fertile plants and shown to have improved efficacy for insect control as well as germline transmissibility, the backcrossing process begins. Even if the event was generated in an elite commercial cultivar and shown not to affect the performance of that cultivar in a negative manner, its full value will be realized when transferred to other elite cultivars. Backcrossing can result in the transfer of a small chromosome segment (that containing the inserted DNA in this case) to a different genotype (the recurrent parent) if sufficient backcrosses are made with sufficient precision. What is sufficient depends on the donor germplasm, the size of backcross populations, the number of genes to be transferred and the ability to 'see' the genome. Molecular markers enable the visualization of individual plant genotypes; the amount of recurrent parent genome and crossovers bordering insertion site(s) can be monitored in each plant of each generation to insure that the final product is almost entirely the original cultivar plus added transgenes. In fact, molecular markers are essential for achievement of 'complete' conversions in a commercially relevant time period. The more marker loci examined in the final backcross product, the greater the confidence that no phenotypic differences will be observed in the absence of selection for the inserted DNA.

To demonstrate that backcross conversion processes have not compromised the agronomic performance of cultivars in production, regardless of molecular marker genotype, isogenic versions of the cultivar ($\pm$ transgene) must be tested in field trials in multiple locations and multiple years. This process can conform to normal field testing and commercial development programs in progressing from small plot to larger 'on-farm' strip trials. Although special regulatory restrictions might apply in the early stages of the development process, the usual large number of phenotypic traits should be measured and compared between the isogenic cultivars. If the crop is hybrid, several different hybrids should be made with each isogenic inbred to test for possible changes in specific and general combining ability undetected by molecular marker analysis. These trials must be conducted in the absence of insect pressure to assure that the performance of the converted transgenic line is maintained. This effort requires careful attention to the application of insecticides and will vary in success with the crop and the insects to be controlled.

Resistance Management

One concern regarding the use of transgenic crops for insect control is the possibility of resistance emerging in insect populations. Resistance to *Bacillus thuringiensis* insecticidal crystal proteins has been observed and studied in both laboratory populations and field populations of insects (see Chapters 2 and 16).

Several observations have emerged from these studies. Cases of resistance observed both in the laboratory and in the field have not been dominant traits; rather they appear to be recessive or co-dominant (additive) and are determined by a small number of loci (McGaughey and Beeman, 1988; Tabashnik *et al.*, 1992). Tolerance to the toxin is typically lost in the absence of selection, suggesting that there may be a fitness cost associated with resistance. Since different cultivars of a given crop species are usually planted in close proximity and because wild host plants species are present in most environments, the opportunity for a continual dilution of any resistance alleles is possible. Another observation is that a chronic high-dose exposure, as occurs with transgenic plants, is different to short-term, sublethal exposure as can occur with *Bacillus thuringiensis* sprays or with populations of resistant insects selected in a laboratory environment. Most laboratory strains selected for resistance are only exposed to the insecticidal crystal protein during a portion of their life cycle, and therefore caution must be used in interpreting whether these strains are truly resistant. Consequently, with a high dose exposure, as occurs with transgenic plants, it is difficult for an insect to avoid exposure by cessation of feeding for a short time. Another consideration is the degree of resistance observed in an insect population given different levels of exposure. Insect populations selected under short-term sublethal doses still cannot survive on transgenic plants expressing chronic high doses (Whalon and Wierenga, 1994). This difference also has implications for selection of resistance. Since resistance is recessive or co-dominant, insects heterozygous for resistance are more susceptible to the insecticidal proteins and are not likely to survive on transgenic plants, thus limiting the prevalence of the resistance alleles.

Another important aspect for resistance management is to have insecticidal proteins with different modes of action. In the case of *Bacillus thuringiensis* insecticidal crystal proteins, resistance is often determined by differences in receptor binding in the insect gut (Van Rie *et al.*, 1990). Since different classes of toxins bind to different receptors, the possibility of reducing the chances for resistance by having plants expressing different insecticidal crystal protein genes has been discussed. In addition to the insecticidal crystal protein genes, new proteins are being discovered with completely different modes of action, which can be used in concert with the insecticidal crystal protein genes, as is being done with chemical pesticides today.

Registration Studies for Commercialization

Regulatory approval of transgenic insect resistant crops is required by many countries. Many that do not currently have a regulatory framework are in the process of formulating one. Countries differ in the laws and agencies regulating transgenic products. Also, the laws, procedures, data requirements, key people and the agencies involved can change with time. The process of regulatory approval can be quite complex. In some countries, in comparison with the approval of chemicals, the process can be fairly straightforward and relatively rapid. The complex demands of multiple agencies, sometimes multiple countries and, usually, multiple issues must be addressed successfully, and normally simultaneously, to prevent the regulatory process from delaying the introduction of increased value to the growers. Scientifically based, data-driven evaluations of risk and benefit continue to provide the best basis for determining safety to the consumer.

Intellectual Property

Genetically engineered plants are complex from the point of view of intellectual property. Numerous materials and methods can be utilized in producing a transgenic commercial crop variety, line or hybrid. Patentable materials used to create the transgenic plant can include: structural genes; proteins; promoters; mRNA leader sequences; transcription terminator sequences; introns; transformation target cells or tissues; machines; vectors; and hosts for vectors. Methods might include: general methods of molecular biology such as chimeric gene construction, synthesis of genes, gene detection; and general methods of transformation such as gene delivery methods. The use of methods and materials results in products which themselves can be patented, e.g., microbes, cells, plants (inbreds, hybrids, varieties), plant parts.

One of the primary sources of risk from intellectual property, both for the parties who do and do not hold the patents, arises from the complexity of intellectual property laws worldwide: the sometimes very long prosecution process, and uncertainty, in the US, about what applications have actually been filed; variation in the scope of claims issuing over time; validity of issued patents, which can be challenged in the courts; and the occasional change in ownership of patents. In addition, intellectual property laws vary by country. Patents are issued throughout the time during which transgenic products are created and developed, and pose difficult-to-assess threats of blockage from commercialization as the investment in commercial development increases. The scope of patents issuing over time can vary from broad to narrow depending on the field of the invention and the date of filing. Patent validity can be challenged. Patents can be bought and sold.

In summary, insect resistant cultivars achieved by genetic engineering entail new non-technical risks for widespread adoption. However, among the set of transgenic cultivars with a given insect resistance phenotype will be those with patent protection, directly or through licenses that will be commercialized; intellectual property itself should not provide a block to bringing value to growers but should enhance the process by improving the potential return on the research investment.

Status of Transgenic Crops in Commercial Development

Several companies are in the process of bringing transgenic plants to the marketplace. Three products have received approval from US governmental agencies to be cultivated and marketed. In potato, CryIIIA has been expressed for control of Colorado potato beetle (Chapter 3). In corn, CryIA(b) has been expressed for the control of European corn borer (Chapter 4). CryIA(c) has been expressed in cotton to control lepidopteran cotton pests (Perlak *et al.*, 1990). These products are currently being grown commercially.

Several products are in advanced stages of development or undergoing the regulatory approval process. CryIA(b) has been expressed for the control of potato tuber moth. CryIH has been expressed in corn for the control of European corn borer and fall armyworm. CryIIA has been expressed in cotton for the control of lepidopteran pests.

Several insecticidal proteins are still in the experimental stages of development. These include cholesterol oxidase in cotton for the control of boll weevil, vegetative insecticidal proteins for the control of corn rootworm, and black cutworm and fall armyworm in maize (see Chapter 7). Patatins, or lipid acyl hydrolases, have been reported to have activity against *Diabrotica* species which are pests of corn. Lectins, α-amylase inhibitors, proteinase inhibitors, chitinases, peroxidases, and plant secondary metabolites are all still in the experimental stages of development (see Chapters 8–14).

Summary

The use of transgenes for increasing host plant resistance to insects is one further advance in agricultural crop production. Non-genetic control measures have not provided a complete solution to the problem of insect predation on agricultural crops. Even integrated pest management strategies, which combine the use of chemicals, resistant germplasm, scouting for insect infestation, crop rotations which interrupt insect life cycles, and modifying planting, harvesting and handling practices have failed to control insect predation adequately. Yield losses due to insects have actually increased slightly over the past two decades in most crops. The use of transgenes to control insect pathogens is a natural extension of plant domestication by man. Transgenic plants resistant to insects fit well with integrated strategies for insect pest management. With transgenes, insect resistance is inherent in the germplasm and consequently does not require scouting and chemical treatments to control insect infestations. Since most resistance genes are single dominant traits, transgenes introduced into elite cultivars can reach the grower rapidly, and breeding transgenes into related germplasm can proceed quickly and precisely through the use of molecular markers. The active principles encoded by insect resistance genes are very well characterized and the modes of action understood before reaching the grower, unlike many resistance traits derived from conventional breeding. This greater knowledge of the final product is enhanced, even mandated, by the process of obtaining regulatory approval from governmental agencies. With transgenic plant technologies, the germplasm pool for resistance genes is as big as nature itself. Agricultural crop production throughout the world is poised to realize the benefits of transgenes for host plant resistance.

References

Bonierbale, M. S., Plaisted, R. L., Pineda, O. and Tanksley, S. D. (1994) QTL analysis of trichome-mediated insect resistance in potato, *Theor. Appl. Genet.* **87**, 973–987.

Christensen, D., Beland, G. and Meghji, M. (1994) Yield loss due to European corn borer in normal and transgenic hybrids, In: *Proceedings of the 48th Annual Corn and Sorghum Research Conference*, Washington: American Seed Trade Association, pp. 43–52.

Christou, P. (1994) Gene transfer to plants via particle bombardment, In: Gelvin, S. and Schilperoort, R. (Eds), *Plant Molecular Biology Manual*, 2nd edn., Dordrecht: Kluwer Academic, pp. A2/1–A2/14.

Dicke, F. F. and Guthrie, W. D. (1988) The most important corn insects, In: Sprague, G. F. and Dudley, J. W. (Eds), *Corn and Corn Improvement*, Agronomy Series No. 18, Madison: American Society of Agronomy, pp. 769–868.

FAO (1991), *FAO Yearbook: Production*, Vol. 45, FAO Statistics Series No. 104, Rome: Food and Agriculture Organization of the United Nations, pp. 61–184.

Felton, G., Donato, K., Broadway, R. and Duffey, S. (1992) Impact of oxidized plant phenolics on the nutritional quality of dietary protein to a noctuid herbivore, *Spodoptera exigua, J. Insect Physiol.* **38**, 277–285.

Hill, D. S. (1987) *Agricultural Insect Pests of Temperate Regions and Their Control*, Cambridge: Cambridge University Press, pp. 187–470.

Hofmann, C., Vanderbruggen, H., Hofte, H., Van Rie, J., Jansens, S. and Van Mellaert, H. (1988) Specificity of *Bacillus thuringiensis* δ-endotoxins is correlated with the presence of high affinity binding sites in the brush border membrane of target insect midguts, *Proc. Natl. Acad. Sci. USA* **85**, 7844–7848.

Höfte, H. and Whiteley, H. R. (1989) Insecticidal crystal proteins of *Bacillus thuringiensis Microbiol. Rev.* **5**, 242–255.

Kogan, M. and Ortman, E. E. (1978) 'Antixenosis – a new term proposed to replace Painter's "Nonpreference" modality of resistance', *ESA Bull.* **24**, 175–176.

Li, J. D., Carroll, J. and Ellar, D. J. (1991) Crystal structure of insecticidal δ-endotoxin from *Bacillus thuringiensis* at 2.5 Å resolution, *Nature* **353**, 815–821.

McGaughy, W. and Beeman R. (1988) Resistance to *Bacillus thuringiensis* in colonies of Indian meal moth and almond moth (Lepidoptera: Pyralidae), *J. Econ. Entomol.* **81**, 28–33.

Nair, S., Bentur, J. S., Prasada Rao, U. and Mohan, M. (1995) DNA markers tightly linked to a gall midge resistance gene (Gm2) are potentially useful for marker-aided selection in rice breeding, *Theor. Appl. Genet.* **91**, 68–73.

Nieto-Lopez, R. M. and Blake, T. K. (1994) Russian wheat aphid resistance in barley: inheritance and linked molecular markers, *Crop Sci.* **34**, 655–659.

Niles, G. A. (1980) Breeding cotton for resistance to insect pests, In: Maxwell, F. G. and Jennings, P. R. (Eds), *Breeding Plants Resistant to Insects*, New York: John Wiley & Sons, pp. 337–370.

Oerke, E.-C., Dehne, H.-W., Schönbeck, F. and Weber, A. (1994) *Crop Production and Crop Protection: Estimated Losses in Major Food and Cash Crops*, Amsterdam: Elsevier Science.

Painter, R. H. (1951) *Insect Resistance in Crop*, New York: The Macmillan Co., p. 520.

Palm, E. W. (1991) Estimated crop losses without the use of fungicides and nematicides and without nonchemical controls, In: Pimentel, D. (Ed.), *CRC Handbook of Pest Management in Agriculture*, Vol. 1, 2nd edn., Boca Raton: CRC Press, pp. 109–113.

Perlak, F. J., Deaton, R. W., Armstrong, T. A., Fuchs, R. L., Sims, S. R., Greenplate, J. T. and Fischoff, D. A. (1990) Insect resistant cotton plants, *Bio/Technology* **8**, 939.

Pimentel, D. (1991) *CRC Handbook of Pest Management in Agriculture*, Vol. 1, 2nd edn., Boca Raton: CRC Press.

Porter, A., Davidson, E. and Liu, J.-W. (1993) Mosquitocidal toxins of Bacilli and their genetic manipulation for effective biological control of mosquitos, *Microbiol. Rev.* **57**, 838–861.

Purcell, J., Isaac, B., Tran, M., Sammons, D., Gillespie, J., Greenplate, J., Solsten, R., Prinsen, M., Pershing, J. and Stonard, R. (1994) Two enzyme classes active in green peach aphid bioassays, *J. Econ. Entomol.* **87**, 15-19.

Sebastian, L. S., Ikeda, R., Huang, N., Imbe, T., Coffman, W. R. and McCouch, S. R. (1996) Molecular mapping of resistance to rice tungro spherical virus and green leafhopper, *Phytopathology* **86**, 25–30.

Singh, D. P. (1986) *Breeding for Resistance to Diseases and Insect Pests*, Berlin: Springer-Verlag, pp. 35–61.

Strickland, J. A., Orr, G. L. and Walsh, T. A. (1995) Inhibition of Diabrotica larval growth by patatin, the lipid acyl hydrolase from potato tubers, *Plant Physiol.* **109**, 667–674.

Smith, C. M. (1989) *Plant Resistance to Insects: A Fundamental Approach*, New York: John Wiley & Sons, pp. 196–197.

Tabashnik, B., Schwartz, J., Finson, N. and Johnson, M. (1992) Inheritance of resistance to

Bacillus thuringiensis in diamondback moth (Lepidoptera: Plutellidae), *J. Econ. Entomol.* **85**, 1046–1055.

Turnipseed, S. G. and Kogan, M. (1987) Integrated control of insect pests, In: *Soybeans: Improvement, Production and Uses*, 2nd edn., Agronomy Series No. 16, Madison: American Society of Agronomy, pp. 779–818.

Van Rie, J., Jansens, S., Hofte, H., Degheele, D. and Van Mellaert, H. (1989) Specificity of *Bacillus thuringiensis* δ-endotoxins: importance of specific receptors on the brush border membrane of the mid-gut of target insects, *Eur. J. Biochem.* **186**, 239–247.

Van Rie, J., McGaughy, W., Johnson, D., Barnett, B. and Van Mellaert, H. (1990) Mechanism of insect resistance to the microbial insecticide *Bacillus thuringiensis, Science* **247**, 72–74.

Walkerpeach, C. and Velten, J. (1994) Agrobacterium-mediated gene transfer to plant cells: cointegrate and binary vector systems, In: Gelvin, S. and Schilperoort, R. (Eds), *Plant Molecular Biology Manual*, 2nd edn., Dordrecht: Kluwer Academic, pp. B1/1–B1/12.

Whalon, M. and Wierenga, J. (1994) *Bacillus thuringiensis* resistant Colorado potato beetle and transgenic plants: some operational and ecological implications for deployment, *Biocontrol Sci. Technol.* **4**, 555–561.

Wilson, F. D., Flint, H. M., Deaton, W. R. and Buehler, R. E. (1994) Yield, yield components and fiber properties of insect-resistant cotton lines containing a *Bacillus thuringiensis* toxin gene, *Crop Sci.* **34**, 38–41.

Young, N. D., Kumar, L., Menancio-Hautea, D., Danesh, D., Talekar, N. S., Shanmugasundarum, S. and Kim, D.-H. (1992) RFLP mapping of a major bruchid resistance gene in mungbean (*Vigna radiata*, L. Wilczek), *Theor. Appl. Genet.* **84**, 839–844.

2

Insect Control with Transgenic Plants Expressing *Bacillus thuringiensis* Crystal Proteins

MARNIX PEFEROEN

Introduction

Bacillus thuringiensis (*Bt*) was discovered in a diseased silk moth population at the beginning of this century by Ishiwata. The bacterium, then recorded as *Bacillus soto*, was lost but subsequently re-isolated in 1909 by Berliner from a diseased Mediterranean flour moth population (Berliner, 1915). Berliner described the presence of an inclusion body or crystal, now considered as the phenotypic characteristic discriminating *Bacillus thuringiensis* from *Bacillus cereus*. Already at the time of discovery Berliner suggested that pathogenic organisms could be used to control insect populations. In the late 1920s the bacterium was mass produced and used to control European corn borer in south-east Europe. The first commercial formulation, Sporéine, was produced in France in 1938. After the Second World War there was a rejuvenated interest in the use of *Bt* for the control of agricultural insect pests. In the 1950s it became clear that the insecticidal activity of the bacterium was predominantly determined by its crystal which is produced during sporulation (Hannay, 1953) and which contains proteins (Hannay and Fitz-James, 1955). *Bt* strains produce other products which have biocidal activities such as phospholipases, immunosuppresive agents, exotoxins (Lüthy, 1980) and vegetative insecticidal proteins (Chapter 7). The increasing number of *Bt* strains isolated from different sources by different people created a need to characterize *Bt* strains. One of the methods still in use today involves the immunological identification of flagella on the surface of the vegetative cells (de Barjac and Bonnefois, 1962), and so far some 45 serotypes have been described (Lecadet *et al.*, 1994). A major step forward in the commercial success of *Bt* was the isolation of HD-1, a strain which proved far more potent than any other strain isolated so far and which still forms the basis of some of today's commercial formulations (Dulmage, 1970). Until 1977 *Bt* was considered to be exclusively toxic to lepidopteran larvae. But then, from a pond in the Negev desert, Goldberg and Margalit (1977) isolated a *Bt* strain, *Bacillus thuringiensis* subsp. *israelensis*, which was specifically toxic to dipteran larvae such as mosquito and black fly larvae. *Bt* strains active against dipteran larvae are now extensively used to control insect disease vectors for malaria and river blindness. In 1981 the first gene encoding a crystal protein was cloned (Schnepf and Whiteley, 1981). In 1983, Krieg *et al.* found a *Bt* strain, *Bacillus thuringiensis* subsp. *tenebrionis*, which proved highly toxic to some coleopteran larvae, including the Colorado potato beetle. This was the trigger for a massive isolation and screening effort for strains and crystal proteins with improved and new activities. *Bt* proved to be a ubiquitous organism and strains have been isolated from a wide range of sources such as forest soil, grain dust, bat dung, sea water, desert sand, and dead and diseased insects (Martin and Travers, 1989; Meadows, 1993). An interesting observation was that *Bt* strains are quite abundant on the phylloplane suggesting that they could help plants to protect themselves from insect feeding damage (Smith and Couche, 1991). Although *Bt* is not an obligate insect pathogen – it can easily be grown on standard microbiological media – it seems that multiplication in soil is rather limited (West *et al.*, 1985). Spores can survive for several years and may occasionally germinate when in contact with nutrients derived from decaying grass or other organic material (West, 1985). The screening effort, mainly conducted at private companies, has led to the collection of perhaps as many as 50 000 isolates which represent several thousand unique strains. Only a fraction of these strains have been analyzed in full detail and one can expect many more interesting discoveries.

Bt Crystal Proteins

Although it was demonstrated 40 years ago that the crystal and its proteins have insecticidal effects, it was the cloning of a crystal protein gene that directed attention toward these genes and proteins (Schnepf and Whiteley, 1981). Transcription of crystal protein genes is normally regulated by sporulation specific sigma factors (Brown and Whiteley, 1988, 1990), although the transcription of Cry3A is controlled by a vegetative sigma factor (Agaisse and Lereclus, 1994). Crystal proteins are accumulated in crystals and as a result are better protected from degradation while exposed to the environment. Different crystal proteins can be packed in one and the same crystal or stored in separate crystals. When different Cry1 type proteins are expressed by a *Bt* bacterium they are packed in a single bipyramidal crystal. The Cry2 type proteins typically form cuboidal crystals, while Cry3 proteins accumulate in rhomboidal crystal. Other crystal structures such as ovoids have been observed.

The first crystal protein genes were fully sequenced in 1985 (Adang *et al.*, 1985; Schnepf *et al.*, 1985). The same crystal protein gene was found in many different strains and it appeared that strains often carry multiple crystal protein genes. This can be explained by the fact that crystal protein genes are often carried by plasmids which can be transferred by conjugation from one strain to another (Gonzales *et al.*, 1982; Jarrett and Stephenson, 1990). In addition, crystal protein genes are sometimes surrounded by inverted repeat sequences which could act as signals for recombination (Bourgouin *et al.*, 1988; Kronstad and Whiteley, 1984; Mahillon *et al.*, 1985). Perhaps, in nature, there is a regular exchange of crystal protein genes between *Bt* bacteria, so that *Bt* should be perceived as a population rather than as a strain. Furthermore, in the environment spores and crystals are very often dissociated so that spores and crystals from different strains may be mixed. From a population perspective it does not matter which of the bacteria carry certain crystal protein genes. The population as a whole, even those bacteria without a crystal protein gene, will benefit from the nutrients provided by the dead insects. For a *Bt* population it is important to have a wide range of crystal proteins so that a wide

Table 2.1 *Bacillus thuringiensis* crystal protein genes

Name	Old name	Access no.	Size (Da)	Activity	Reference
Cry1Aa1	CryIA(a)	M11250	133 500	Lepidoptera	Schnepf *et al.* (1985)
Cry1Ab1	CryIA(b)	M13898	130 615	Lepidoptera	Wabiko *et al.* (1986)
Cry1Ac1	CryIA(c)	M11068	131 000	Lepidoptera	Adang *et al.* (1985)
Cry1Ad1	CryIA(d)	M73250	131 000	Lepidoptera	Payne and Sick (1993b)
Cry1Ae1	CryIA(e)	M65252	131 000	Lepidoptera	Lee and Aronson (1991)
Cry1Ba1	CryIB	X06711	136 500	Lep./Coleop.	Brizzard and Whiteley (1988)
Cry1Bb1	ET5	L32020	136 500	Lepidoptera	Donovan and Tan (1994)
Cry1Bc1	CryIB(c)	Z46442	138 000	Lepidoptera	Bishop *et al.* (unpublished)
Cry1Ca1	CryIC	X07518	132 000	Lepidoptera	Honée *et al.* (1988)
Cry1Cb1	CryIC(b)	M97880	130 500	Lepidoptera	Kalman *et al.* (1993)
Cry1Da1	CryID	X54160	129 500	Lepidoptera	Höfte *et al.* (1990)
Cry1Db1	PrtB	Z22511	129 000	Unknown	Lambert (unpublished)
Cry1Ea1	CryIE	X53985	133 236	Lepidoptera	Visser *et al.* (1990)
Cry1Eb1	CryIE(b)	M73253	130 500	Lepidoptera	Payne and Sick (1993a)
Cry1Fa1	CryIF	M63897	130 500	Lepidoptera	Chambers *et al.* (1991)
CryFb1	PrtD	Z22512	131 000	Unknown	Lambert (unpublished)
Cry1Ga1	PrtA	Z22510	129 500	Unknown	Lambert (unpublished)
Cry1Ha1	PrtC	Z22513	130 000	Unknown	Lambert (unpublished)
Cry1Ia1	CryV	X62821	81 200	Lep./Coleop.	Tailor *et al.* (1992)
Cry1Ib1	CryV	U07642	80 000	Unknown	Shin *et al.* (unpublished)
Cry1Ja1	ET4	L32019	129 500	Lepidoptera	Donovan and Tan (1994)
Cry1Jb1	ET1	U31527	131 000	Lepidoptera	Von Tersch and Gonzalez (1994)
Cry1Ka1	CryI	U28801		Unknown	Koo (unpublished)
Cry2Aa1	CryIIA	M31738	70 500	Lep./Dip.	Donovan *et al.* (1989)
Cry2Ab1	CryIIB	M23724	70 500	Lepidoptera	Widner and Whiteley (1989)
Cry2Ac1	CryIIC	X57252	69 000	Lepidoptera	Wu *et al.* (1991)
Cry3Aa1	CryIIIA	M22472	72 500	Coleoptera	Herrnstadt *et al.* (1987)
Cry3Ba1	CryIIIB	X17123	74 300	Coleoptera	Sick *et al.* (1990)
Cry3Bb1	CryIIIBb	M89794	72 500	Coleoptera	Donovan *et al.* (1992)
Cry3Ca1	CryIIID	X59797	73 000	Coleoptera	Lambert *et al.* (1992b)
Cry4Aa1	CryIVA	Y00423	134 545	Diptera	Ward and Ellar (1987)
Cry4Ba1	CryIVB	X07423	126 000	Diptera	Chungjatupornchai *et al.* (1988)
Cry5Aa1	CryVA(a)	L07025	153 500	Nematoda	Sick *et al.* (1994)
Cry5Ab1	CryVA(b)	L07026	143 000	Nematoda	Narva *et al.* (1991)
Cry5Ba1	PS86Q3	U19725	139 000	Coleoptera	Payne and Michaels (1995)
Cry6Aa1	CryVIA	L07022	52 500	Nematoda	Narva *et al.* (1993)
Cry6Ba1	CryVIB	L07024	44 000	Nematoda	Narva *et al.* (1991)
Cry7Aa1	CryIIIC	M64478	126 500	Coleoptera	Lambert *et al.* (1992b)
Cry7Ab1	CryIIICb	U04367	126 500	Coleoptera	Payne and Fu (1994)
Cry8Aa1	CryIIIE	U04364	128 500	Coleoptera	Foncerrada *et al.* (1992)
Cry8Ba1	CryIIIG	U04365	130 000	Coleoptera	Michaels *et al.* (1993)
Cry8Ca1	CryIIIF	U04366	127 500	Coleoptera	Ogiwara *et al.* (1995)
Cry9Aa1	CryIG	X58120	128 500	Lepidoptera	Smulevitch *et al.* (1991)
Cry9Ba1	CryIX	X75019	127 000	Lepidoptera	Shevelev *et al.* (1993)
Cry9Ca1	CryIH	Z37527	129 800	Lepidoptera	Lambert *et al.* (1996)
Cry10Aa1	CryIVC	M12662	75 000	Diptera	Thorne *et al.* (1986)
Cry11Aa1	CryIVD	M31737	71 500	Diptera	Donovan *et al.* (1988)
Cry11Ba1	Jeg80	X86902		Diptera	Delecluse (unpublished)
Cry12Aa1	CryVB	L07027	139 500	Nematoda	Narva *et al.* (1991)
Cry13Aa1	CryVC	L07023	89 000	Nematoda	Schnepf *et al.* (1992)
Cry14Aa1	CryVD	U13955	132 800	Coleoptera	Payne and Narva (1994)
Cry15Aa1	34kDa	M76442	38 000	Lepidoptera	Brown and Whiteley (1992)
Cyt1Aa1	CytA	X03182	27 340	Cytotoxin	Waalwijk *et al.* (1985)
Cyt2Aa1	CytB	Z14147	29 000	Cytotoxin	Koni and Ellar (1995)

range of food sources can be exploited. Perhaps this continuous remodeling explains the ecological success of *Bt*, with new combinations of crystal proteins providing access to new niches (Lambert and Peferoen, 1992).

In 1989, Höfte and Whiteley proposed a nomenclature and classification scheme for the 42 crystal protein genes sequences known at that time. Based on their insecticidal activity the crystal protein genes were ordered in four groups: Lepidoptera-specific (type I); Lepidoptera- and Diptera-specific (type II); Coleoptera-specific (type III); and Diptera-specific (type IV). Further subdivisions within the groups were determined by sequence homology and the 42 crystal protein sequences were found to represent 14 different crystal protein genes. There were several inconsistencies in the original classification and the ongoing discovery of crystal proteins with very different amino acid sequences and biocidal activities urged the development of a new classification system. The new system ranks crystal proteins on the basis of their amino acid sequence only. The 96 crystal protein sequences reported by mid-1995 are now classified into 17 groups (Crickmore *et al.*, 1996). Table 2.1 shows 54 crystal protein genes with less than 96% amino acid sequence identity.

Many crystal protein genes encode proteins of some 130 to 140 kDa with a trypsin resistant core of some 60 kDa. The insecticidal activity of the crystal protein resides in this trypsin resistant protein derived from the N-terminal half of the crystal protein. Probably, the C-terminal half of the protein is important for the crystallization of the protoxin, often in a bipyramidal shape. Some crystal proteins, like the Cry3 proteins, seem to consist of the N-terminal half of these large 130–140 kDa proteins only, as if their genes are naturally truncated. Deletion studies whereby gene fragments were produced to define the shortest gene fragment encoding a fully toxic protein, have shown that a block of amino acids highly conserved between most crystal protein genes delineates the C-terminus of the active toxin (Adang *et al.*, 1985; Höfte *et al.*, 1986; MacIntosh *et al.*, 1990). Sequence alignments show the presence of up to five conserved regions within the toxic moieties of crystal proteins and an additional three in the C-terminal half.

The variation of crystal proteins identified since the first cloned crystal protein gene in 1981 is beyond expectations. So far, new crystal proteins have mostly been found through bioassay screening. Perhaps more sophisticated screening methods, such as antibody (Höfte *et al.*, 1988) and polymerase chain reaction screening (Carozzi *et al.*, 1991), can lead to the discovery of crystal proteins which cannot be easily identified in activity assays.

Structure–Function Relationship

The elucidation of the three-dimensional structure of Cry3A by Li *et al.* (1991) and of the toxic moiety of Cry1Aa by Grochulski *et al.* (1995) basically shows a three domain structure (Figure 2.1). Domain I of some 250 N-terminal amino acids is a helical bundle with six alpha helices surrounding a central alpha helix and is very likely the part of the toxin involved in membrane insertion and pore formation. This domain shows some striking resemblances with the membrane insertion folds of colicin and diphtheria toxin (Parker and Pattus, 1993). Domain I of the Cry1Ac and Cry3Bb protein on its own is able to insert itself in artificial lipid bilayers which clearly indicates that domain I is responsible for pore formation (Walters *et al.*, 1993; Von Tersch *et al.*, 1994). Synthetic peptides corresponding to helices of

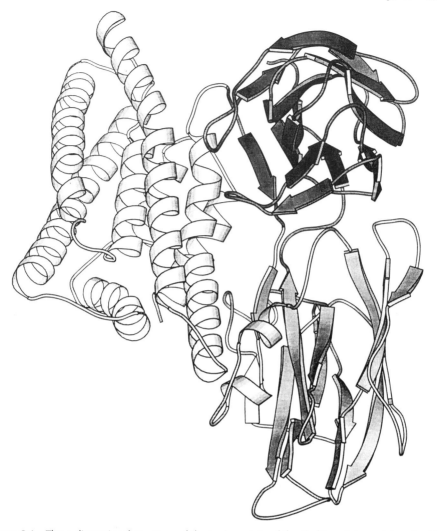

Figure 2.1   Three-dimensional structure of the toxic moiety of the CryIA protein as determined by Grochulski *et al.* (1995) (reprinted with kind permission of P. Grochulski)

domain I interact with phospholipid vesicles, planar lipid bilayers and insect midgut membranes (Cummings *et al.*, 1994; Gazit and Shai, 1995; Gazit *et al.*, 1994). Domain II, corresponding to the central 200 amino acids of the Cry toxins, consists of three beta sheets and probably is the part of the protein interacting with target molecules in insect midgut epithelial cells. This domain is similar to the vitelline membrane outer layer protein I of the hen egg; a protein which has transferase-like activity and which can bind to endothelial cells (Shimizu *et al.*, 1994). Studies whereby hybrid crystal proteins were produced have shown that the activity spectrum of one crystal protein could be transferred to another crystal protein simply by exchanging regions containing or being part of domain II (Ge *et al.*, 1989; Masson *et al.*, 1994; Schnepf *et al.*, 1990). Also, binding studies with hybrid proteins showed that binding is determined at least in part by domain II (Lee *et al.*, 1992). Lu *et al.* (1994) have shown that the six amino acids which form the loop at the bottom of

25

sheet 2 in domain II of Cry1Aa are essential for its binding to the midgut epithelial cells (Grochulski *et al.*, 1995). The C-terminal 150 amino acids of the toxin are folded in a beta sandwich with the last 25 amino acids covered by the two sheets. The function of domain III is unknown although a role in stabilizing overall structure (Li *et al.*, 1991), defining specificity (Honée *et al.*, 1991; Bosch *et al.*, 1994), ion channel formation (Chen *et al.*, 1993) or binding (Aronson *et al.*, 1995) has been postulated. The structure determined for Cry3A and for Cry1Aa is probably the general structure for the majority of the toxic moieties of crystal proteins since the five conserved blocks of amino acids form the core of the structure. A thorough understanding of the structure–function relationship could allow the design of crystal proteins with improved or new activities.

Mechanism of Action

Cytological studies have demonstrated that δ-endotoxins act by destroying the midgut epithelium, leading to starvation, paralysis, septicemia and death (Manthavan *et al.*, 1989). Mechanism-of-action studies have revealed the major steps between the ingestion of the crystal and the disruption of the gut. In general terms, upon ingestion, crystal proteins are dissolved from the crystals and are proteolitically activated to a trypsin resistant core fragment. This protein passes through the pores in the peritrophic membrane and binds to a membrane protein in the brush border of the midgut epithelial cells and inserts into the membrane. This insertion leads to the formation of pores and lysis of the cells. Crystal proteins operate by this general scheme but are characterized by a unique activity spectrum. Figure 2.2 presents the relative insecticidal activities of four crystal proteins when tested against five insects and illustrates the unique activity spectrum of crystal proteins. It is inter-

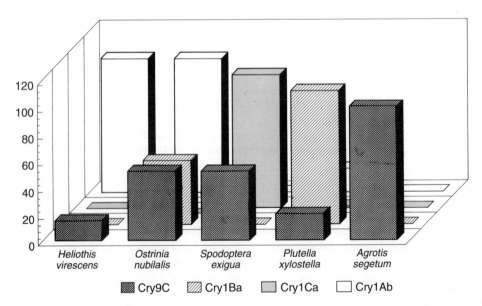

Figure 2.2 Relative insecticidal activity of four crystal proteins tested against five lepidopteran larvae

esting to note that sequence homology is a poor indicator for insecticidal spectrum. Cry1Aa is more than 90% identical with Cry1Ab and Cry1Ac but only Cry1Aa is toxic to silkmoth larvae. Cry3A and Cry7A, with only some 37% amino acid identity, are both toxic to Colorado potato beetle larvae. This extreme specificity is still the focus of attention in mechanism-of-action studies.

When *Bt* crystals are ingested by insects, the first step in the mode of action is that crystal proteins are dissolved from the crystal. In the Cry3A crystal, the proteins are arranged around large solvent channels (Li *et al.*, 1991). The pH in the gut of lepidopteran larvae varies between pH 9 and 12 and lepidopteran-specific crystal proteins can only be solubilized above pH 9.5 (Knowles and Dow, 1993). In Colorado potato beetle larvae the pH of the midgut fluid is around 6.0 (Slaney *et al.*, 1992). However, *in vitro*, Cry3A – a crystal protein toxic to Colorado potato beetle – is soluble at pH below 3.5 and above pH 9.5. Solubilization is therefore pH-dependent; other factors such as detergents or the reducing potential of the gut to break up intermolecular bridges, could however play a role (Bietlot *et al.*, 1990). Crystal proteins are protoxins which are probably not active as such. The lepidopteran specific protoxins are typically proteins of some 130 kDa. Proteases with a trypsin-like activity, either from the insect or from the *Bacillus* itself, gradually remove some 600 amino acids from the C-terminus of the protein and some 30 to 50 amino acids from the N-terminus (Choma *et al.*, 1990). This generates a protein of some 60 kDa which corresponds to the N-terminal half of the original crystal protein. This 60-kDa protein is trypsin-resistant and is the active core of the crystal protein (Höfte and Whiteley, 1989). Furthermore, crystal proteins such as Cry3A, which do not have this C-terminal part of 600 amino acids, undergo some proteolytic processing at their N-terminus (McPherson *et al.*, 1988). DNA associated with crystal proteins could play a role in the crystallization and in proteolytic processing (Bietlot *et al.*, 1993). A most dramatic illustration of the role of solubilization and proteolytic activation for insecticidal activity is presented by Cry7Aa. Cry7Aa, formerly described as CryIIIC, is a crystal protein of some 130 kDa whose N-terminal half has some 32–37% sequence identity with coleopteran-specific protein such as Cry3A, Cry3Ba and Cry3C (Lambert *et al.*, 1992a). Cry7Aa was tested as a crystal preparation against some lepidopteran and coleopteran larvae but no insecticidal activity could be demonstrated. However, when Cry7Aa was dissolved and trypsinized before oral administration it proved toxic to Colorado potato beetle larvae. It was proposed that the Cry7Aa crystal protein, like other 130 kDa crystal proteins, is not soluble in the slightly acidic conditions of the Colorado potato beetle gut but requires the alkaline condition in the gut of Lepidoptera. When Cry7Aa crystal protein is dissolved in an alkaline buffer and mixed with gut juice from Colorado potato beetle larvae, it is processed to a trypsin-resistant core fragment which is highly toxic to these larvae. On the other hand Cry7Aa can be dissolved in the alkaline gut fluid of tobacco horn worm larvae but then the dissolved crystal protein is totally degraded by the gut proteases, which at least in part explains why it has no activity in these larvae. A similar situation is observed with Cry1Ba, which as a protoxin only has activity against certain lepidopteran larvae, while *in vitro* solubilization of the protoxin reveals the latent activity against Colorado potato beetle larvae (Bradley *et al.*, 1995). From an ecological point of view one can speculate on the relevance of producing a crystal with a 'silent' insecticidal activity. Probably the Colorado potato beetle is not the natural target of the Cry7Aa or Cry1Ba crystal proteins.

Differential proteolytic activation of one and the same crystal protein can lead to toxins with a different activity spectrum. A Cry1Ab protein was found to be activated into a lepidopteran toxin by midgut juice of lepidopteran larvae, while gut juice of dipteran larvae removed a few more amino acids from its C-terminus, destroying the lepidopteran activity but revealing a toxicity for mosquito larvae (Haider *et al.*, 1986).

Following the processing of the protoxin into the trypsin-resistant core fragment, the δ-endotoxin passes through the pores of the peritrophic membrane – a sieve in the insect gut – and interacts with the gut epithelial cells (Bravo *et al.*, 1992b). In the late 1980s it was shown that δ-endotoxins can interact with brush border membrane vesicles prepared from the insect midgut with a high affinity (Hofmann *et al.* 1988; Wolfersberger *et al.*, 1987). Specific binding of crystal proteins to the gut epithelium has been described for a wide range of lepidopteran insects (Bravo *et al.*, 1992a; Denolf *et al.*, 1993; Eschriche *et al.*, 1994; Ferré *et al.*, 1991; Van Rie *et al.*, 1989, 1990a,b), some dipteran larvae (Ravoahangimalala and Charles, 1995; Ravoahangimalala *et al.*, 1993) and some coleopteran larvae (Belfiore *et al.*, 1994; Slaney *et al.*, 1992). Many lepidopteran larvae are susceptible to more than one crystal protein. It was shown that different crystal proteins may have different binding sites in one and the same insect. In general Cry1A type proteins seem to compete for the same binding site while most other crystal proteins seem to have a unique receptor. Initially there appeared to be a correlation between binding and toxicity. For example, Cry1Ba is not toxic to *Manduca sexta* larvae and does not show specific binding to their midgut. In contrast, the Cry1Ba protein is highly toxic to *Pieris brassicae* and binds specifically to the brush border of the midgut epithelial cells (Hofmann *et al.*, 1988). *Heliothis virescens* is susceptible to Cry1Aa, Cry1Ab and Cry1Ac with Cry1Aa as the least toxic and Cry1Ac as the most toxic protein, corresponding to an increasing number of binding sites (Van Rie *et al.*, 1990b). However, there are insects in which increased binding does not correspond to increased toxicity (Wolfersberger, 1990) or in which crystal proteins bind without being toxic (Garczynski *et al.*, 1991). There are however no crystal proteins which are toxic but do not bind to the brush border of midgut epithelial cells. Upon closer examination, it appears that the interaction between a crystal protein, or its activated form, and the midgut epithelial cells is really a two-step process. Initial binding is a reversible process characterized by the affinity and the number of accessible binding sites. At a later stage binding can become irreversible probably due to the insertion of at least part of the crystal protein into the cell membrane. This insertion seems essential for the pore formation leading to cell lysis and establishes a correlation between the irreversible binding and toxicity of crystal proteins (Chen *et al.*, 1995; Liang *et al.*, 1995; Rajamohan *et al.*, 1995).

In recent years, the search for the binding molecules in the cell membrane of midgut cells has led to the identification of amino peptidases (Knight *et al.*, 1994; Sangadala *et al.*, 1994; Valaitis *et al.*, 1995) and a cadherin-like protein (Vadlamudi *et al.*, 1995) as possible receptors for the activated crystal proteins. It is unclear whether the role of the receptor is just to concentrate the δ-endotoxins at the cell membrane or whether the receptor actively takes part in the membrane insertion process either by modifying δ-endotoxins for membrane insertion or by becoming part of the pore itself. It is also not clear whether a single δ-endotoxin molecule can form a pore or whether some oligomerization is required (Walters *et al.*, 1994). The pores formed by δ-endotoxins have been reported to be potassium-selective (Sacchi *et al.*, 1986), non-selective (Knowles and Ellar, 1987), permeable to cations and small molecules (Hendrickx *et al.*, 1989) and cation- and anion-selective (Schwartz *et al.*, 1993). Perhaps these differences in observations can be explained by the fact that pore formation was studied with different cell types, different toxins and with different assay systems. Knowles and Dow (1993) postulate that upon pore formation in the columnar cells of the midgut epithelium, there is a rapid influx of ions which leads to depolarization of the apical membrane which in turn causes a rise of intracellular pH. The potassium pump stops and the columnar cells swell and lyse osmotically, leading to disruption of the integrity of the gut epithelium, starvation, and insect death. It appears that the specific action of activated crystal proteins is related to their pore-forming capacity which in turn is dependent on receptor binding.

Bt Plants

Engineering plants with crystal protein genes from *Bacillus thuringiensis* was one of the first projects in plant biotechnology. The project seemed technically feasible and commercially relevant. The use of *Bt* sprays had demonstrated their specificity and safety, the few *Bt* crystal proteins known at that time proved to be very active against certain important agronomic insect pests, the crystal proteins were encoded by single genes, and discovery programs indicated that *Bt* was an excellent source of proteins with improved and new pesticidal activities. Yet it has taken ten years to develop the technology to the point that seeds containing *Bt* crystal protein genes are becoming available to the growers. In the first experiments *Bt* crystal protein genes were cloned in a T-DNA vector and introduced into tobacco and tomato plants using *Agrobacterium tumefaciens* (Adang *et al.*, 1987; Barton *et al.*, 1987; Fischhoff *et al.*, 1987; Vaeck *et al.*, 1987). Vectors containing the entire coding sequence for the Cry1Aa, Cry1Ab or Cry1Ac protoxins were used to transform tobacco plants by Barton *et al.* (1987), Vaeck *et al.* (1987) and Adang *et al.* (1987), respectively. Different promoters were used but levels of crystal protein in the regenerated transgenic plants were at most a few ng/mg total plant protein and no significant insecticidal activity related to *Bt* was observed. There was some confusion concerning the possible toxic effects on plant cells by protoxins and Northern blot analysis revealed crystal protein gene mRNA species which were too short to encode toxic proteins (Adang *et al.*, 1987; Barton *et al.*, 1987). Since at that time it was already known that the insecticidal activity of crystal proteins resides in its N-terminal half, plants were also transformed with truncated crystal protein genes. Tobacco plants engineered with gene fragments encoding toxic fragments of Cry1Aa (Barton *et al.*, 1987) and Cry1Ab (Vaeck *et al.*, 1987) killed *Manduca sexta* larvae feeding from their leaves. Tomato plants with gene fragments encoding truncated forms of Cry1Ac proteins (Fischhoff *et al.*, 1987) proved protected from feeding damage by *Heliothis virescens* and *Helicoverpa zea* larvae. Cotton plants engineered with truncated crystal proteins controlled *Heliothis virescens* larvae (Perlak *et al.*, 1990). Using a truncated *cry1Ab* gene under the control of a wound-stimulated promoter (Velten *et al.*, 1984) a number of potato varieties were transformed in order to engineer resistance against potato tuber moth larvae (*Phthorimaea operculella*) (Peferoen *et al.*, 1990). Potato tuber moth is one of the most important pests of potato in storage and in the field, in tropical and subtropical areas, but also in the Andes and Himalayas. Larvae infesting the potato tubers not only cause a direct damage but open up the tubers to pathogens. Out of 166 transgenic potato lines, 24 proved highly insecticidal for potato tuber moth larvae. Tests whereby tubers were infested with neonate tuber moth larvae and stored for nearly two months demonstrated that transgenic tubers were fully protected from insect feeding damage. Field evaluations of transgenic tomato (Delannay *et al.*, 1989) and tobacco plants (Warren *et al.*, 1992) expressing truncated crystal protein genes showed substantial levels of insecticidal activity against important pest insects. Typically, in these experiments levels of crystal protein were at most 0.01% of the plant protein. Despite these promising results, it became clear that for many crops, in order to achieve total protection from agronomically important pests, levels of crystal protein expression had to be improved.

Levels of expression with chimeric crystal protein genes were typically lower than those obtained with other chimeric transgenes in plants indicating that the coding region of crystal protein genes inhibits efficient expression in plants. Crystal protein genes are A + T-rich genes while typical plant genes have a high G + C content. This means that codon usage is very different from a typical plant gene so that translation could be inefficient which in turn could lead to instability of the *cry* gene transcript. A + T-rich regions could cause erroneous splicing or premature 3′ end formation and AU motifs can destabilize mRNA transcripts (Koziel *et al.*, 1993b; van Aarssen *et al.*, 1995). Murray *et al.* (1991) examined expression of three full-length and truncated crystal protein genes and observed a high number of truncated transcripts. They concluded that the truncated transcripts were due to the instability of the crystal protein gene messengers rather than premature termination. Several research groups modified the crystal protein coding region, while preserving the amino acid sequence, in order to eliminate motifs which were speculated to limit expression levels.

Perlak *et al.* (1991) compared expression levels of two crystal proteins under the control of a (CaMV)35S promoter with different types of modifications in the coding region. First, in a *cry1Ab* and a *cry1Ac* gene, 62 of 1743 bases were changed to eliminate potential polyadenylation signals and ATTTA sequences in A + T-rich regions. In a second series of modifications, 390 of 1845 bases were changed to further reduce potential polyadenylation signals, to eliminate all ATTTA motifs, to increase overall G + C levels and to replace the bacterial codons with plant-preferred codons. Both with the partially and fully modified genes a larger number of insecticidal tobacco and tomato plants was obtained than with the wild type truncated genes. Levels of crystal protein were highest with the fully modified genes and were up to 0.3% of the plant proteins. No single region of the crystal protein genes was identified as being the major block to expression. Transcript levels were increased but not to the level which could be expected on the basis of the increase in protein levels. It was concluded that the modifications improved the translational efficiency in plants. In cotton, the modified genes also resulted in an increase in crystal protein levels with high levels of cotton bollworm control (Perlak *et al.*, 1990). In the field, cotton lines with a truncated modified *cry1Ab* gene showed high levels of protection from pink bollworm, cotton leaf perforator and beet armyworm larvae (Wilson *et al.*, 1992). In a similar approach, a coleopteran specific crystal protein gene was modified, introduced into potato, and high levels of resistance to Colorado potato beetles were obtained (Perlak *et al.*, 1993).

In a study on transcript formation of truncated *cry1Ab* genes, van Aarssen *et al.* (1995) demonstrated that initiation of transcription and the stability of the *cry* messenger was comparable to other transgenes such as *bar* or *nptII*. A region in the middle of the *cry1Ab* coding sequence was found to interfere with transcript accumulation. In this region, 63 translationally neutral substitutions were made in a stretch of some 110 nucleotides and the 28 N-terminal codons, not essential for toxicity, were deleted. Tobacco plants transformed with this *cry1Ab6* gene showed a 50-fold higher level of *cry1Ab* transcript compared to the unmodified plants. Transcript levels could be further increased by removal of two 5′ splice sites leading to the formation of full-length transcripts only. These experiments indicated that the expression problem was primarily situated in the nucleus where low levels of full-length *cry1Ab* transcripts were being produced. The *cry1Ab6* gene, under the control of a wound-stimulated TR2′ or a constitutive 35S promoter, was engineered into potato plants (Jansens *et al.*, 1995). Up to 90% of the transgenic plants showed high levels of control of potato tuber moth larvae, a dramatic increase compared to the approx. 15% obtained with the non-modified *cry1Ab* gene (Peferoen *et al.*, 1990). Furthermore, tunneling in leaves with the *cry1Ab6* gene was significantly lower than tunneling in leaves with the unmodified *cry1Ab* gene. In leaves of *cry1Ab6* plants only small tunnels were found which clearly indicates that larvae could not continue their development, while in plants with the unmodified *cry1Ab* gene tunnels of larger sizes were also observed. In a storage test 3 kg tubers of each potato line were exposed to 600 potato tuber moth eggs on day 0. At the end of the 66-day trial, potato tuber moth had completed two generations and 381 pupae with a total weight of 22 g were recovered from the control tubers. In the tubers from the *cry1Ab6* line no pupae were found and tubers appeared undamaged (Figure 2.3). Such transgenic tubers have also been evaluated in potato tuber storage houses in Portugal, and their protection from infestation by tuber moth larvae was confirmed. Overall, these results demonstrate that modifications to a specific area in the *cry1Ab* coding region can drastically improve levels of expression in plants.

Recently, an interesting alternative strategy to improve expression levels of *Bt* crystal protein genes was developed by targeting the gene to the chloroplasts

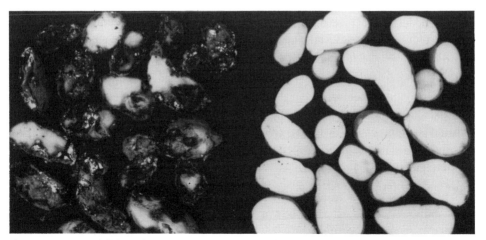

Figure 2.3   Control (left) and transgenic potato tubers after storage and exposure for 66 days to potato tuber moth larvae

Figure 2.4 Tunneling of European corn borer larvae in stalks of control corn (lower) and corn expressing the *cry9C* gene

(McBride *et al.*, 1995). A chimeric gene with an unmodified coding sequence for the Cry1Ac protoxin was integrated into the plastid genome by homologous recombination and was amplified to some 10 000 copies per cell producing Cry1Ac protoxin at 3–5% of the soluble protein.

Potato and cotton plants protected from Colorado potato beetle and the *Heliothis* complex, respectively, are expected to be marketed in 1996.

Bt Corn

European corn borer (*Ostrinia nubilalis*) is a major pest of corn in Europe and was introduced in North America in the beginning of this century. In the US Corn Belt the insect has up to three generations with first generations feeding primarily on leaves and later generations tunneling into the stalk. Besides physiological disruption of normal plant development the larvae may also cause broken stalks and dropped ears, and opens the plant to pathogens. Chemical control is only effective before larvae enter the stalks and requires extensive scouting. Some natural resistance to first generation European corn borer (ECB) larvae has been identified and hybrids with this natural host plant resistance have been developed. Nevertheless, annual yield loss by ECB infestation is estimated at US$800–900 million (Fox, 1995). The development of corn transformation systems allowed the introduction of foreign genes and the production of large numbers of transgenic corn lines (D'Halluin *et al.*, 1992; Fromm *et al.*, 1990; Gordon-Kamm *et al.*, 1990). Koziel *et al.* (1993a) modified a truncated *cry1Ab* gene by increasing G + C levels to some 65% and replacing bacterial codons with maize-preferred codons. The modified gene was engineered in maize by particle bombardment of immature maize embryos. Besides the constitutive (CaMV)35S promoter the *cry1Ab* gene was also expressed under the control of the phosphoenolpyruvate carboxylase (PEPC) and a pollen specific promoter. Levels of Cry1Ab protein expressed in leaves with the *cry1Ab* gene under the control of the PEPC promoter proved less variable between lines than with the (CaMV)35S promoter, ranging from 1.0 to 1.8 μg/mg plant protein and increasing up to 4 μg/mg plant protein. Field trials with corn lines expressing high levels of Cry1A proteins showed high levels of protection from foliar feeding and stalk tunneling by first and second generations of European corn borer larvae, respectively (Armstrong *et al.*, 1995; Koziel *et al.*, 1993a). Recently, corn lines were engineered with the *cry9C* gene, a gene encoding a crystal protein which is highly toxic to an unusual broad range of lepidopteran larvae, including important corn pests such as European corn borer, black cutworm (*Agrotis ypsilon*) and South-western corn borer (*Diatraea grandiosella*) (Lambert *et al.*, 1996). The corn plants tested in greenhouse and field showed high levels of protection from feeding damage by European corn borer and black cutworm larvae and very low levels of tunneling were observed in the stalks (Figure 2.4). An issue often neglected is that currently, plant engineering in general and corn transformation in particular is still a lengthy and expensive process. Transformation is based on random integration of the trans-genes in the genome whereby the expression pattern of the endogenous and the transgenes can be dramatically influenced. Hundreds of transgenic events must be generated in order to select those lines which not only have the phenotype as encoded by the transgene but also have the expected agronomic performance in the field. Hybrid corn seed containing a *Bt* gene to reduce yield loss by European corn borer damage is scheduled for market introduction in 1996.

Bt Resistance

The major concern related to the use of crops expressing insecticidal proteins is the possibility that insects develop resistance to these proteins (McGaughey and Whalon, 1992). Insects have the capacity to develop resistance to all kinds of insecti-cides with more than 500 insect species reported to be resistant to one or another type of insecticide (Georghiou and Lagunes-Tejeda, 1991). In insects, resistance is a pre-adaptive phenomenon which develops by selection of rare individuals in a popu-lation which can survive a certain insecticide treatment.

    Bt resistance has been studied mostly in insects cultured in the lab and exposed to *Bt* crystal protein. Such laboratory selection experiments can indicate the reper-toire of resistance mechanisms available in a certain population and are essential to study inheritance of resistance genes. However, laboratory populations usually have a smaller genetic basis and may not contain all resistance genes as found in certain field populations. There are in fact several examples of selection experiments with *Bt* proteins on laboratory populations of diamondback moths which failed to select *Bt* resistant insects, while this very insect was the first to develop resistance to *Bt* applied as a biopesticide in the field. Also, the laboratory is a very protective environment compared to the field. Resistance mechanisms which are linked to certain fitness costs may only become selected in contained conditions. Laboratory selection experiments do not predict if resistance will develop in the field or which resistance mechanisms will be selected.

    The first substantial levels of resistance to *Bt* were observed in Indian meal moth populations from grain bins regularly treated with *Bt* formulations (McGaughey,

1985). On additional exposures in the laboratory to formulated *Bt*, levels of resistance could be further increased. When the selection pressure was removed, resistance levels decreased. Resistance proved to be partially recessive and to involve few loci (McGaughey and Beeman, 1988). A study on the resistance mechanism showed that resistance was caused by a 50-fold lower affinity of the target site on the midgut brush border for the Cry1Ab protein (Van Rie *et al.*, 1990b), one of the crystal proteins in the biopesticidal product used to control and further select the insects. The insects resistant to the Cry1Ab protein proved more susceptible to Cry1Ca, a protein which the insects had not been exposed to. The increase in susceptibility for the Cry1Ca protein was related to an increase in binding sites for the Cry1Ca protein on their midgut epithelial cells (Van Rie *et al.*, 1990b).

Heliothis virescens has been used in several laboratory selection experiments resulting in colonies with different levels of resistance to *Bt* proteins (Gould *et al.*, 1992; MacIntosh *et al.*, 1991; Stone *et al.*, 1989). In one population binding of the crystal proteins was changed but could not explain the level of resistance observed (MacIntosh *et al.*, 1991). In another laboratory selected population cross-resistance to a Cry2Aa protein was observed (Gould *et al.*, 1992). Binding of crystal proteins proved unaltered. Starting from a population of Colorado potato beetles collected from a field treated with formulated Cry3A products, continued selection in the laboratory produced a strain resistant to the Cry3A protein (Whalon *et al.*, 1993). Almond moth (*Cadra cautella*) (McGaughey and Beeman, 1988), cottonwood leaf beetle (*Chrysomela scripta*) (Bauer *et al.*, 1994), cabbage looper (*Trichoplusia ni*) (Estada and Ferré, 1994), cotton leafworm (*Spodoptera littoralis*) (Müller-Cohn *et al.*, 1994), beet armyworm (*Spodoptera exigua*) (Moar *et al.*, 1995) and European corn borer (*Ostrinia nubilalis*) (Bolin *et al.*, 1995) populations resistant to certain *Bt* proteins have been obtained by selection in the laboratory.

So far, diamondback moth is the only insect which has developed resistance to *Bt* applied in the field (Tabashnik *et al.*, 1990). From different islands in Hawaii, populations were collected from areas which had been treated with different regimes of *Bt* sprays. Diamondback moth populations from *Bt* treated areas were less susceptible to *Bt* than the untreated laboratory population. Populations from heavily treated fields proved more resistant to *Bt* than insects from the less intensively treated areas. Furthermore, continued selection in the laboratory resulted in even higher levels of resistance (Tabashnik *et al.*, 1991). Resistance proved to be determined by a change at the Cry1Ac binding site whereby the binding of Cry1Ac proteins to the midgut epithelial cells was dramatically reduced (Tabashnik *et al.*, 1994a). A diamondback moth population collected from a field in the Philippines frequently treated with *Bt* sprays also showed a high level of resistance to Cry1Ab, one of the proteins in the sprays. The same insect population proved fully susceptible to other crystal proteins such as Cry1Ba and Cry1Ca (Ferré *et al.*, 1991). Again, resistance was caused by a change in the target site for the Cry1Ab protein, while the receptors for Cry1Ba and Cry1Ca were fully functional. Resistance related to an altered target site has also been observed in diamondback moth populations in Florida (Tang *et al.*, 1996). It therefore appears that a change in the target site is one of the preferred mechanisms of resistance to *Bt* crystal proteins used by insects. It is clear that almost at every step in the mechanism of action there may be a mechanism which would inhibit or block the activity of crystal proteins. Perhaps insects can change the characteristics of their midgut fluids so that crystal proteins become precipitated; or insects may have gut proteases which can destroy crystal proteins; or the peritrophic membrane can trap the crystal proteins on their passage to the midgut epithelium. Insects may also have mechanisms which would allow a rapid repair of damage inflicted by crystal proteins. It is unclear whether such mechanisms are available and whether they can become selected in a field population.

Development of resistance is dependent on many different parameters. A high frequency of the resistance gene in a population which is being exposed to an insecticide will of course lead to resistance more rapidly than when present at low frequencies. Typically, it is assumed that in unexposed populations resistance genes occur at fairly low frequencies. There are no data available on initial *Bt* resistance gene frequencies in insect populations, but in many computer simulations a frequency of 10^{-3} is postulated. Cross-resistance to other crystal proteins may also jeopardize control strategies when different crystal proteins are being used (Gould, 1994). In many insect populations selected for *Bt* resistance, no cross-resistance has been observed (Estada and Ferré, 1994; Ferré *et al.*, 1991; Tabashnik *et al.*, 1993; Van Rie *et al.*, 1990b). There are some cases where some levels of cross-resistance have been observed (Gould *et al.*, 1992; Moar *et al.*, 1995; Müller-Cohn *et al.*, 1994; Tabashnik *et al.*, 1994b). Often resistance to one of the Cry1A crystal proteins leads to resistance to other Cry1A proteins as well. This is no surprise since Cry1A proteins share binding sites and as resistance is often based on a modification in the receptor, cross-resistance seems inevitable. The genetics of the resistance will also to a great extent determine its development under selection. Recessive genes only exert their phenotypic effect when in a homozygous state and spread less rapidly through the population than dominant resistance genes. In diamondback moth, resistance to *Bt* crystal proteins seems autosomally inherited, involving few loci and is partially or completely recessive (Tabashnik *et al.*, 1992). In *Heliothis virescens*, resistance was autosomally inherited and was additive (Gould *et al.*, 1992; Sims and Stone, 1991). Little is known about frequencies of resistance genes in insect populations which have not been exposed to selective agents. Fitness costs associated with resistance maintains resistance gene frequencies at low levels. In many of the *Bt* resistant populations the level of resistance decreases when the population is no longer exposed to the crystal protein, indicating that resistance to *Bt* has a fitness cost (Groeters *et al.*, 1993; McGaughey and Beeman, 1988; Tabashnik *et al.*, 1991, 1994a). Insects with a high number of generations will show a more rapid development of resistance than insects with only one generation in a growing season (Georghiou and Taylor, 1986). In populations with a low influx of genes, resistance may become more rapidly fixed, while in populations with a continuous influx of novel genes, resistance may be continuously diluted. Resistance genes can also be transmitted from one population to another. The way insecticides are applied also has a dramatic effect on the resistance development. A topical application of *Bt* or expression of *Bt* crystal proteins in a plant imposes different selection pressures on an insect population.

Resistance Management

Management of resistance to *Bt* products can benefit from the extensive theoretical and empirical background obtained with other insecticides. However, *Bt* products are far more specific than traditional insecticides and have no direct toxic effects upon specialist and generalist natural enemies (Johnson and Gould, 1992). *Bt* sprays have a shorter lifespan than synthetic insecticides. *Bt* proteins expressed by the crop can be present during the entire growing season and only insects feeding on the plant tissue are exposed to the *Bt* protein. It is often assumed that resistance might develop more rapidly to plants expressing *Bt* crystal proteins than to plants sprayed with a formulated *Bt* biopesticide. In 1988 companies developing *Bt* biopesticidal sprays and transgenic plants formed the *Bt* Management Working Group to promote research on the judicial use of *Bt* products.

The goal of resistance management is to keep the frequency of resistance genes in a population below the level which would require an increased insecticide usage in order to maintain the same levels of pest control. Resistance management strategies are mostly based on computer simulations which estimate the impact of certain resistance management tactics on resistance development (Gould, 1994; Mallet and Porter, 1992). Resistance management tactics for *Bt* plants are different from resistance management strategies for *Bt* sprays, since *Bt* sprays act upon an insect population in a very different way than *Bt* plants. *Bt* sprays have a fairly short persistence in the field and also their coverage of the field is incomplete. This means that in a field sprayed with a *Bt* product there is a whole range of dosages. At some spots in the field or on the crop there is no *Bt* while at other sites there is sufficient *Bt* to kill even the higher instar larvae and perhaps larvae which have some level of resistance. With *Bt* plants, depending on the promoters used to direct the expression of the crystal protein gene, a high level of the crystal protein can be achieved for the entire growing season in the entire crop. It is also possible to target the expression of the crystal protein genes to certain tissues of the crop so that only certain parts of the plant are fully protected. The expression can even be triggered either by the feeding of the insect itself (Peferoen *et al.*, 1990) or by the application of a chemical (Williams *et al.*, 1992).

There are different strategies which at least in theory should retard the spread of resistance genes through a population. One of the strategies envisaged with *Bt* plants is the expression of a dose high enough to kill homozygous resistant insects and thereby changing the crop to a non-host. This has never been achieved with a sprayable product, but could be possible with *Bt* plants since in plants *Bt* proteins can be expressed at levels which are 500-fold higher than the LC_{50} (Koziel *et al.*, 1993a). A Colorado potato beetle population resistant to a formulated *Bt* spray could not survive on potato plants expressing the same protein (Whalon and Wierenga, 1994). The ultra-high-dose strategy is not really being aimed at; it is rather something which is hoped for. In most *Bt* plant engineering programs, transgenic plants which provide the highest levels of *Bt* crystal proteins and the highest levels of insect control are selected. It remains to be seen whether for certain insects a crop can be changed into a non-host, driving the insects away from the crop.

Another strategy aims at removing most of the resistance genes from the population. Most of the resistance genes in a population are present in a heterozygous condition. In most cases *Bt* resistance is partially recessive and it is expected that plants can be engineered so that they produce levels which consistently kill all heterozygous resistant insects. Although this strategy would eliminate most resistance genes from the population, when applied on its own, it would still lead to the development of resistance. *Bt* plants killing susceptible and heterozygous resistant insects would permit survival of homozygous resistant insects and a rapid fixation of resistance in the population. A regular influx of wild type genes could dilute the homozygous resistance to its heterozygous form, so that these insects can again be controlled by the *Bt* plants. This could be accomplished by maintaining plants in the vicinity of the crop which do not express *Bt* crystal proteins. In these refuge areas, susceptible insects could survive and mate with insects which have been selected by the *Bt* plants. Offspring of such crosses would be heterozygous for the resistance gene and effectively controlled by the *Bt* plants. The refuge strategy could continuously dilute the resistance developing in an insect population (Alstad and Andow, 1995; Roush, 1994). Shelton reported (at the 28th Annual Meeting of the Society for Invertebrate Pathology, Cornell University, Ithaca, New York, 1995) on the retardation of resistance development to *Bt* expressing Brassica plants in a diamondback moth population by maintaining 5–20% of the plants as a refuge allowing the survival of susceptible insects. The high-dose–refuge strategy is perceived by most experts in this area as one of the most effective resistance management strategies. Probably in some cases, insect pests feed on plants other than the *Bt* crop and mate with insects selected on the *Bt* crop so that there is no need for the creation of additional refugia. In other cases, refugia will have to be developed in the vicinity of the crop and for each pest–crop complex. The number of susceptible insects needed and the size of the refuge area will have to be determined (Gould, 1994). Implementation will require the commitment by extension people, crop consultants and farmers (Kennedy and Whalon, 1995).

Another valuable strategy is to target the insect pest with multiple insecticides, acting at different targets. This can be accomplished with *Bt* crystal proteins since there are several crystal proteins which act against the same insect. Many of these crystal proteins bind to different receptors and therefore are less prone to cross-resistance (Van Rie *et al.*, 1989, 1990a). There are well documented cases where insects have developed resistance to a certain crystal protein by changing their receptor for the crystal protein, while being fully susceptible to other crystal proteins (Ferré *et al.*, 1991; Tabashnik *et al.*, 1993; Van Rie *et al.*, 1990b). The multiple target strategy exploits the fact that while, in a population, insects homozygous for one resistance gene are rare, insects homozygous for multiple resistance genes are extremely rare. When using multiple crystal proteins, even insects homozygous for one or two resistance genes but heterozygous for another resistance gene would still be controlled by crops expressing multiple *Bt* crystal proteins. The prime condition is that there is very low cross-resistance for the different crystal proteins. The combination of a *Bt* crystal protein with another insecticidal protein would also provide a valuable multiple target strategy. It is however essential that each of the insecticidal proteins used on its own has a very high insecticidal activity. An important aspect of the multiple target strategy is that the tactic is incorporated into the crop which makes it more robust and less prone to errors (Kennedy and Whalon, 1995). This strategy can also be combined with other resistance management strategies and will increase the effectiveness of the resistance management (Roush, 1994). Planting mixtures of *Bt* seeds and *Bt*-free seeds and expression of *Bt* proteins in certain organs of the plants only, has been proposed as a resistance management tactic but the effectiveness is debated. One of the problems envisaged is that these strategies would allow survival of a large number of heterozygously resistant insects.

It is expected that each pest–crop complex may require a specific implementation of certain resistance management strategies. Virtually all strategies proposed today are based on theoretical models. Experience with *Bt* crops grown under different agronomic conditions is essential to define what is required in terms of resistance management.

Concluding Remarks

Bacillus thuringiensis has been a source of a wide variety of crystal proteins each with its own spectrum of activity. Although it may become more difficult, new crystal proteins will be found. Knowledge of the structure of δ-endotoxin combined with information on the function of certain structures within the molecule is expected to lead to the design of crystal proteins with improved or new activities. In some applications, *Bt* plants will compete with existing pest control strategies and may generate substantial economic and ecological benefits. In other cases, like *Bt* corn, the technology provides solutions to problems that could not be managed effectively before and thus allow growers to harvest more of the yield potential of the crop. We are at the very beginning of the commercial use of *Bt* crops and much remains to be learned on how to deploy such crops in a sustainable manner. Obviously, *Bt* plants are not a stand-alone product. *Bt* plants are an additional tool to control insect pests and should be integrated with other pest management strategies.

References

Adang, L. F., Staver, M. J., Rocheleau, T. A., Leighton, J., Barker, R. F. and Thompson, D. V. (1985) Characterized full-length and truncated plasmid clones of the crystal protein of *Bacillus thuringiensis* subsp. *kurstaki* HD-73 and their toxicity to *Manduca sexta, Gene* **36**, 289–300.

Adang, M. J., Firoozabady, E., Klein, J., Deboer, D., Sekar, V., Kemp, J. D., Murray, E., Rocheleau, T. A., Rashka, K., Staffield, G. *et al.* (1987) Application of a *Bacillus thuringiensis* crystal protein for insect control. In: Arntzen, C. J. and Ryan, C. (Eds), *Molecular Strategies for Crop Protection*, New York: Alan R. Liss, pp. 345–353.

Agaisse, H. and Lereclus, D. (1994) Expression in *Bacillus subtilis* of the *Bacillus thuringiensis cryIIIA* toxin gene is not dependent on a sporulation-specific sigma factor and is increased in a *spo0* mutant, *J. Bacteriol.* **176**, 4734–4741.

Alstad, D. N. and Andow, D. A. (1995) Managing the evolution of insect resistance to transgenic plants, *Science* **268**, 1894–1896.

Armstrong, C. L., Parker, G. B., Pershing, J. C., Brown, S. M., Sanders, P. R., Duncan, D. R., Stone, T., Dean, D. A., DeBoer, D. L., Hart, J. *et al.* (1995) Field evaluation of European corn borer control in progeny of 173 transgenic corn events expressing an insecticidal protein from *Bacillus thuringiensis, Crop Sci.* **35**, 550–557.

Aronson, A. I., Wu, D. and Zhang, C. (1995) Mutagenesis of specificity and toxicity regions of a *Bacillus thuringiensis* protoxin gene, *J. Bacteriol.* **177**, 4059–4065.

Barton, K., Whiteley, H. and Yang, N.-S. (1987) *Bacillus thuringiensis* δ-endotoxin in transgenic *Nicotiana tabacum* provides resistance to lepidopteran insects, *Plant Physiol.* **85**, 1103–1109.

Bauer, L. S., Koller, C. N., Miller, D. L. and Hollingworth, R. M. (1994) Laboratory selection of cottonwood leaf beetle, *Chrysomela scripta*, to *Bacillus thuringiensis* var. *tenebrionis* delta-endotoxin, *Abstracts of the XXVIIth Annual Meeting of the Society for Invertebrate Pathology, Montpellier, France*, p. 68.

Belfiore, C. J., Vadlamudi, R. K., Osman, Y. A. and Bulla, L. A. Jr (1994) A specific binding protein from *Tenebrio molitor* for the insecticidal toxin of *Bacillus thuringiensis* subsp. *tenebrionis, Biochem. Biophys. Res. Commun.* **200**, 359–364.

Berliner, E. (1915) Uber die Schlaffsucht der Mehlmottenraupe (*Ephestia kuehniella* Zell), und ihren Erreger *Bacillus thuringiensis* n. sp., *Zeitschrift für angewandtes Entomologie* **2**, 29–56.

Bietlot, H. P. L., Vishnubhatla, I., Carey, P. R., Pozsgay, M. and Kaplan, H. (1990) Characterization of the cysteine residues and disulfide linkages in the protein crystal of *Bacillus thuringiensis, Biochem. J.* **267**, 309–315.

Bietlot, H. P. L., Schernthaner, J. P., Milne, R. E., Clairmont, F. R., Bhella, R. S. and Kaplan, H. (1993) Evidence that the CryIA crystal protein from *Bacillus thuringiensis* is associated with DNA, *J. Biol. Chem.* **268**, 8240–8245.

Bolin, P. C., Hutchison, W. D. and Andow, D. A. (1995) Selection for resistance to *Bacillus thuringiensis* CryIA(c) endotoxin in a Minnesota population of European corn borer, *Abstracts of the 28th Annual Meeting of the Society for Invertebrate Pathology, Cornell University, Ithaca, New York, USA*, p. 8.

Bosch, D., Schipper, B., van der Kleij, H., de Maagd, R. A. and Stiekema, W. J. (1994) Recombinant *Bacillus thuringiensis* proteins with new properties: possibilities for resistance management, *Bio/Technology* **12**, 915–918.

Bourgouin, C., Delecluse, A., Ribier, J., Klier, A. and Rapoport, G. (1988) A *Bacillus thuringiensis* subsp. *israelensis* gene encoding a 125-kilodalton larvicidal polypeptide is associated with inverted repeat sequences, *J. Bacteriol.* **170**, 3575–3583.

Bradley, D., Harkey, M. A., Kim, M.-K., Biever, K. D. and Bauer, L. S. (1995) The insecticidal CryIB crystal protein of *Bacillus thuringiensis* has dual specificty to coleopteran and lepidopteran larvae, *J. Invert. Pathol.* **65**, 162–173.

Bravo, A., Hendrickx, K., Jansens, S. and Peferoen, M. (1992a) Immunocytochemical analysis of specific binding of *Bacillus thuringiensis* insecticidal crystal proteins to lepidopteran and coleopteran midgut membranes, *J. Invert. Pathol.* **60**, 247–253.

Bravo, A., Jansens, S. and Peferoen, M. (1992b) Immunocytochemical localization of *Bacillus thuringiensis* insecticidal crystal proteins in intoxicated insects, *J. Invert. Pathol.* **60**, 237–246.

Brizzard, B. L. and Whiteley, H. R. (1988) Nucleotide sequence of an additional crystal protein gene cloned from *Bacillus thuringiensis* subsp. *thuringiensis, Nucl. Acid Res.* **16**, 2723–2724.

Brown, K. L. and Whiteley, H. R. (1988) Isolation of a *Bacillus thuringiensis* RNA polymerase capable of transcribing crystal protein genes, *Proc. Natl. Acad. Sci. USA* **85**, 4166–4170.

Brown, K. L. and Whiteley, H. R. (1990) Isolation of the second *Bacillus thuringiensis* RNA polymerase that transcribes from a crystal protein gene promoter, *J. Bacteriol.* **172**, 6682–6688.

Brown, K. L. and Whiteley, H. R. (1992) Molecular characterization of two novel crystal protein genes from *Bacillus thuringiensis* subsp. *thompsoni, J. Bacteriol.* **174**, 549–557.

Carozzi, N. B., Kramer, V. C., Warren, G. W., Evola, S. and Koziel, M. G. (1991) Prediction of insecticidal activity of *Bacillus thuringiensis* strains by polymerase chain reaction product profiles, *Appl. Environ. Microbiol.* **57**, 3057–3061.

Chambers, J. A., Jelen, A., Gilbert, M. P., Jany, C. S., Johnson, T. B. and Gawron-Burke, C. (1991) Isolation and characterization of a novel insecticidal crystal protein gene from *Bacillus thuringiensis* subsp. *aizawai, J. Bacteriol.* **173**, 3966–3976.

Chen, X. J., Lee, M. K. and Dean, D. H. (1993) Site-directed mutations in a highly conserved region of *Bacillus thuringiensis* δ-endotoxin affect inhibition of short circuit current across *Bombyx mori* midguts, *Proc. Natl. Acad. Sci. USA* **90**, 9041–9045.

Chen, X. J., Curtiss, A., Alcantara, E. and Dean, D. H. (1995) Mutations in domain I of *Bacillus thuringiensis* δ-endotoxin CryIAb reduce the irreversible binding of toxin to *Manduca sexta* brush border membrane vesicles, *J. Biol. Chem.* **270**, 6412–6419.

Choma, C. T., Surewicz, W. K., Carey, P. R., Pozsgay, M., Raynor, T. and Kaplan, H. (1990) Unusual proteolysis of the protoxin and toxin of *Bacillus thuringiensis* – structural implications, *Eur. J. Biochem.* **189**, 523–527.

Chungjatupornchai, W., Höfte, H., Seurinck, J., Angsuthanasombat, C. and Vaeck, M. (1988) Common features of *Bacillus thuringiensis* toxins specific for Diptera and Lepidoptera, *Eur. J. Biochem.* **173**, 9–16.

39

Crickmore, N., Zeigler, D. R., Feitelson, J., Schnepf, E., Lambert, B., Lereclus, D., Baum, J. and Dean, D. H. (1996) Revision of the nomenclature for *Bacillus thuringiensis* cry genes (in preparation).

Cummings, C. E., Armstrong, G., Hodgman, T. C. and Ellar, D. J. (1994) Structural and functional studies of a synthetic peptide mimicking a proposed membrane inserting region of a *Bacillus thuringiensis* δ-endotoxin, *Mol. Memb. Biol.* **11**, 87–92.

de Barjac, H. and Bonnefois, A. (1962) Essai de classification biochimique et sérologique de 24 souches de Bacillus du type B. thuringiensis, *Entomophaga* **1**, 5–31.

Delannay, X., LaVallee, B. J., Proksch, R. K., Fuchs, R. L., Sims, S. R., Greenplate, J. T., Marrone, P. G., Dodson, R. B., Augustine, J. J., Layton, J. G. and Fischhoff, D. A. (1989) Field performance of transgenic tomato plants expressing the *Bacillus thuringiensis* var. *kurstaki* insect control protein, *Bio/Technology* **7**, 1265–1269.

Denolf, P., Jansens, S., Peferoen, M., Degheele, D. and van Rie, J. (1993) Two different *Bacillus thuringiensis* δ-endotoxin receptors in the midgut brush border membrane of the European corn borer, *Ostrinia nubilalis* (Hübner) (Lepidoptera: Pyralidae), *Appl. Environ. Microbiol.* **59**, 1828–1837.

D'Halluin, K., Bonne, E., Bossut, M., De Beuckeleer, M. and Leemans, J. (1992) Transgenic maize plants by tissue electroporation, *The Plant Cell* **4**, 1495–1505.

Donovan, W. P. and Tan, Y. (1994) *Bacillus thuringiensis* cryET4 and cryET5 genes and proteins toxic to lepidopteran insects, US Patent Number 5 322 687.

Donovan, W. P., Dankocsik, C. and Gilbert, M. P. (1988) Molecular characterization of a gene encoding a 72-kilodalton mosquito-toxic crystal protein from *Bacillus thuringiensis* subsp. *israelensis*, *J. Bacteriol.* **170**, 4732–4738.

Donovan, W. P., Gonzalez, J. M., Gilbert, M. P. and Dankocsik, C. (1989) Amino acid sequence and entomocidal activity of the P2 crystal protein, *J. Biol. Chem.* **264**, 4740.

Donovan, W. P., Rupar, M. J., Slaney, A. C., Malvar, T., Gawron-Bruke, M. C. and Johnston, T. B. (1992) Characterization of two genes encoding *Bacillus thuringiensis* insecticidal crystal proteins toxic to Coleoptera species, *Appl. Environ. Microbiol.* **58**, 3921–3927.

Dulmage, H. T. (1970) Insecticidal activity of HD-1, a new isolate of *Bacillus thuringiensis* var. *alesti*, *J. Invert. Pathol.* **15**, 232–239.

Escriche, B., Martinez-Ramirez, A. C., Real, M. D., Silva, F. J. and Ferré, J. (1994) Occurrence of three different binding sites for *Bacillus thuringiensis* δ-endotoxins in the midgut brush border membrane of the potato tuber moth, *Phthorimaea operculella* (Zeller), *Arch. Insect Biochem. Physiol.* **26**, 315–327.

Estada, U. and Ferré, J. (1994) Binding of insecticidal crystal proteins of *Bacillus thuringiensis* to the midgut brush border of the cabbage looper, *Trichoplusia ni* (Hübner) (Lepidoptera: Noctuidae), and selection for resistance to one of the crystal proteins, *Appl. Environ. Microbiol.* **60**, 3840–3846.

Ferré, J., Real, M. D., Van Rie, J., Jansens, S. and Peferoen, M. (1991) Resistance to the *Bacillus thuringiensis* bioinsecticide in a field population of *Plutella xylostella* is due to a change in a midgut membrane receptor, *Proc. Natl. Acad. Sci. USA* **88**, 5119–5123.

Fischhoff, D. A., Bowdish, K. S., Perlak, F. J., Marrone, P. G., McCormick, S. M., Niedermeyer, J. G., Dean, D. A., Kusano-Kretzmer, K., Mayer, E. J., Rochester, D. E. *et al.* (1987) Insect tolerant transgenic tomato plants, *Bio/Technology* **5**, 807–813.

Foncerrada, L., Sick, A. J. and Payne, J. M. (1992) Novel coleopteran-active *Bacillus thuringiensis* isolate and a novel gene encoding a coleopteran-active toxin, Patent EP 0 498537.

Fox, J. L. (1995) EPA okays *Bt* corn; USDA eases plant testing, *Bio/Technology* **13**, 1035–1036.

Fromm, M. E., Morrish, F., Armstrong, C., Williams, R., Thomas, J. and Klein, T. (1990) Inheritance and expression of chimeric genes in the progeny of transgenic maize plants, *Bio/Technology* **8**, 833–839.

Garczynski, S. F., Crim, J. W. and Adang, M. J. (1991) Identification of putative insect brush border membrane-binding molecules specific to *Bacillus thuringiensis* δ-endotoxin by protein blot analysis, *Appl. Environ. Microbiol.* **57**, 2816–2820.

Gazit, E. and Shai, Y. (1995) The assembly and organization of the α5 and α7 helices from the pore-forming domain of *Bacillus thuringiensis* δ-endotoxin, *J. Biol. Chem.* **270**, 2571–2578.

Gazit, E., Bach, D., Kerr, I. D., Sansom, M. S. P., Chejanovsky, N. and Shai, Y. (1994) The α5 segment of *Bacillus thuringiensis* δ-endotoxin: *in vitro* activity, ion channel formation and molecular modeling, *Biochem. J.* **304**, 895–902.

Ge, A. Z., Shivarova, N. I. and Dean, D. H. (1989) Location of the *Bombyx mori* specificity domain on a *Bacillus thuringiensis* δ-endotoxin protein, *Proc. Natl. Acad. Sci.* **86**, 4037–4041.

Georghiou, G. P. and Lagunes-Tejeda, A. (1991) *The Occurrence of Resistance to Pesticides in Arthropods*, Rome: Food and Agriculture Organization of the United Nations.

Georghiou, G. P. and Taylor, C. E. (1986) Factors influencing the evolution of resistance, In: *Pesticide Resistance: Strategies and Tactics for Management*, Washington: National Academic Press, pp. 157–169.

Goldberg, L. J. and Margalit, J. (1977) A bacterial spore demonstrating rapid larvicidal activity against *Anopheles serengetii, Uranotaenia unguiculata, Culex univittatus, Aedes aegypti* and *Culex pipiens, Mosquito News* **37**, 355–358.

Gonzalez, J. M., Brown, B. and Carlton, B. (1982) Transfer of *Bacillus thuringiensis* plasmids coding for δ-endotoxin among strains of *B. thuringiensis* and *B. cereus, Proc. Natl. Acad. Sci.* **79**, 6951–6955.

Gordon-Kamm, W. J., Spencer, T. M., Mangano, M. L., Adams, T. R., Daines, R. J., Start, W. G., O'Brien, J. V., Chambers, S. A., Adams, W. R., Willetts, N. G. *et al.* (1990) Transformation of maize cells and regeneration of fertile transgenic plants, *The Plant Cell* **2**, 603–618.

Gould, F. (1994) Potential problems with high-dose strategies for pesticidal engineered crops, *Biocontrol Sci. Technol.* **4**, 451–461.

Gould, F., Martinez-Ramirez, A., Anderson, A., Ferré, J., Silva, F. J. and Moar, W. (1992) Broad-spectrum resistance to *Bacillus thuringiensis* toxins in *Heliothis virescens, Proc. Natl. Acad. Sci.* **89**, 7986–7990.

Grochulski, P., Masson, L., Borisova, S., Pustzai-Carey, M., Schwartz, J.-L., Brousseau, R. and Cygler, M. (1995) *Bacillus thuringiensis* CryIA(a) insecticidal toxin: crystal structure and channel formation, *J. Mol. Biol.* **254**, 447–464.

Groeters, F. R., Tabashnik, B. E., Finson, N. and Johnson, M. W. (1993) Resistance to *Bacillus thuringiensis* affects mating success of the diamondback moth (Lepidoptera: Plutellidae), *J. Econ. Entomol.* **86**, 1035–1039.

Haider, M. Z., Knowles, B. and Ellar, D. J. (1986) Specificity of *Bacillus thuringiensis* var. *colmeri* insecticidal delta-endotoxin by differential processing of the protoxin by larval gut proteases, *Eur. J. Biochem.* **156**, 531–540.

Hannay, C. L. (1953) Crystalline inclusions in aerobic spore-forming bacteria, *Nature* **172**, 1004.

Hannay, C. L. and Fitz-James, P. (1955) The protein crystals of *Bacillus thuringiensis* Berliner, *Can. J. Microbiol.* **1**, 674–710.

Hendrickx, K., De Loof, A. and Van Mellaert, H. (1989) Effects of *Bacillus thuringiensis* delta-endotoxin on the permeability of brush border membrane vesicles from tobacco hornworm (*Manduca sexta*) midgut, *Compar. Biochem. Physiol.* **95C**, 241–245.

Herrnstadt, C., Gilroy, T. E., Sobieski, D. A., Bennett, B. D. and Gaertner, F. H. (1987) Nucleotide sequence and deduced amino acid sequence of a coleopteran-active delta-endotoxin gene from *Bacillus thuringiensis* subsp. *san diego, Gene* **57**, 37–46.

Hofmann, C., Vanderbruggen, H., Höfte, H., Van Rie, J., Jansens, S. and Van Mellaert, H. (1988) Specificity of *Bacillus thuringiensis* δ-endotoxins is correlated with the presence of high-affinity binding sites in the brush border membrane of target insect midguts, *Proc.*

Natl. Acad. Sci. **85**, 7844–7848.

Höfte, H. and Whiteley, H. R. (1989) Insecticidal crystal proteins of *Bacillus thuringiensis*, *Microbiol. Rev.* **53**, 242–255.

Höfte, H., Soetaert, P., Jansens, S. and Peferoen, M. (1990) Nucleotide sequence and deduced amino acid sequence of a new Lepidoptera-specific crystal protein gene from *Bacillus thuringiensis*, *Nucl. Acid Res.* **18**, 5545.

Höfte, H., Seurinck, J., Van Houtven A. and Vaeck, M. (1987) Nucleotide sequence of a gene encoding an insecticidal protein of *Bacillus thuringiensis* var. *tenebrionis* toxic against Coleoptera, *Nucl. Acid Res.* **15**, 7183.

Höfte, H., Van Rie, J., Jansens, S., Van Houtven, A., Vanderbruggen, H. and Vaeck, M. (1988) Monoclonal antibody analysis and insecticidal spectrum of three types of lepidopteran-specific insecticidal crystal proteins of *Bacillus thuringiensis*, *Appl. Environ. Microbiol.* **54**, 2010–2017.

Höfte, H., De Greve, H., Seurinck, J., Jansens, S., Mahillon, J., Ampe, C., Vandekerckhove, J., Vanderbruggen, H., Van Montagu, M., Zabeau, M. and Vaeck, M. (1986) Structural and functional analysis of a cloned delta endotoxin of *Bacillus thuringiensis* Berliner 1715, *Eur. J. Biochem.* **161**, 273–280.

Honée, G., VanderSalm, T. and Visser, B. (1988) Nucleotide sequence of crystal protein gene isolated from *B. thuringiensis* subspecies entomocidus 60.5 coding for a toxin highly active against *Spodoptera* species, *Nucl. Acid Res.* **16**, 6240.

Honée, G., Convents, D., Van Rie, J., Jansens, S., Peferoen, M. and Visser, B. (1991) The C-terminal domain of the toxic fragment of a *Bacillus thuringiensis* crystal protein determines receptor binding, *Mol. Microbiol.* **5**(11), 2799–2806.

Jansens, S., Cornelissen, M., De Clercq, R., Reynaerts, A. and Peferoen, M. (1995) *Phthorimaea operculella* (Lepidoptera: Gelechiidae) resistance in potato by expression of the *Bacillus thuringiensis* CryIA(b) insecticidal crystal protein, *J. Econ. Entomol.* **88**, 1469–1476.

Jarrett, P. and Stephenson, M. (1990) Plasmid transfers between strains of *Bacillus thuringiensis* infecting *Galleria mellonella* and *Spodoptera littoralis*, *Appl. Environ. Microbiol.* **56**, 1608–1614.

Johnson, M. T. and Gould, F. (1992) Interaction of genetically engineered host-plant resistance and natural enemies of *Heliothis virescens* (Lepidoptera: Noctuidae) in tobacco, *Environ. Entomol.* **21**, 586–597.

Kalman, S., Kiehne, K. L., Libs, J. L. and Yamamoto, T. (1993) Cloning of a novel *cryIC*-type gene from a strain of *Bacillus thuringiensis* subsp. *galleriae*, *Appl. Environ. Microbiol.* **59**, 1131–1137.

Kennedy, G. G. and Whalon, M. E. (1995) Managing pest resistance to *Bacillus thuringiensis* endotoxins: constraints and incentives to implementation, *J. Econ. Entomol.* **88**, 454–460.

Knight, P. J. K., Crickmore, N. and Ellar, D. J. (1994) The receptor for *Bacillus thuringiensis* CryIA(c) delta-endotoxin in the brush border membrane of the lepidopteran *Manduca sexta* is aminopeptidase N, *Mol. Microbiol.* **11**(3), 429–436.

Knowles, B. H. and Dow, J. A. T. (1993) The crystal δ-endotoxins of *Bacillus thuringiensis*: models for their mechanism of action on the insect gut, *BioEssays* **15**, 469–476.

Knowles, B. H. and Ellar, D. J. (1987) Colloid-osmotic lysis is a general feature of the mechanism of action of *Bacillus thuringiensis* δ-endotoxins with different insect specificities, *Biochim. Biophys. Acta* **924**, 509–518.

Koni, P. A. and Ellar, D. J. (1995) Cloning and characterization of a novel *Bacillus thuringiensis* cytolytic delta-endotoxin, *J. Mol. Biol.* **229**, 319–327.

Koziel, M. G., Beland, G. L., Bowman, C., Carozzi, N. B., Crenshaw, R., Crossland, L., Dwason, J., Desai, N., Hill, M., Kadwell, S. *et al.* (1993a) Field performance of elite transgenic maize plants expressing an insecticidal protein derived from *Bacillus thuringiensis*, *Bio/Technology* **11**, 194–200.

Koziel, M. G., Carozzi, N. B., Currier, T. C., Warren, G. W. and Evola, S. V. (1993b) The insecticidal crystal proteins of *Bacillus thuringiensis*: past, present and future uses, *Bio-technol. Genet. Engin. Rev.* **11**, 171–228.

Krieg, A., Huger, A. M., Langenbruch, G. A. and Schnetter, W. (1983) *Bacillus thuringiensis* var. *tenebrionis*: ein neuer gegenüber Larven von Coleopteren wirksamer Pathotyp, *J. Appl. Entomol.* **96**, 500–508.

Kronstad, J. W. and Whiteley, H. R. (1984) Inverted repeat sequences flank the *Bacillus thuringiensis* crystal protein gene, *J. Bacteriol.* **160**, 95–102.

Lambert, B. and Peferoen, M. (1992) Insecticidal promise of *Bacillus thuringiensis*. Facts and mysteries about a successful biopesticide, *BioScience* **42**, 112–122.

Lambert, B., Buysse, L., Decock, C., Jansens, S., Piens, C., Saey, B., Seurinck, J., Van Auden-hove, K., Van Rie, J., Van Vliet, A. and Peferoen, M. (1996) A *Bacillus thuringiensis* insecticidal crystal protein with a high activity against members of the family Noctuidae, *Appl. Environ. Microbiol.* **62**, 80–86.

Lambert, B., Van Audenhove, K., Theunis, W., Agouda, R., Jansens, S., Seurinck, J. and Peferoen, M. (1992a) Nucleotide and deduced amino acid sequence of cryIIID, a novel coleopteran-active crystal protein gene from a *Bacillus thuringiensis* subsp. *kurstaki* isolate, *Gene* **110**, 131–132.

Lambert, B., Höfte, H., Annys, K., Jansens, S., Soetaert, P. and Peferoen, M. (1992b) Novel *Bacillus thuringiensis* insecticidal crystal protein with a silent activity against coleopteran larvae, *Appl. Environ. Microbiol.* **58**, 2536–2542.

Lecadet, M.-M, Frachon, E., Dumanoir, V. C. and de Barjac, H. (1994) An updated version of the *Bacillus thuringiensis* strains classification according to H-serotypes, *Abstracts of the XXVIIth Annual Meeting of the Society for Invertebrate Pathology, Montpellier, France*, p. 345.

Lee, C.-S. and Aronson, A. I. (1991) Cloning and analysis of δ-endotoxin genes from *Bacillus thuringiensis* subsp. *alesti*, *J. Bacteriol.* **173**, 6635–6638.

Lee, M. K., Milne, R. E., Ge, A. Z. and Dean, D. H. (1992) Location of a *Bombyx mori* receptor binding region on a *Bacillus thuringiensis* δ-endotoxin, *J. Biol. Chem.* **267**, 3115–3121.

Li, J., Carroll, J. and Ellar, D. J. (1991) Crystal structure of insecticidal δ-endotoxin from *Bacillus thuringiensis* at 2.5 Å resolution, *Nature* **353**, 815–821.

Liang, Y., Patel, S. and Dean, D. H. (1995) Irreversible binding of *Bacillus thuringiensis* δ-endotoxin to *Lymantria dispar* brush border membrane vesicles is directly correlated to toxicity, *J. Biol. Chem.* **270**, 24719–24727.

Lu, H., Rajamohan, F. and Dean, D. H. (1994) Identification of amino acid residues of *Bacillus thuringiensis* δ-endotoxin CryIA(a) associated with membrane binding and toxicity to *Bombyx mori*, *J. Bacteriol.* **176**, 5554–5559.

Lüthy, P. (1980) Insecticidal toxins of *Bacillus thuringiensis*, *FEMS Microbiol. Lett.* **8**, 1–7.

MacIntosh, S. C., McPherson, S. L., Perlak, F. J., Marrone, P. G. and Fuchs, R. L. (1990) Purification and characterization of *Bacillus thuringiensis* var. *tenebrionis* insecticidal proteins produced in *E. coli*, *Biochem. Biophys. Res. Commun.* **170**, 665–672.

MacIntosh, S. C., Stone, T. B., Jokerst, R. S. and Fuchs, R. L. (1991) Binding of *Bacillus thuringiensis* proteins to a laboratory selected line of *Heliothis virescens*, *Proc. Natl. Acad. Sci.* **88**, 8930–8933.

Mahillon, J., Seurinck, J., Van Rompuy, V., Delcour, J. and Zabeau, M. (1985) Nucleotide sequence and structural organization of an insertion sequence element (IS231) from *Bacillus thuringiensis* strain Berliner 1715, *EMBO J.* **4**, 3895–3899.

Mallet, J. and Porter, P. (1992) Preventing insect adaptation to insect-resistant crops: are seed mixtures or refugia the best strategy? *Proc. R. Soc. London Series B* **250**, 165–169.

Manthavan, S., Sudha, P. M. and Pechimuthu, S. M. (1989) Effect of *Bacillus thuringiensis* on the midgut cells of *Bombyx mori* larvae: a histopathological and histochemical study, *J. Invert. Pathol.* **53**, 217–227.

Martin, P. A. W. and Travers, R. S. (1989) Worldwide abundance and distribution of *Bacillus thuringiensis* isolates, *Appl. Environ. Microbiol.* **55**, 2437–2442.

Masson, L., Mazza, A., Gringorten, L., Baines, D., Aneliunas, V. and Brousseau, R. (1994) Specificity domain localization of *Bacillus thuringiensis* insecticidal toxins is highly dependent on the bioassay system, *Mol. Microbiol.* **14**, 851–860.

McBride, K. E., Svab, Z., Schaaf, D. J., Hogan, P. S., Stalker, D. M. and Maliga, P. (1995) Amplification of a chimeric *Bacillus* gene in chloroplasts leads to an extraordinary level of an insecticidal protein in tobacco, *Bio/Technology* **13**, 362–365.

McGaughey, W. H. (1985) Insect resistance to the biological insecticide *Bacillus thuringiensis*, *Science* **229**, 193–195.

McGaughey, W. H. and Beeman, R. W. (1988) Resistance to *Bacillus thuringiensis* in colonies of Indian meal moth and almond moth (lepidoptera: Pyralidae), *J. Econ. Entomol.* **81**, 28–33.

McGaughey, W. H. and Whalon, M. E. (1992) Managing insect resistance to *Bacillus thuringiensis* toxins, *Science* **258**, 1451–1455.

McPherson, S. A., Perlak, F. J., Fuchs, R. L., Marrone, P. G., Lavrik, P. B. and Fischhoff, D. A. (1988) Characterization of the coleopteran-specific protein gene of *Bacillus thuringiensis* var. *tenebrionis*, *Bio/Technology* **6**, 61–66.

Meadows, M. P. (1993) *Bacillus thurngiensis* in the environment: ecology and risk assessment. In: Entwhistle, P. F., Croy, J. S., Bailey, M. J. and Higgs, S. (Eds), *Bacillus thuringiensis an Environmental Biopesticide: Theory and Practice*, New York: John Wiley & Sons, pp. 195-220.

Michaels, T. E., Narva, K. E. and Foncerrada, L. (1993) Process for controlling scarab pests with *Bacillus thuringiensis* isolates, Patent WO 93/15206.

Moar, W. J., Pusztai-Carey, M., Van Faassen, H., Bosch, D., Frutos, R., Rang, C., Luo, K. and Adang, M. (1995) Development of *Bacillus thuringiensis* CryIC resistance by *Spodoptera exigua* (Hübner) (Lepidoptera: Noctuidae), *Appl. Environ. Microbiol.* **61**, 2086–2092.

Müller-Cohn, J., Chafaux, J., Buisson, C., Golois, N., Sanchis, V. and Lereclus, D. (1994) *Spodoptera littoralis* resistance to the *Bacillus thuringiensis* CryIC toxin and cross resistance to other toxins, *Abstracts of the XXVIIth Annual Meeting of the Society for Invertebrate Pathology Montpellier, France*, p. 66.

Murray, E. E., Rocheleau, T., Eberle, M., Stock, C., Sekar, V. and Adang, M. (1991) Analysis of unstable RNA transcripts of insecticidal crystal protein genes of *Bacillus thuringiensis* in transgenic plants and electroporated protoplasts, *Plant Mol. Biol.* **16**, 1035–1050.

Narva, K. E., Payne, J. M., Schwab, G. E., Hickle, L. A., Galasan, T. and Sick. A. (1991) Novel *Bacillus thuringiensis* microbes active against nematodes, and genes encoding novel nematode-active toxins cloned from *Bacillus thuringiensis* isolates, Patent EP 0462721.

Narva, K. E., Schwab, G. E., Galasan, T. and Payne, J. M. (1993) Gene encoding a nematode-active toxin cloned from a *Bacillus thuringiensis* isolate, US Patent 5 236 843.

Ogiwara, K., Hori, H., Minami, M., Takeuchi, K., Sato, R., Ohba, M. and Iwahana, H. (1995) Nucleotide sequence of the gene encoding a novel delta-endotoxin from *Bacillus thuringiensis* serovar *japonensis* strain Buibui specific to scarabeid beetles, *Curr. Microbiol.* **30**, 227–235.

Parker, M. W. and Pattus, F. (1993) Rendering a membrane protein soluble in water: a common packing motif in bacterial protein toxins, *Trends Biochem. Sci.* **18**, 391–395.

Payne, J. M. and Fu, J. M. (1994) Coleopteran-active *Bacillus thuringiensis* isolates and genes encoding coleopteran-active toxins, US Patent 5 286 486.

Payne, J. M. and Michaels, T. E. (1995) *Bacillus thuringiensis* isolates selectively active against certain coleopteran pests, US Patent 5 427 786.

Payne, J. M. and Narva, K. E. (1994) Process for controlling corn rootworm larvae, Patent WO 94/16079.

Payne, J. M. and Sick, A. J. (1993a) Genes encoding lepidopteran-active toxins and transformed hosts, US Patent 5 206 166.

Payne, J. M. and Sick, A. J. (1993b) *Bacillus thuringiensis* isolate active against lepidopteran pests, and genes encoding novel lepidopteran-active toxins, US Patent 5 246 852.

Peferoen, M., Jansens, S., Reynaerts, A. and Leemans, J. (1990) Potato plants with engineered resistance against insect attack, In: Vayda, M. E. and Park, W. C. (Eds), *Molecular and Cellular Biology of the Potato*, Wallingford: CAB International, pp. 193–204.

Perlak, F. J., Deaton, R. W., Armstrong, T. A., Fuchs, R. L., Sims, S. S., Greenplate, J. T. and Fischhoff, D. A. (1990) Insect resistant cotton plants, *Bio/Technology* **8**, 939–943.

Perlak, F. J., Fuchs, R. L., Dean, D. A., McPherson, S. L. and Fischhoff, D. A. (1991) Modification of the coding sequence enhances plant expression of insect control protein genes, *Proc. Natl. Acad. Sci.* **88**, 3324–3328.

Perlak, F. J., Stone, T. B., Muskopf, Y. M., Peterson, L. J., Parker, G. B., McPherson, S. A., Wyman, J., Love, S., Reed, G., Biever, D. and Fischhoff, D. A. (1993) Genetically improved potatoes: protection from damage by Colorado potato beetles, *Plant Mol. Biol.* **22**, 313–321.

Rajamohan, F., Alcantara, E., Lee, M. K., Chen, X. J., Curtiss, A. and Dean, D. H. (1995) Single amino acid changes in domain II of *Bacillus thuringiensis* CryIAb δ-endotoxin affect irreversible binding to *Manduca sexta* midgut membrane vesicles, *J. Bacteriol.* **177**, 2276–2282.

Ravoahangimalala, O. and Charles, J.-F. (1995) *In vitro* binding of *Bacillus thuringiensis* var *israelensis* individual toxins to midgut cells of *Anopheles gambiae* larvae (Diptera: Culicidae), *FEBS Lett.* **362**, 111–115.

Ravoahangimalala, O., Charles, J.-F. and Schoeller-Raccaud, J. (1993) Immunological localization of *Bacillus thuringiensis* serovar *israelensis* toxins in midgut cells of intoxicated *Anopheles gambiae* larvae (Diptera: Culicidae), *Res. Microbiol.* **144**, 271–278.

Roush, R. T. (1994) Managing pests and their resistance to *Bacillus thuringiensis*: can transgenic crops be better than sprays?, *Biocontrol Sci. Technol.* **4**, 501–516.

Sacchi, V. F., Parenti, P, Hanozet, G. M., Giordana, B., Lüthy, P. and Wolfersberger, M. G. (1986) *Bacillus thuringiensis* toxin inhibits K + -gradient-dependent amino acid transport across the brush-border membrane of *Pieris brassicae* midgut cells, *FEBS Lett.* **204**, 213–218.

Sangadala, S., Walters, F. S., English, L. H. and Adang, M. J. (1994) A mixture of *Manduca sexta* aminopeptidase and phosphatase enhances *Bacillus thuringiensis* insecticidal CryIA(c) toxin binding and ^{86}Rb^{+}-K^{+} efflux *in vitro*, *J. Biol. Chem.* **269**, 10088–10092.

Schnepf, H. E. and Whiteley, H. R. (1981) Cloning and expression of the *Bacillus thuringiensis* crystal protein gene in *Escherichia coli*, *Proc. Natl. Acad. Sci.* **78**, 2893–2897.

Schnepf, H. E., Wong, H. C. and Whiteley, H. R. (1985) The amino acid sequence of a crystal protein from *Bacillus thuringiensis* deduced from the DNA base sequence, *J. Biol. Chem.* **260**, 6264–6272.

Schnepf, H. E., Tomczak, K., Ortega, J. P. and Whiteley, H. R. (1990) Specificity-determining regions of lepidopteran-specific insecticidal proteins produced by *Bacillus thuringiensis, J. Biol. Chem.* **265**, 20923–20930.

Schnepf, H. E., Schwab, G. E., Payne, J. M., Narva, K. E. and Foncerrada, L. (1992) Novel nematode-active toxins and genes which code therefor, Patent WO 92/19739.

Schwartz, J.-L., Garneau, L., Savaria, D., Masson, L., Brousseau, R. and Rousseau, E. (1993) Lepidopteran-specific crystal toxins from *Bacillus thuringiensis* from cation- and anion-selective channels in planar lipid bilayers, *J. Memb. Biol.* **132**, 53–62.

Shevelev, A. B., Svarinsky, M. A., Karasin, A. I., Kogan, Y. N., Chestukhina, G. G. and Stepanov, V. M. (1993) Primary structure of *cryX*, the novel delta-endotoxin-related gene from *Bacillus thuringiensis* spp., *FEBS Lett.* **336**, 79–82.

Shimizu, T., Vassylyev, D. G., Kiod, S., Doi, Y. and Morikawa, K. (1994) Crystal structure of vitelline membrane outer layer protein I (VMO-I): a folding motif with homologous Greek key structures related by an internal three-fold symmetry, *EMBO J.* **13**, 1003–1010.

Sick, A., Gaertner, F. and Wong, A. (1990) Nucleotide sequence of a coleopteran-active toxin gene from a new isolate of *Bacillus thuringiensis* subsp. *tolworthi, Nucl. Acid Res.* **18**, 1305.

Sick, A. J., Schwab, G. E. and Payne, J. M. (1994) Genes encoding nematode-active toxins cloned from *Bacillus thuringiensis* isolates, Patent EP 046721.

Sims, S. R. and Stone, T. B. (1991) Genetic basis of tobacco budworm resistance to an engineered *Pseudomonas fluorescens* expressing the δ-endotoxin of *Bacillus thuringiensis kurstaki, J. Invert. Pathol.* **57**, 206–210.

Slaney, A. C., Robbins, H. L. and English, L. (1992) Mode of action of *Bacillus thuringiensis* toxin CryIIIA: an analysis of toxicity in *Leptinotarsa decemlineata* (Say) and *Diabrotica undecimpunctata Howardi* Barber, *Insect Biochem. Mol. Biol.* **22**, 9–18.

Smith, R. A. and Couche, G. A. (1991) The phylloplane as a source of *Bacillus thuringiensis* variants, *Appl. Environ. Microbiol.* **57**, 311–315.

Smulevitch, S. V., Osterman, A. L., Shevelev, A. B., Kaluger, S. V., Karasin, A. I., Kadyrov, R. M., Zagnitko, O. P., Chestukhina, G. G. and Stepanov, V. M. (1991) Nucleotide sequence of a novel delta-endotoxin gene *cryIg* of *Bacillus thuringiensis* ssp. *galleriae, FEBS Lett.* **293**, 25–28.

Stone, T. B., Sims, S. R. and Marrone, P. G. (1989) Selection of tobacco budworm for resistance to a genetically engineered *Pseudomonas fluorescens* containing the δ-endotoxin of *Bacillus thuringiensis* subsp. *kurstaki, J. Invert. Pathol.* **53**, 228–234.

Tabashnik, B. E., Cushing, N. L., Finson, N. and Johnson, M. W. (1990) Field development of resistance to *Bacillus thuringiensis* in diamondback moth (Lepidoptera: Plutellidae). *J. Econ. Entomol.* **83**, 1671–1676.

Tabashnik, B. E., Finson, N. and Johnson, M. W. (1991) Managing resistance to *Bacillus thuringiensis*: lessons from the diamondback moth (Lepidoptera: Plutellidae), *J. Econ. Entomol.* **84**, 49–55.

Tabashnik, B. E., Schwartz, J. M., Finson, N. and Johnson, M. W. (1992) Inheritance of resistance to *Bacillus thuringiensis* in diamondback moth (Lepidoptera: Plutellidae), *J. Econ. Entomol.* **85**, 1046–1055.

Tabashnik, B. E., Finson, N., Johnson, M. W. and Moar, W. J. (1993) Resistance to toxins from *Bacillus thuringiensis* subsp. *kurstaki* causes minimal cross-resistance to *B. thuringiensis* subsp. *aizawai* in the diamondback moth (Lepidoptera: Plutellidae), *Appl. Environ. Microbiol.* **59**, 1332–1335.

Tabashnik, B. E., Finson, N., Groeters, F. R., Moar, W. J., Johnson, M. W., Luo, K. and Adang, M. J. (1994a) Reversal of resistance to *Bacillus thuringiensis* in *Plutella xylostella, Proc. Natl. Acad. Sci. USA* **91**, 4120–4124.

Tabashnik, B. E., Finson, N., Johnson, M. W. and Heckel, D. G. (1994b) Cross-resistance to *Bacillus thuringiensis* toxin CryIF in the diamondback moth (*Plutella xylostella*), *Appl. Environ. Microbiol.* **60**, 4627–4629.

Tailor, R., Tippett, J., Gibb, G., Pells, S., Pike, D., Jordon, L. and Ely, S. (1992) Identification and characterization of a novel *Bacillus thuringiensis* delta-endotoxin entomocidal to coleopteran and lepidopteran larvae, *Mol. Microbiol.* **6**, 1211–1217.

Tang, J. D., Shelton, A. M., Van Rie, J., De Roeck, S., Moar, W. J., Roush, R. T. and Peferoen, M. (1996) Toxicity of *Bacillus thuringiensis* spore and crystal to resistant diamondback moth (*Plutella xylostella*), *Appl. Environ. Microbiol.* **62**, 564–569.

Thorne, L., Garduno, F., Thompson, T., Decker, D., Zounes, M. A., Wild, M., Walfield, A. M. and Pollock, T. J. (1986) Structural similarity between the lepidoptera- and diptera-specific insecticidal endotoxin genes of *Bacillus thuringiensis* subsp. 'kurstaki' and 'israelensis', *J. Bacteriol.* **166**, 801–811.

Vadlamudi, R. K., Weber, E., Ji, I., Ji, T. H. and Bulla, L. A. Jr (1995) Cloning and expression of a receptor for an insecticidal toxin of *Bacillus thuringiensis, J. Biol. Chem.* **270**, 5490–5494.

Vaeck, M., Reynaerts, A., Höfte, H., Jansens, S., De Beuckeleer, M., Dean, C., Zabeau, M.,

Van Montagu, M. and Leemans, J. (1987) Transgenic plants protected from insect attack, *Nature* **327**, 33–37.

Valaitis, A. P., Lee, M. K., Rajamohan, F. and Dean, D. H. (1995) Brush border membrane aminopetidase-N in the midgut of the gypsy moth serves as the receptor for the Cry1A(a) δ-endotoxin of *Bacillus thuringiensis*, *Insect Biochem. Mol. Biol.* **25**(10), 1143–1151.

van Aarssen, R., Soetaert, P., Stam, M., Dockx, J., Gosselé, V., Seurinck, J., Reynaerts, A. and Cornelissen, M. (1995) *cry*IA(b) transcript formation in tobacco is inefficient, *Plant Mol. Biol.* **28**, 513–524.

Van Rie, J., Jansens, S., Höfte, H., Degheele, D. and Van Mellaert, H. (1989) Specificity of *Bacillus thuringiensis* δ-endotoxins – importance of specific receptors on the brush border membranes of the mid-gut of target insects, *Eur. J. Biochem.* **186**, 239–247.

Van Rie, J., Jansens, S., Höfte, H., Degheele, D. and Van Mellaert, H. (1990a) Receptors on the brush border membrane of the insect midgut as determinants of the specificity of *Bacillus thuringiensis* delta-endotoxins, *Appl. Environ. Microbiol.* **56**, 1378–1385.

Van Rie, J., McGaughey, W. H., Johnson, D. E., Barnett, B. D. and Van Mellaert, H. (1990b) Mechanism of insect resistance to the microbial insecticide *Bacillus thuringiensis*, *Science* **247**, 72–74.

Velten, J., Velten, L., Hain, R. and Schell, J. (1984) Isolation of a dual plant promoter fragment from the Ti plasmid of Agrobacterium tumefaciens, *EMBO J.* **12**, 2723–2730.

Visser, B., Munsterman, E., Stoker, A. and Dirkse, W. G. (1990) A novel *Bacillus thuringiensis* gene encoding a *Spodoptera exigua*-specific crystal protein, *J. Bacteriol.* **172**, 6783–6788.

Von Tersch, M. A. and Gonzalez, J. M. (1994) *Bacillus thuringiensis* cryET1 toxin gene and protein toxic to lepidopteran insects, US Patent 5 356 623.

Von Tersch, M. A., Slatin, S. L., Kulesza, C. A. and English, L. H. (1994) Membrane-permeabilizing activities of *Bacillus thuringiensis* coleopteran-active toxin CryIIIB2 and CryIIIB2 domain I peptide, *Appl. Environ. Microbiol.* **60**, 3711–3717.

Waalwijk, C., Dullemans, A. M., vanWorkum, M. E. S. and Visser, B. (1985) Molecular cloning and the nucleotide sequence of the Mr28,000 crystal protein gene of *Bacillus thuringiensis* subsp. *israelensis*, *Nucl. Acid Res.* **13**, 8207–8217.

Wabiko, H., Raymond, K. C. and Bulla, L. A. Jr (1986) *Bacillus thuringiensis* entomocidal protoxin gene sequence and gene product analysis, *DNA* **5**, 305–314.

Walters, F. S., Slatin, S. L., Kulesza, C. A. and English, L. H. (1993) Ion channel activity of N-terminal fragments from CryIA(c) delta-endotoxin, *Biochem. Biophys. Res. Commun.* **196**, 921–926.

Walters, F. S., Kulesza, C. A., Phillips, A. T. and English, L. H. (1994) A stable oligomer of *Bacillus thuringiensis* delta-endotoxin, CryIIIA, *Insect Biochem. Mol. Biol.* **24**, 963–968.

Ward, E. S. and Ellar, D. J. (1987) Nucleotide sequence of a *Bacillus thuringiensis* var. *israelensis* gene encoding a 130 kDa delta-endotoxin, *Nucl. Acid Res.* **15**, 7195.

Warren, G. W., Carozzi, N. B., Desai, N. and Koziel, M. G. (1992) Field evaluation of transgenic tobacco containing a *Bacillus thuringiensis* insecticidal protein gene, *J. Econ. Entomol.* **5**, 1651–1659.

West, A. W. (1985) Persistence of *Bacillus thuringiensis* and *Bacillus cereus* in soil supplemented with grass or manure, *Plant and Soil*, **83**, 389–398.

West, A. W., Burges, H. D., Dixon, T. J. and Wyborn, C. H. (1985) Survival of *Bacillus thuringiensis* and *Bacillus cereus* spore inocula in soil: effects of pH, moisture, nutrient availability and indigenous microorganisms, *Soil Biol. Biochem.* **17**, 657–665.

Whalon, M. E. and Wierenga, J. M. (1994) *Bacillus thuringiensis* resistant Colorado potato beetle and transgenic plants: some operational and ecological implications for deployment, *Biocontrol Sci. Technol.* **4**, 555–561.

Whalon, M. E., Miller, D. L., Hollingworth, R. M., Grafius, E. J. and Miller, J. R. (1993) Selection of a Colorado potato beetle (Coleoptera: Chrysomelidae) strain resistant to *Bacillus thuringiensis*, *J. Econ. Entomol.* **86**, 1516–1521.

Widner, W. R. and Whiteley, H. R. (1989) Two highly related insecticidal crystal proteins of

Bacillus thuringiensis subsp. *kurstaki* possess different host range specificities, *J. Bacteriol.* **171**, 965–974.

Williams, S., Friedrich, L., Dincher, S., Carozzi, N., Kessmann, H., Ward, E. and Ryals, J. (1992) Chemical regulation of *Bacillus thuringiensis* δ-endotoxin expression in transgenic plants, *Bio/Technology* **10**, 540–543.

Wilson, F. D., Flint, H. M., Deaton, W. R., Fischhoff, D. A., Perlak, F. J., Armstrong, T. A., Fuchs, R. L., Berberich, S. A., Parks, N. J. and Stapp, B. R. (1992) Resistance of cotton lines containing a *Bacillus thuringiensis* toxin to pink bollworm (Lepidoptera: Gelechiidae) and other insects, *J. Econ. Entomol.* **4**, 1516–1521.

Wolfersberger, M. G. (1990) The toxicity of two *Bacillus thuringiensis* δ-endotoxins to gypsy moth larvae is inversely related to the affinity of binding sites on midgut brush border membranes for the toxins, *Experientia* **46**, 475–477.

Wolfersberger, M. G., Lüthy, P., Maurer, A., Parenti, P., Sacchi, V., Giordana, B. and Hanozet, G. (1987) Preparation and partial characterization of amino acid transporting brush border membrane vesicles from the larval midgut of the cabbage butterfly (*Pieris brassicae*), *Compar. Biochem. Physiol.* **86**(a), 301–308.

Wu, D., Cao, X. L., Bai, Y. Y. and Aronson, A. I. (1991) Sequence of an operon containing a novel δ-endotoxin gene from *Bacillus thuringiensis*, *FEMS Microbiol. Lett.* **81**, 31–36.

3

The Development of a Comprehensive Resistance Management Plan for Potatoes Expressing the Cry3A Endotoxin

JENNIFER FELDMAN and TERRY STONE

Introduction

In May 1995, NewLeaf Russet Burbank potatoes became the first genetically modified, insect-resistant crop to receive full federal approval for commercialization. NewLeaf potato plants express the Cry3A protein derived from *Bacillus thuringiensis* subsp. *tenebrionis* (Cry3A), which is selectively active against certain Coleopteran insects including the Colorado potato beetle (CPB). Three federal agencies evaluated various aspects of NewLeaf potatoes, including their food quality (FDA), potential for becoming a plant pest (USDA), and the human, environmental and non-target safety of the Cry3A protein itself (EPA). The EPA was the lead agency in this process, and their evaluation was as rigorous as that conducted for conventional pesticides. However, unlike their reviews of previous microbial or chemical insecticides, the EPA also considered the risk of insect resistance development to NewLeaf potatoes as part of their regulatory assessment.

In 1992, EPA formed the Pesticide Resistance Management Workgroup (PRMW) within the Office of Pesticide Programs, to begin consideration of resistance as one element of environmental risk in the evaluation of herbicides, fungicides, and insecticides. During the regulatory review process from 1993–95, Monsanto Co. and its seed potato division, NatureMark, worked closely with the PRMW to identify research needs and generate the data necessary to devise workable resistance prevention and management strategies for NewLeaf potatoes. NatureMark, Monsanto Co., outside academic experts, and representatives from the potato industry, including potato growers, met with the EPA repeatedly during the three-year review period to keep them abreast of progress toward the development of a comprehensive management plan. In July 1994, NatureMark and Monsanto submitted a written document to the EPA outlining these resistance management plans. The strategies described in this document were based largely on research results from multiple field and laboratory experiments carried out by outside academic and government cooperators.

On March 1, 1995, following internal review and prior to granting the registration, the EPA convened a Scientific Advisory Panel (SAP) meeting to facilitate an open discussion and review of the plan by individuals from academia, industry, regulatory agencies and the general public. Members of the SAP consisted of several entomologists, plant breeders, toxicologists and other biologists. The panelists found the plan to be detailed, thorough, and appropriate to delay or prevent the widespread development of resistance by the CPB to the Cry3A protein produced in NewLeaf potatoes (US EPA, 1995). EPA registration of the Cry3A protein produced in NewLeaf potatoes was finalized on May 5, 1995. This action signified the final federal regulatory approval necessary for the sale of these seed potatoes to growers.

Below is a transcript of the resistance management plan submitted to the EPA in support of EPA registration No. 524-474. Similar plans have been developed for *Bacillus thuringiensis* (*Bt*)-expressing corn and cotton. Additional research has contributed to the understanding of CPB biology, NewLeaf potatoes and the potential for resistance development since the plan was submitted. However, as it is the first comprehensive management plan to be developed for an insect control product before market introduction, this document is significant from a historical perspective. Moreover, it truly represents a collaborative effort between industry, academia, government and growers. NatureMark is committed to preserving the long-term durability of NewLeaf potatoes, and will continue to work with these groups to gain additional knowledge and refine management strategies. Flexibility and continued cooperation will be necessary to accomplish this. The development and implementation of effective management strategies is an ongoing and dynamic process, but the information contained in this document forms the basis for sound tactics to delay the evolution of resistance in the CPB to NewLeaf potatoes.

Strategies to Maximize the Utility and Durability of Colorado Potato Beetle Resistant Potatoes

Monsanto and NatureMark (an operating unit of Monsanto) have developed an environmentally compatible alternative for Colorado potato beetle (CPB) control through the genetic modification of potato plants. These CPB resistant potatoes (tradename NewLeaf) produce the insect control protein derived from the common soil bacterium *Bacillus thuringiensis* subsp. *tenebrionis* (Cry3A). The protein produced by NewLeaf potatoes is identical to that found in nature and in commercial Cry3A formulations which have been registered with the EPA since 1988.

Results from three years of field experiments conducted throughout the primary potato growing regions have demonstrated that NewLeaf potatoes are protected season-long from all foliage feeding CPB life stages, including summer adults. Moreover, because of the selectivity of the Cry3A protein, beneficial insects are unaffected. Growers planting these potatoes will not require chemical insecticide applications to control CPB. This substantial reduction in insecticide use will enhance biological control and support the implementation of other alternative management strategies for non-target potato pests such as aphids and leafhoppers.

To achieve these benefits, it is important that NewLeaf potatoes be implemented and managed properly. In this respect, these plants are no different than any other crop protection product that has been used over the last century. To successfully maximize the long-term use of these potatoes, two interconnected management components are required. First is the development of integrated pest management techniques that allow farmers to optimize the utility of the plants for potato pest control. In essence, this is the development of a total insect management package that will be centered around NewLeaf potatoes. Second, the development and implementation of strategies to prevent the evolution of insect resistance to the Cry3A protein are needed to ensure the durability of the potatoes as a pest control tool.

The CPB has demonstrated an ability to develop resistance to a wide variety of insecticides. Field resistance to Cry3A has not been reported, but laboratory selection has been achieved following repeated exposure (Whalon *et al.*, 1993). To address the potential for CPB to develop resistance to the Cry3A protein expressed in NewLeaf potatoes, Monsanto and NatureMark scientists have held extensive consultations in recent years with leading pest and resistance management researchers from academia, government, and Extension. In collaboration with these experts, field and laboratory studies have been conducted to evaluate nearly every suggested strategy for managing CPB resistance to the Cry3A protein. Continuing experiments, combined with existing information, will provide the basis for sound, practical, management programs built on an understanding of potato production and agronomic practices. As an outcome of this research effort, the following have been identified as potential resistance management strategies:

(1) Agronomic and other pest management practices that promote multiple tactics for insect control, including cultural, biological and chemical factors.

(2) Monitoring of CPB populations for susceptibility to the Cry3A protein.

(3) High-dose expression of the Cry3A protein in potatoes to control CPB heterozygous for resistance alleles.

(4) Refugia as hosts for Cry3A susceptible insects provided through non-CPB resistant potatoes.

(5) Development of novel CPB control proteins with a distinct mode of action from the NewLeaf insect control protein.

The following is an overview of research progress to date on each of the above strategies with plans for their implementation when appropriate. In addition to these plans, continuing efforts are being made to educate potato growers as to the most effective ways to integrate these potatoes within their current production practices. This cooperative effort between growers, academia, Extension, NatureMark and Monsanto will help ensure that the benefits of NewLeaf potatoes are fully realized and sustained.

Resistance Management through Integrated Pest Management

Current control of CPB relies heavily upon the use of chemical insecticides. In some areas, where CPB insensitivity has reduced the efficacy of currently registered insecticides, the cost of control can exceed $200/acre each season (Ferro and Boiteau, 1992). Additional CPB management options which include crop rotation, vacuum suction (Boiteau *et al.*, 1992), propane flaming (Moyer, 1992; Moyer *et al.*, 1991), polyethylene-lined trenches (Roush, 1993; Wyman, 1993) and trap plots (Roush, 1993; Roush and Tingey, 1992) are often not practical, effective, economic or easily implemented throughout the season (Roush, 1993; Wyman, 1993).

Microbial Cry3A formulations containing the insecticidal protein have been commercially available for CPB control since the late 1980s (Zehnder and Gelernter, 1989). These formulations are variably effective due to poor spray timing, inadequate plant coverage, short residual activity, and an inability to control large larvae and adults (Ferro and Lyon, 1991; Ferro and Gelernter, 1989; Zehnder and Gelernter, 1989). In contrast, NewLeaf potatoes produce the Cry3A protein throughout the potato foliage, and at a level high enough to control all CPB life stages throughout the growing season (Perlak *et al.*, 1993). Such consistently sustained control of CPB is not possible with currently available microbial, chemical, or physical control methods. In addition, populations of predaceous and parasitic insects can increase unhindered by the reduced application of broad spectrum chemical insecticides, since no additional applications are needed to control the CPB. These beneficial insects can then aid in the control of CPB, as well as non-target potato insect pests such as aphids and leafhoppers, and the diseases they transmit. The combination of NewLeaf potatoes and beneficial insects provides a safe and environmentally compatible foundation for the implementation of other potato pest and resistance management practices.

The Impact of NewLeaf Potatoes on Integrated Pest Management Practices

Studies were conducted in 1992, 1993, and 1994 to determine the effect of pest control practices on arthropod population dynamics. As predicted, the populations of generalist predators were significantly enhanced in NewLeaf potato plots that were protected from CPB damage but were not treated with broad spectrum insecticides (Figure 3.1) (Studies were conducted by Gary Reed, Jeff Wyman, Jeff Stewart and Casey Hoy in multiple locations). Aphid populations in these plots remained stable throughout the season as a result of this improved biological activity (Figure 3.2).

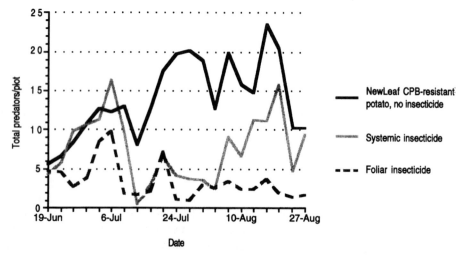

Figure 3.1 Total predatory arthropods found in potato plots with experimental pest management regimes (Hermiston, Oregon, 1992) (Courtesy of Gary Reed)

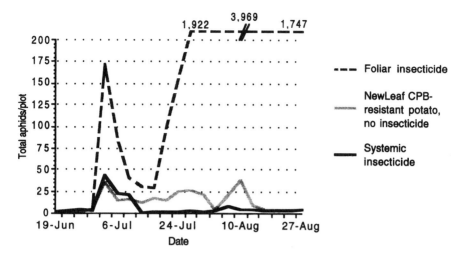

Figure 3.2 Total aphids found in potato plots with experimental pest management regimes (Hermiston, Oregon, 1992) (Courtesy of Gary Reed)

NewLeaf potatoes should be incorporated into integrated pest management (IPM) programs as an integral, but not a stand-alone, measure. Growers should continue to utilize existing, accepted pest management practices that are designed to reduce pesticide inputs and delay resistance development. These practices include:

(1) Cultural controls such as crop rotation, trap crops, non-crop barriers, and overwintering habitat destruction to delay and reduce the colonizing CPB population.

(2) Physical controls such as propane flamers, crop vacuums, and trench traps to reduce the overwintering CPB population.

(3) Biological controls including predators, parasites, and fungal and microbial control agents.

(4) Selective chemical controls, when necessary, for pests other than CPB.

It is anticipated that non-chemical tactics for CPB management will increase the potential for biological control by allowing beneficial arthropod predators and parasites to increase in the agroecosystem. These natural enemies can contribute to pest and resistance management of CPB and other potato insects.

Resistance Monitoring

The first step in resistance management is to establish an estimate of the target pest's baseline susceptibility to the pesticide. Once this baseline is established, regular monitoring is performed to detect changes in susceptibility which may indicate the preliminary stages of resistance in an individual population. In 1992, NatureMark initiated a program for monitoring the susceptibility of CPB to the Cry3A protein expressed in NewLeaf potatoes (Everich *et al.*, 1994). The objectives of this ongoing program, both immediate and long-term, are twofold:

(1) To establish baseline susceptibility to the Cry3A protein in geographically diverse CPB populations.

(2) To monitor CPB populations both before and after exposure to foliar Cry3A and NewLeaf potatoes so that resistance management programs can be modified if necessary.

In cooperation with participating entomologists across North America, nine CPB populations from both experimental and commercial farms were sampled in 1992. In 1993, beetles from each of 19 commercial potato farms were collected and tested for susceptibility to the Cry3A protein. The pesticide treatment history, including the total number of prior Cry3A applications, was obtained for each population.

Galen Dively at University of Maryland found that LC_{50}s ranged from 1.8 (confidence interval 0.7–2.8) $\mu g/ml$ in a Wisconsin population to 8.2 (confidence interval 4.3–12.6) $\mu g/ml$ in a Massachusetts population in 1992. In 1993, LC_{50} values ranged from 1.01 (0.4–1.6) $\mu g/ml$ in a North Carolina population to 5.22 (3.21–7.54) $\mu g/ml$ in a Maryland population (Everich *et al.*, 1994). These differences indicate that a wide range in susceptibility to Cry3A exists among geographically distinct populations of CPB. The 4.5- and 5.2-fold range exhibited in 1992 and 1993, respectively, is typical of the range in susceptibility of other insects to *Bt* protein and does not signify that resistance has developed to the protein in these populations. In a similar study, Stone and Sims (1993) identified an 8-fold range in susceptibility for tobacco budworm and a 16-fold range in susceptibility for corn earworm to the *Bt* subsp. *kurstaki* protein. Because a range in susceptibility to *Bt* protein naturally exists, direct population comparisons cannot be used to identify the presence of resistance within a population. Each geographically distinct population should be monitored over time to detect and confirm shifts in susceptibility.

The CPB populations that were sampled in 1993 are being tested again in 1994, and will be evaluated repeatedly in the coming years for susceptibility to Cry3A. Additional CPB populations will also be sampled. A discriminating dose assay, which utilizes one concentration of Cry3A protein high enough to identify resistant individuals, will be developed so that an even larger number of CPB populations can be efficiently tested.

Colorado Potato Beetle High-Dose Strategy

NewLeaf potatoes express soluble Cry3A protein throughout the foliage at a level of 0.1–0.2% total leaf protein. This concentration is estimated to be approximately 50–100 times the dose required to kill neonate CPB larvae (Perlak *et al.*, 1993). Assuming that resistance to Cry3A is the result of a single major gene that is inherited as a recessive or co-dominant trait (MacIntosh *et al.*, 1991; McGaughey and Beeman, 1988; Sims and Stone, 1991), the 'high-dose' hypothesis predicts that all homozygous susceptible (SS) and heterozygous resistant (SR) individuals will be killed (Figure 3.3). Therefore, only homozygous resistant (RR) beetles will have the

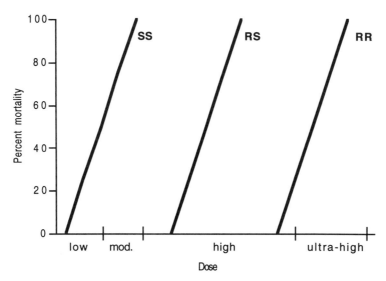

Figure 3.3 The effect of increasing dose on survival of homozygous susceptible (SS), heterozygous resistant (SR) and homozygous resistant (RR) insects

ability to survive on NewLeaf plants. These insects will be extremely rare and will most likely mate with susceptible insects giving rise to heterozygous progeny (Gould, 1986).

In the field, NewLeaf potatoes are protected from feeding by all stages of CPB, including summer adults. Unlike many insecticides, this protection extends throughout the season (Figure 3.4) (Hoy, personal communication). Both foliar and systemic insecticides gradually decay as they are exposed to sunlight, high temperatures, and

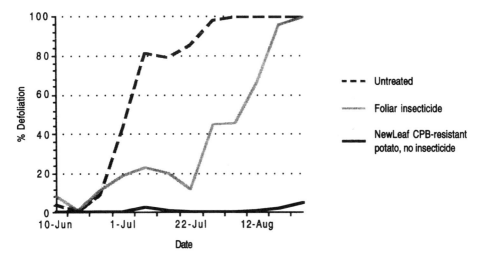

Figure 3.4 Foliar defoliation caused by Colorado potato beetle feeding in potatoes with experimental pest management regimes (Wooster, Ohio, 1993)

irrigation water. Immediately following application, they may be present at levels high enough to kill most or all individuals. However, at some point during the decay period, the dose 'discriminates' between truly susceptible individuals and those with some tolerance to the pesticide. In this fashion, externally applied insecticides select very keenly for resistance by allowing survival and subsequent proliferation of resistance genes, while eliminating susceptible genotypes. In contrast, NewLeaf potatoes produce Cry3A protein at a consistently high level throughout the season. It is hypothesized that CPB exposed to NewLeaf potatoes may never experience this discriminating dose, and survival will be limited to those CPB individuals with genes conferring resistance to a very high dose of Cry3A. As seen below, these individuals will be rare or nonexistent.

Estimating Gene Frequency

Estimating the frequency of resistant alleles in natural insect populations may give some indication of the potential rate of resistance development. It is assumed that the high-dose strategy will provide no benefit over a low-dose strategy if resistant alleles are more common than 1×10^{-3}, or if 1 in 1 000 000 individuals are homozygous resistant (Roush, 1994).

In support of the above, in 1992, 22 000 egg masses (approximately 440 000 individuals) were collected from natural CPB populations and placed on Nature-Mark's NewLeaf potatoes. Larvae were permitted to feed on the plants after egg hatch, and survivors were to be counted at the end of their larval development. No CPB larvae survived beyond the first instar stage. These results support an actual resistance gene frequency of 1×10^{-3} or less (Roush, unpublished data).

Evaluating the Survival of Resistant CPB on NewLeaf Potatoes

Researchers at Michigan State University have subjected CPB to selection in the laboratory with microbial Cry3A formulations, and have demonstrated 200–400-fold resistance to these preparations (Whalon *et al.*, 1993). These resistant insects were challenged with NewLeaf potatoes to investigate their potential to survive on plants that express the Cry3A protein at a very high level. Both resistant and susceptible larval and adult CPB were fed on NewLeaf potatoes for a 96-hour period. Mortality and weight gain of each life stage were measured.

Highly resistant CPB were unable to complete development on CPB resistant plants. Resistant adults laid no eggs after feeding on NewLeaf plants, and virtually all resistant first instar larvae died before the end of the 96-hour feeding period. First instar larvae gained no weight during this time. Both feeding and weight gain in larger larvae were also significantly reduced, suggesting that they probably would have succumbed if the assay had continued for a longer period (Wierenga *et al.*, 1996). Although these insects demonstrated a high level of resistance to gradually increasing doses of microbial Cry3A in the laboratory, they were unable to survive and reproduce on high Cry3A protein expressing potatoes. The authors concluded that NewLeaf plants represent a true high-dose strategy for resistance management.

Refugia

One way to ensure that individual insects with high level (homozygous) resistance do not multiply in number, is to increase the chance that they mate with homozygous susceptible genotypes. This in effect 'dilutes out' resistance genes that occur through chance mutation, and prevents their build-up in the population. As discussed in the previous section on 'Colorado Potato Beetle High-Dose Strategy', only homozygous resistant (RR) insects are predicted to survive high-dose Cry3A, while homozygous susceptible (SS) and heterozygous resistant (SR) individuals will not. In order to ensure a continual influx of susceptible genes into the population, an untreated refuge for their survival should be provided.

There are several potential strategies for incorporating a refuge for Cry3A susceptible CPB into the cropping system. These are:

(1) *Within-field*
   Seed mix: Standard (non-transformed) seed potatoes are randomly mixed with NewLeaf seed potatoes at a predetermined ratio so that CPB resistant plants are immediately adjacent to susceptible plants.
   Refuge strips: Standard potatoes are grown in adjacent rows within the field or at the field edge.

(2) *Between-field*
   Adjacent fields: Not all potato varieties will contain the Cry3A protein and not all fields will be planted to NewLeaf potatoes, therefore, Cry3A-expressing fields will always exist as a spatial mosaic. CPB can move freely between fields with different treatment histories.
   Non-crop hosts: Nightshade and other plant species support CPB in potato production areas. These unselected populations can serve as a susceptible refuge.

(3) *Between-year*
   Potatoes are typically rotated with other crops as a standard management practice. NewLeaf potatoes may be rotated with standard cultivars or non-solanaceous hosts to ensure that CPB are not exposed solely to potatoes expressing the Cry3A protein.

A number of factors will dictate the choice and ultimate success of various refugia strategies. These include, but are not limited to, the following:

(1) The extent of CPB movement from plant to plant and patch to patch. If too little adult movement takes place, susceptible insects may never come in contact with potentially resistant individuals. However, if too much movement takes place such that larvae or adults readily move between NewLeaf and standard potato plants, all individuals will eventually undergo selection with the Cry3A protein. In this circumstance, the value of a refuge may be negated.

(2) Mate and host finding behavior. Colorado potato beetle adults move within and between fields in an effort to locate host plants and/or mates. Mating may occur following diapause in early spring, after completing larval development in mid-season, or before entering the soil in the fall. At each of these phenological stages, CPB host/mate seeking behavior and movement may differ. The success of refugia strategies is strongly influenced by insect movement between

standard and CPB resistant plants, as this is when potentially resistant genotypes may mate with susceptibles. Consequently, complex behavioral patterns
must be considered when designing the size and proximity of refuge plantings.

(3) Industry acceptance. In order for resistance management strategies to be
readily adopted, they must not compromise pest control efficacy or agronomic
quality. If plants in the refuge are severely defoliated, growers will treat them
with conventional pesticides to prevent yield losses. This would negate the
value of the refuge by eliminating susceptible CPB, and would eliminate the
benefit of transgenic, non-chemical insect control. Even economically acceptable strategies will require the voluntary cooperation on the part of potato
growers and pest managers.

An advantage to the seed mix strategy is that it ensures compliance by commercial
producers when seed is sold premixed. However, premixing seed may expose seed
growers to unacceptable risks which they are unwilling to assume. In order to
ensure freedom from disease, seed lots currently retain a distinct identity throughout
numerous years of field multiplication, as per state seed certification requirements.
Seed mixing may jeopardize certification status by altering this line identity. In addition, since potatoes are a large seeded crop, effective mixing is difficult. Inefficiencies
in mixing may lead to small-scale control failures and legal questions of liability.

Evaluating the Potential for Seed Mixes and Other Refugia Strategies

Studies to date have been designed to investigate various refugia options in order to
determine the most effective and practical methods of implementation. Since the
seed mix option would require the most direct involvement by NatureMark and its
seed producing cooperators, much of the current cooperative research concerns the
evaluation of this strategy. Results from numerous studies conducted by leading
entomologists have shown the following:

(1) Adult CPB appear to 'graze' on multiple plants and contact both NewLeaf and
standard potato plants in mixed plantings (Beverly Burden, personal
communication). As a result, the seed mix does not provide an entirely
unselected refuge for CPB adults.

(2) Eggs are laid at random and are deposited with equal frequency on NewLeaf
and standard potato plants in mixed plantings. However, egg laying is suppressed on all plants after feeding on NewLeaf potatoes (Beverly Burden and
Rick Roush, personal communication).

(3) Small larvae are relatively stationary and rarely move off their original plant.

(4) Large larvae move to adjacent plants, particularly as host plants become defoliated (Burden and Wyman, unpublished).

(5) Physiological tolerance to the Cry3A protein is correlated with avoidance
behavior. Both of these traits appear to be heritable (Hoy and Head, 1995).

The overall effect of feeding suppression which is observed in seed mixes may be
desirable to a grower from the standpoint of plant protection. However, in order for
a refuge successfully to provide a barrier to resistance development, it must harbor
insects which do not undergo pesticide-mediated selection. Large larvae from stan-

dard (refuge) plants may acquire a dose of Cry3A protein as a consequence of inter-plant movement. If these insects feed on both NewLeaf and standard potato foliage, they may experience a moderate dose of Cry3A protein, which may compromise the efficacy of the high-dose strategy. In addition, tolerant insects may have a greater likelihood of moving off NewLeaf plants and surviving on standard plants in mixed seed settings. This could gradually lead to, rather than delay, resistance.

Ongoing studies will determine the optimal method for implementing refugia strategies. Long-term studies using isolated CPB populations that are subjected to various management strategies will provide real-life evidence of their potential for managing resistance.

Education and Communication

Even the most sophisticated resistance management strategies will not be successful if they are not communicated to, and adopted by, commercial potato producers. Providing growers with up-to-date information on how to maximize the longevity and value of NewLeaf potatoes is an important role for the NatureMark and Monsanto product development teams. In cooperation with potato specialists, we are currently developing integrated crop management strategies which incorporate both agronomic and pest control components. An effective pest management strategy is a 'local strategy', and involves using cultural options such as crop rotation and over-wintering habitat destruction which must be tailored to individual production areas.

To communicate pest and resistance management strategies effectively, a multi-level approach is being developed. Regional strategies will be presented to NewLeaf growers in the same ways that information is presented today: using technical bulletins; mailings of recent research results and presentations at local grower meetings, state potato organization meetings, and training seminars. Research results have already been presented at both academic and potato industry meetings in the majority of potato producing states. The organizations that represent growers, such as the National Potato Council and individual state potato commissions, have offered assistance in distributing Integrated Crop Management information to their members. At the local level, NatureMark will have the unique opportunity to provide information to growers when they contract to produce NewLeaf seed potatoes.

NatureMark and Monsanto will continue to conduct cooperative research with University Extension agents and crop consultants to maximize the use of NewLeaf potatoes. These results will be shared with growers through established information-al channels to keep them abreast of the latest management strategies. This multi-level approach to communicating the pest and resistance management options possible with NewLeaf potatoes will help maximize the value and durability of a very important new tool for controlling CPB.

Strategies for the Future

Multiple gene and alternate gene strategies hold potential for substantially delaying or halting resistance development. Potatoes with other genes for resistance to CPB are currently under development. In addition, Monsanto is actively searching for additional insect active proteins and other mechanisms for transgenic control of the Colorado potato beetle. These new products can be managed in conjunction with existing potatoes to maximize their resistance management potential. They may be deployed with NewLeaf potatoes in fields, between fields, between regions, or over time.

NewLeaf potatoes hold tremendous promise for controlling CPB as they are highly effective and will allow growers to substantially reduce their pesticide inputs, thus opening new windows for biological control of potato pests. Therefore, it is in the best interest of industry, academia, USDA Extension and the agricultural community to preserve the longevity of this environmentally sound pest control product through proper use and good management. To that end, Monsanto and Nature-Mark have developed a series of strategies which will help to effectively manage the potential for CPB resistance. Over the next few years, as additional information is gathered, details of this program and the incorporation of NewLeaf potatoes into existing pest management programs will be further developed and refined.

References

Boiteau, G., Misener, C., Singh, R. P. and Bernard, G. (1992) Evaluation of a vacuum collector for insect pest control in potato, *Am. Pot. J.* **69**, 157–166.

Everich, R. C., Dively, G. P. and Linduska, J. J. (1994) Monitoring, selection, and characterization of *Bt* susceptibility in Colorado potato beetle, PhD Dissertation. University of Maryland, College Park, MD.

Ferro, D. N. and Boiteau, G. (1992) Management of major insect pests of potato, In: Rowe, R. C. (Ed.), *Plant Health Management in Potato Production*, St Paul: Am. Phytopath. Soc. Press, pp. 209–234.

Ferro, D. N. and Gelernter, W. D. (1989) Toxicity of a new strain of *Bacillus thuringiensis* to Colorado potato beetle (Coleoptera: Chrysomelidae), *J. Econ. Entomol.* **82**, 750–755.

Ferro, D. N. and Lyon, S. M. (1991) Colorado potato beetle (Coleoptera: Chrysomelidae) larval mortality: operative effects of *Bacillus thuringiensis* subsp. *san diego, J. Econ. Entomol.* **84**, 806–809.

Gould, F. (1986) Simulation models for predicting durability of insect-resistant germ plasm: a deterministic diploid, two locus model, *Environ. Entomol.* **15**, 1–10.

Hoy, C. W. and Head, G. (1995) Correlation between behavioral and physiological responses to transgenic potatoes containing *Bacillus thuringiensis* d-endotoxin in *Leptinotarsa decemlineata* (Coleoptera: Chrysomelidae), *J. Econ. Entomol.* **88**, 480–486.

MacIntosh, S. C., Stone, T. B., Jokerst, R. S. and Fuchs, R. L. (1991) Binding of *Bacillus thuringiensis* proteins to a laboratory-selected line of *Heliothis virescens, Proc. Natl. Acad. Sci. USA* **88**, 28–33.

McGaughey, W. H. and Beeman, R. W. (1988) Resistance to *Bacillus thuringiensis* in colonies of the Indian meal moth and the Almond moth (Lepidoptera: Pyralidae), *J. Econ. Entomol.* **81**, 28–33.

Moyer, D. D. (1992) Fabrication and operation of a propane flamer for Colorado potato beetle control, *Cornell Coop. Extension Bull.*

Moyer, D., Kujawski, R., Derksen, R., Moeller, R., Sieczka, J. B. and Tingey, W. M. (1991) Development of a propane flamer for Colorado potato beetle control, Mimeo, Riverhead, New York: Cornell Cooperative Extension, Suffolk County.

Perlak, F., Stone, T. B., Muskopf, Y. M., Petersen, L. J., Parker, G. B., McPherson, S. A., Wyman, J., Love, S., Biever, D., Reed, G. and Fischhoff, D. (1993) Genetically improved potatoes: protection from damage by Colorado potato beetles, *Plant Molec. Biol.* **22**, 313–321.

Roush, R. T. (1993) Transgenic host plant resistance and insect management in potatoes, In: Monsanto Company (1993), Registration of the plant pesticide *Bacillus thuringiensis* subsp. tenebrionis (B.t.t.) Colorado potato beetle control protein and exemption from the requirement of a tolerance for the B.t.t. protein, 21 volumes, Submitted to the Environmental Protection Agency, September 10, 1993.

Roush, R. T. (1994) Managing pests and their resistance to *Bacillus thuringiensis*: can transgenic crops be better than sprays? *Biocontrol. Sci. Tech.* **4**, 501–516.

Roush, R. T. and Tingey, W. M. (1992) Evolution and management of resistance in the Colorado potato beetle, Leptinotarsa decemlineata, In: Denholm, I., Devonshire, A. L. and Holloman, D. W. (Eds), *Resistance '91: Achievements and Developments in Combating Pesticide Resistance*, Essex, UK: Elsevier Applied Science, pp. 61–74.

Sims, S. R. and Stone, T. B. (1991) Genetic basis of tobacco budworm resistance to an engineered *Pseudomonas fluorescens* expressing the delta-endotoxin of *Bacillus thuringiensis kurstaki.*, *J. Invert. Pathol.* **57**, 206–210.

Stone, T. B. and Sims, S. R. (1993) Geographic susceptibility of *Heliothis virescens* and *Helicoverpa zea* (Lepidoptera: Noctuidae) to *Bacillus thuringiensis*, *J. Econ. Entomol.* **86**, 989–994.

United States Environmental Protection Agency (1995) Analysis of SAP and public comments on pesticide resistance management for the CryIIIA delta endotoxin in potatoes and the Pesticide Resistance Management Workgroup's recommendations, EPA memorandum to Monsanto Co., May 2, 1995. Docket No. OPP00401.

Whalon, M. E., Miller, D. I., Hollingworth, R. M., Grafius, E. J. and Miller, J. R. (1993) Selection of a Colorado potato beetle strain resistant to *Bacillus thuringiensis*, *J. Econ. Entomol.* **86**, 226–233.

Wierenga, J., Norris, D. L. and Whalon, M. E. (1996) Stage-specific mortality of transgenic potatoes to the Colorado potato beetle, *J. Econ. Entomol.* **89**, 1047–1052.

Wyman, J. A. (1993) Impacts of transgenic host plant resistance to Colorado potato beetle on potato culture in the United States, In: Monsanto Company (1993), Registration of the plant pesticide *Bacillus thuringiensis* subsp. tenebrionis (B.t.t.) Colorado potato beetle control protein and exemption from the requirement of a tolerance for the B.t.t. protein, 21 volumes, submitted to the Environmental Protection Agency, September 10, 1993.

Zehnder, G. W. and Gelernter, W. D. (1989) Activity of the M-ONE formulation of a new strain of *Bacillus thuringiensis* against Colorado potato beetle (Coleoptera: Chrysomelidae): relationship between susceptibility and insect life stage, *J. Econ. Entomol.* **82**, 756–761.

Acknowledgments

The authors wish to acknowledge the numerous individuals whose research contributed to this document, and to the successful implementation of resistance management plans. Special thanks to Rick Roush, Jeff Wyman, Galen Dively, Casey Hoy, Gary Reed and Jeff Stewart.

4

Transgenic Maize Expressing a *Bacillus thuringiensis* Insecticidal Protein for Control of European Corn Borer

NADINE B. CAROZZI and MICHAEL G. KOZIEL

Overview of Crop Damage from European Corn Borer

With the ability to introduce new genes into plants, many groups have evaluated numerous strategies to reduce insect damage in a variety of crops through the use of transgenic plants expressing insecticidal proteins. One of the most important field crops is corn (maize) and there has been considerable effort aimed at producing insect resistant transgenic maize plants. Transgenic maize resistant to European corn borer is among the first insect resistant crops to be commercialized. Maize is susceptible to several insect pests which can cause significant damage during the growing season and have the potential to produce substantial yield losses. European corn borer (ECB), *Ostrinia nubilalis*, causes extensive damage throughout the maize growing regions in the United States. In the central United States there are typically two generations of ECB. Early planted maize is most susceptible to the first generation. Eggs laid on the underside of maize leaves hatch in 3–5 days. Larvae feed on leaf tissue and bore into the leaf whorl and later into the stalk. The first generation of ECB causes primarily leaf and leaf whorl damage on young pre-anthesis maize plants. Second generation ECB populations feed on numerous plant structures before tunneling into the stalk. The moths are attracted to fresh pollen, green silks and immature kernels. Tissues upon which young second generation ECB can establish include silks, pollen that has accumulated in the leaf axils, ear sheaths, and collar tissue (Showers *et al.*, 1989). Larvae also feed directly on kernels before tunneling into the cob or ear shank. Once in the stalk they continue to develop and feed on interior stalk tissue. ECB larvae can hollow out significant portions of the stalk resulting in reduced water and nutrient flow, poor ear development and stalk breakage. Larvae pupate within the stalk and emerge as adult moths. In warmer growing regions the number of ECB generations can exceed two and reach as many as five per growing season. Corn fields with heavy infestations have a characteristic appearance. Stalks are bent over or completely broken, often with the ears severed and dropped to the ground. Strong autumn winds and rains will increase stalk breakage and further increase the number of dropped ears in weakened plants in the last few weeks prior to harvest. ECB feeding can result in stunted plants as well as

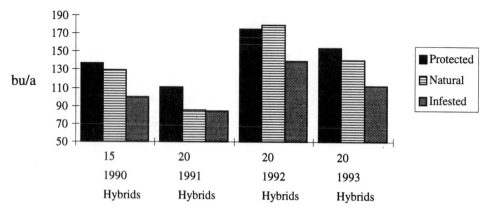

Figure 4.1 Yield losses resulting from natural and artificial ECB infestation of commercial maize hybrids from 1990 through 1993 at Bloomington, Illinois; bu/a, bushels per acre (Christensen *et al.*, 1993)

the obvious loss of ears from severed stalks. The majority of the yield loss results from the physiological disruption of normal plant nutritive uptake caused by stalk tunneling and stalk lodging. It has been estimated that a single borer can cause 3–7% yield loss per plant (Lynch, 1980). In the US corn belt ECB populations have reached economic threshold levels in each of the last three years. In 1994 an estimated $1 billion was lost to ECB in the US alone. Ciba Seeds conducted a series of experiments over a four-year period in which the effects of ECB damage were measured in artificially and naturally infested plots and in plots protected from damage by chemical insecticides (Christensen *et al.*, 1993). A summary of results from 1990–93 shows the yield reduction resulting from ECB infestation (Figure 4.1).

Natural infestations vary considerably in intensity from year to year but can be as bad as intense artificial infestation (1991). Yield losses can exceed 25% in years of heavy infestation. In addition, plants with ECB stalk damage have a higher incidence of stalk rot caused by fungi such as *Fusarium moniliforme* and *Gibberella zern.* Efficient control of ECB can lower damage from diseases that invade plants through ECB entry holes.

Transgenic Maize for Control of European Corn Borer

Effective control of ECB has been very difficult to achieve due to its feeding habits. Chemical sprays are not fully effective because ECB spends little time feeding on the surface of leaves and is therefore inaccessible to the sprays. Once they have penetrated the stalk there is little that can be done to control them. Presently farmers may spray for ECB several times a year and hire field scouts to help determine the best timing for these sprays based on the presence of new ECB generations. Traditional plant breeding has met with only limited success in controlling ECB damage, particularly from second generation ECB. Continuous expression of an insecticidal protein in tissues of maize on which ECB feeds offers an ideal solution to the problems of controlling ECB damage. Certain endotoxin proteins produced by *Bacillus thuringiensis* (*Bt*), a Gram-positive spore forming bacterium, possess potent insecticidal activities. In particular, the CryIA(b) endotoxin is very effective against ECB with an LC_{50} of 4–10 ng/cm^2. The insecticidal activities of the *Bt* δ-endotoxins are characterized by a narrow spectrum of activity. That is, while frequently possessing a potent insecticidal activity, a given δ-endotoxin is typically known to act upon only a few insects. This narrow spectrum of insecticidal activity is accompanied by a lack of toxicity toward vertebrates, making these proteins ideal candidates for use in transgenic plants to confer insect resistance against one or a few pests of a particular crop. Transgenic plants provide a continual dosage of the active protein in contrast to microbial sprays which have a short half-life under field conditions. Chronic exposure may be more effective in controlling an insect than a single acute dose strategy since if the single dose is not well timed or is not lethal, the insect can recover from the exposure.

Engineering δ-Endotoxins of *Bacillus thuringiensis* for Expression in Maize

Several groups have introduced genes encoding *Bt* δ-endotoxins, or insecticidal fragments of δ-endotoxins, into plants with the intent of expressing the δ-endotoxin at insecticidal levels (for review, see Koziel *et al.*, 1993b). The native genes from *Bt* encoding the δ-endotoxins are not expressed in maize. It has been speculated that the high A + T content of the native endotoxin genes result in a number of fortuitous processing signals recognized by plant cells. These fortuitous processing signals could initiate polyadenylation, intron splicing, mRNA instability or, most likely, some combination of these. Expression of high levels of the δ-endotoxins has been effected in plants by introducing synthetic genes encoding δ-endotoxins or insecticidal fragments of δ-endotoxins (Koziel *et al.*, 1993a; Perlak *et al.*, 1991). These synthetic genes all contain a higher G + C content than the native gene, thus removing the fortuitous processing signals. We engineered a maize optimized synthetic gene encoding CryIA(b) derived from *Bacillus thuringiensis* var. *kurstaki* HD-1 (Koziel *et al.*, 1993a) by using the most preferred codon from maize for each amino acid (Murray *et al.*, 1989) and increasing the G + C content from 37% in the native gene to 65% in the synthetic gene. Our synthetic gene encodes the N-terminal 648 amino acids of the *cryIA(b)* gene. In addition to designing a gene for optimal expression in maize it is also critical to engineer it such that the insecticidal protein will be expressed in plant tissues that are most susceptible to insect feeding. The maize optimized *cryIA(b)* gene was placed under the control of the maize phosphoenolpyruvate carboxylase (PEPC) promoter (Hudspeth and Grula, 1989) which expresses primarily in green tissues; a maize pollen-specific promoter maize (Estruch *et al.*, 1994) and a maize pith-preferred promoter to provide high expression in the stalk (Kramer and Koziel, 1995). Transgenic maize lines also contain the *cryIA(b)* gene under control of a maize promoter from a metallothionein-like promoter (de Frammond, 1991). All promoter combinations tested have proven capable of providing protection from extremely heavy infestation with ECB in the field.

Transformation of Maize

In the past several years tremendous progress has been made in the efficiency of maize transformation (Fromm *et al.*, 1990; Gordon-Kamm *et al.*, 1990; Koziel *et al.*, 1993a). As a result of these transformation successes combined with the ability to design and synthesize genes capable of being expressed in maize cells, transgenic maize plants expressing insecticidal levels of a given δ-endotoxin are being routinely tested in the field for their ability to withstand insect feeding damage (Armstrong *et al.*, 1995; Koziel *et al.*, 1993a). We have used particle bombardment to introduce foreign DNA into either immature maize embryos or type I callus of various elite maize genotypes. Early transformation success was dependent upon use of non-elite lines. Breeding programs for conversion of non-elite transformed lines into elite commercial lines can take several years and in certain cases the transgenic loci may be closely linked to undesirable phenotypic characteristics which reduce yield. The occurrence of this yield drag is minimized when transformation can be done directly into commercial lines. We have been able to manipulate elite commercial genotypes in tissue culture and render them amenable to transformation and regeneration. The time it takes to introduce a transgenic hybrid to the market is greatly decreased if the original transformation is done with an elite line.

Results of Field Trials with European Corn Borer Resistant Transgenic Maize

Efficacy of Transgenic Lines

Our first insecticidal transgenic maize field test in 1992 was designed to evaluate two maize lines in elite commercial genotypes that were expressing high levels of CryIA(b). Maize optimized *cryIA(b)* vectors containing either the CaMV 35S promoter or the maize PEPC and pollen promoters were introduced into maize via Biolistics. Immature embryos were plated and bombarded using the PDS-1000He Biolistic device (BIO-RAD, Hercules, CA) as previously described (Koziel *et al.*, 1993a). The selectable marker for the transformation experiments was the *bar* gene which confers resistance to the herbicide phosphinothricin (Thompson *et al.*, 1987). Several different transformation events expressed detectable levels of CryIA(b) protein by ELISA and also exhibited strong insecticidal activity in ECB bioassays. Based on strong insecticidal activity, two lines, Event 171 and Event 176, were chosen for further analysis. Plants from each event were crossed with several different elite maize inbred lines and T1 progeny field tested (Koziel *et al.*, 1993a). Event 171 contained the synthetic *cryIA(b)* gene under control of the CaMV 35S promoter and Event 176 contained two versions of the synthetic *cryIA(b)* gene, one under control of the maize PEPC promoter and the other under control of a maize pollen promoter. The PEPC promoter is expressed in essentially all green tissues of maize while the pollen promoter is specific for pollen. When plants in the field reached about 40 cm of extended leaf height, infestation with laboratory reared ECB larvae was started on both the transgenic and non-transgenic controls. About 300 neonate larvae were applied per week to each plant for four consecutive weeks to simulate first generation corn borer infestations. Starting two weeks after the initial ECB infestation each plant was rated weekly for four weeks for ECB1 foliar damage using a modified 1-9 Guthrie rating scale (Guthrie *et al.*, 1960) where a rating of 1 was reserved for plants with no visible damage. The mean ECB1 damage rating was determined for each line. As plants reached anthesis, 300 larvae per week per plant were applied for another four consecutive weeks to simulate second generation infestation. Approximately 50 days following the first infestation of second generation ECB, tunneling damage was measured in a 92 cm section of stalk, 46 cm above and 46 cm below the primary ear node. Maize lines expressing CryIA(b) protein provided season-long protection from repeated ECB infestations. Protection from both first and second brood injury was achieved despite the abnormally high infestation rate of 2400 larvae per plant. All transgenic plants from both events were superior to non-transgenic inbreds for protection from both foliar feeding damage and internal stalk damage. When averaged over all hybrid lines, the lines containing the PEPC and pollen promoter *cryIA(b)* genes had better performance than the 35S promoter *cryIA(b)* lines for foliar and internal stalk damage. The average leaf damage rating and mean tunnel length in cm for the best transgenic line of Event 176 was 1.6 and 1.7, respectively, compared to the control inbred with a foliar damage rating of 7.2 and tunneling damage of 59.3 cm.

Comparisons Between Field Efficacy and CryIA(b) Levels in Leaf and Pith

Subsequent to our 1992 field test we have generated over 200 independent trans-genic maize events from microprojectile bombardment of various maize tissues, including immature embryos and Type I callus. Many of these transgenic lines have been evaluated in field tests. New lines tested include plants transformed with either the truncated or the full length maize optimized gene expressed by various com-binations of promoters derived from maize including the PEPC, pollen-specific, and pith-preferred promoter. A pith promoter was isolated from maize to generate lines with high pith expression thought to be important for good control of stalk boring ECB. Many of the new events containing the synthetic gene expressed using a variety of maize promoters provide excellent protection against first and second brood ECB attack. All transgenic plants examined to date expressing CryIA(b) at detectable levels are superior to lines containing a natural ECB resistance allele as assessed by both foliar damage and tunneling damage.

In 1994 we field tested several new CryIA(b) lines that had different levels of expression and different patterns of expression in critical ECB target tissues. In the transgenic lines that we studied, levels of CryIA(b) in leaves are stable throughout development with a slight increase occurring at anthesis. The combinations of tissue-specific promoters that we have used have allowed us to generate lines with high levels of CryIA(b) in green tissue, pith, and pollen. Table 4.1 shows the CryIA(b) protein concentrations in ng CryIA(b) per mg soluble protein in both leaf and pith tissue and relates this to the ECB stalk tunneling damage. A non-transformed control line, 684, shows the most tunneling damage with a mean of 87.6 cm. Lines with high leaf and pith expression, such as 852 and 934, withstood insect attack and showed little tunneling damage. Line 845 which contains lower levels of CryIA(b) in both leaf and pith also provided excellent control against ECB tunnel-ing damage. Excellent ECB protection can be obtained even in plants expressing approximately 10 ng CryIA(b) per mg soluble protein in leaf and pith. This is due, at least in part, to the high specific activity of the CryIA(b) protein against ECB. While a plant may be producing a lower level of CryIA(b) when compared with other plants, as long as that level is above the level needed for control of ECB and is in the tissues upon which ECB feeds, the plant is protected. Lines 592, 225, and 684 all had high levels of CryIA(b) expression in the leaf but less than detectable expression of CryIA(b) in pith tissues. These lines were very susceptible to damage from second

Table 4.1 Second generation ECB damage

Line	CryIA(b) Leaf conc. (ng/mg protein)		CryIA(b) Pith conc. (ng/mg protein)		Mean tunnel length (cm)	
	Mean	Range	Mean	Range	Mean	Range
852	819 ± 157	613–984	730 ± 238	487–956	0.75 ± 1	0–2
845	12 ± 4	7–16	9 ± 3	6–12	2 ± 2	0–4
934	166 ± 44	103–204	213 ± 150	118–447	8 ± 3	5–11
592	26 ± 7	17–33	< 5	—	27 ± 23	13–61
225	662 ± 180	517–924	< 5	—	57 ± 24	41–85
684	114 ± 12	104–130	< 5	—	86 ± 5	80–90
Control	0		0		87.6 ± 4	

generation ECB stalk tunneling. As expected from the high level of CryIA(b) in the leaves, the foliar damage on all these transgenic CryIA(b) lines was very low, less than 2 on the Guthrie scale. Although high levels of CryIA(b) expression in leaves may be an indication of a highly expressing line it alone cannot be taken as an indicator of a line with good season-long protection against ECB. High levels of CryIA(b) in leaves provides good protection against first generation ECB that feed primarily on leaf and leaf whorl tissue but not against the extensive stalk damage cause by subsequent generations of ECB. The maize PEPC promoter provides expression in green tissue including leaf and the rind surrounding the stalk but does not always provide expression in pith. Events expressing high levels of CryIA(b) will sometimes have expression in pith tissue as is seen in Event 176 (Koziel *et al.*, 1993a) but other events such as 592, 225, and 684 that have no detectable pith expression are not able to provide season-long protection against ECB. Transgenic lines that contain *cryIA(b)* genes under the control of the PEPC, pollen, and pith promoters, afford an insurance of generating lines with good pith expression for stalk protection. Expression of CryIA(b) in other ECB target tissues such as pollen and silks may play at least a contributing role in producing a line with superior ECB resistance. Results in Table 4.1 point to the importance of producing at least 10 ng CryIA(b) per mg soluble protein in pith. The importance of CryIA(b) expression in other tissues such as pollen and silks is still being evaluated. Recent studies with transgenic lines containing the *cryIA(b)* gene under control of a maize metallothionein-related promoter show that this promoter can provide a high level of expression in both leaf and pith resulting in excellent ECB protection.

Conversion and Testing of Transgenic Maize Lines for Commercialization

The transgenic locus of Event 176 containing two versions of the maize-optimized *cryIA(b)* gene under control of the maize PEPC and a maize pollen promoter has now been moved into many elite maize inbred lines. The original transformed maize event was first field tested in 1992 and by 1995 inbred maize lines had been converted with it and hybrids made from those inbreds. Transgenic ECB resistant inbreds converted with Ciba Seeds' transformation Event 176 were analyzed for standard traits in the absence of European corn borers. Commercial hybrid maize lines containing the maize optimized *cryIA(b)* gene show no yield penalty for expressing foreign genes when compared with isogenic hybrids not containing the transgenes. Traits such as yield, moisture, and stalk strength were found to be identical to the normal inbred within small variations not statistically significant, indicating that there were no deleterious genes in the recipient genotype tightly linked to the inserted DNA. Neither the expression of CryIA(b) nor the insertion of foreign genes into the Event 176 transgene insertion locus has a deleterious impact on yield or other physiological processes. Figure 4.2 shows the reduced tunneling in Event 176 versus the isogenic ECB susceptible hybrid measured in small plot trials under natural and artificial ECB infestations. When compared to ECB infested plants, the CryIA(b) expressing hybrids consistently produce a significantly higher yield than their non-transgenic counterparts (Figure 4.3). The average US yield increase over two years of testing is about 14%. *Bt* maize provides season-long control with 95% reduction in ECB feeding damage achieved under field conditions with heavy insect pressure. Additionally, since the insecticidal protein is present in the plant during the entire growing season, ECB control is achieved regardless of the number and timing of moth flights.

Following proper engineering of a maize optimized *cryIA(b)* gene, transformation of elite maize hybrids, and conversion of other elite genotypes by backcrossing using molecular markers, the result is a set of new insect resistant maize hybrids with outstanding performance versus the best conventional hybrids available (Figure 4.4).

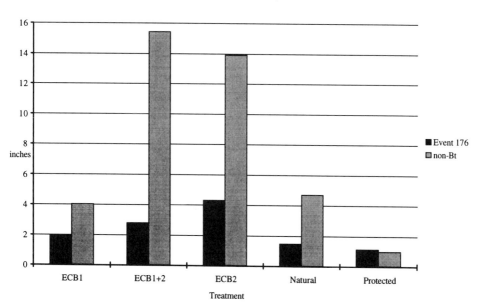

112 Day Hybrid: Small Plot Tunneling Damage at 8 Locations in 1993-1994

Figure 4.2 ECB larval tunneling in *Bt* (Event 176) and normal isogenic hybrids under artificial infestations of first (ECB1) and second (ECB2) generation ECB; natural infestation; and chemically protected plants in small plot trials (Evola, 1996)

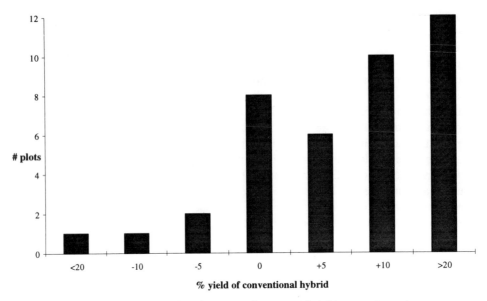

Figure 4.3 Degree of superior yield performance of Event 176 hybrids versus isogenic conventional hybrids in 1994 strip tests under natural ECB infestation (Evola, 1996)

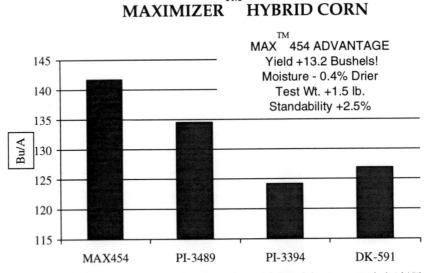

Figure 4.4 Yield, moisture at harvest, test weight and standability of the Event 176 hybrid (Ciba Seeds Max 454) versus the mean of three competitor hybrids of the same maturity

Overview of Registration Process for ECB Resistant Maize Hybrids

In August 1995, Event 176 became the first transgenic maize line to be registered with the United States Environmental Protection Agency for commercialization. As part of this registration process, it has passed several safety tests for both health and environmental impact. It is currently sold under the trade name of Ciba Seeds Maximizer™ Hybrid Corn with KnockOut™ Built-In Corn Borer Control.

In the US, regulatory oversight is provided by three agencies of the federal government: the Environmental Protection Agency (EPA), the Food and Drug Administration (FDA), and the US Department of Agriculture (USDA). The EPA review of the data package included: product (novel proteins) chemistry studies (molecular weight, N-terminal sequencing, post-translational modifications and breakdown products); comparison of bacterial and plant produced insecticidal proteins; quantification of novel proteins in plant parts at various developmental stages; mammalian safety tests including *in vitro* digestibility studies and acute oral toxicity studies in mice; and tests for effects on non-target organisms (quail, earthworm, ladybeetle, Daphnia, honeybee, dynamics of non-target insect populations). The EPA review found no health issues, no safety issues, and no environmental issues. The only issue was resistance management due to a concern that *Bt* maize might jeopardize the use of *Bt* microbials by organic farmers. The EPA registration of Event 176 was granted with certain terms and conditions. The registration is for five years, renewable based on development of an effective resistance management strategy.

Ciba Seeds will develop a long-term resistance management plan within three years, which includes 'structured refugia' or an effective alternative and implement this plan within five years. Representative distribution areas will be monitored annually for resistant ECB and findings reported to the EPA. Customers will be instructed to contact Ciba Seeds if unexpected levels of ECB damage occur, and incidents of unexpected ECB damage will be investigated. Cases of 'confirmed resistance' as defined by controlled ECB studies showing a significant increase in LC_{50} in a standard CryIA(b) diet bioassay and significant damage to Maximizer™ leaf tissue will be reported to the EPA. For cases of 'confirmed resistance' immediate mitigation measures will be implemented including notification of local customers and extension agents, and recommending methods to reduce local ECB populations. Other action items will include reporting information for each state to the EPA annually, and providing grower education to increase awareness of resistance management and promote responsible product use. Ciba Seeds will also implement an EPA-approved research program on ECB biology (including ECB movement, survival in maize and overwintering in non-maize hosts); the feasibility of 'structured refugia' for resistance management; a diagnostic assay that can identify resistant ECB; the effects of Maximizer™ maize on non-ECB pests; and the biology of ECB resistance.

The FDA (Center for Food Safety and Nutrition and the Center for Veterinary Medicine) raised no concerns after reviewing information supporting conclusions that the nutritional composition of Event 176 maize and the levels of natural toxins were unchanged and that the novel gene products raised no safety or allergenicity issues. The USDA reviewed all data submitted to the EPA and FDA, as well as the agronomic performance of Event 176 maize, which was found to be indistinguishable from isogenic maize except for ECB resistance. Overwintering capability was not altered, nor was the nutritional composition.

In summary, evaluations of Event 176 maize have shown efficacious biological control of ECB. In addition, Event 176 plants are indistinguishable from non-transformed plants in the absence of ECB infestation, yet are superior to non-transgenic plants in the presence of ECB.

Resistance Management

There is concern that the widespread use of maize expressing *Bt* δ-endotoxins to control insect damage might lead to development of resistance of the insect pest to the δ-endotoxins expressed in the maize lines. Such concerns are based largely on predictions from computer models. Small changes in various parameters have a dramatic effect on the predictions obtained from these models. Cases where resistance to *Bt* δ-endotoxins has occurred in the laboratory and in nature are rare. Resistance of this kind has typically developed, or been cultivated, using isolated populations of insects and selecting with sublethal concentrations of endotoxin for short periods of the insect's life cycle (acute exposure). Transgenic plants producing high levels of δ-endotoxin present a different type of selection pressure, a high chronic dose exposure. No reports of resistance to chronic high-dose exposure of δ-endotoxins are known. Although strategies have been proposed to prevent, or delay, the onset of resistance (McGaughey and Whalon, 1992), which one will be most effective in the field is unclear at this point in time. It is likely that development of resistance will be heavily influenced by both the insect and crop in question and the resistance management strategy must be tailored to each crop–insect combination. ECB usually has only two generations per year in the majority of the US corn belt and would make resistance seem a much smaller issue in maize for ECB than for other crop–insect combinations where there are more generations per year.

Because of the desire to maintain the efficacy of the *Bt* endotoxins as insecticidal agents, we are actively evaluating several potential strategies for preventing or delaying occurrence of resistance. The expression of more than one active compound such as two δ-endotoxins which bind different receptors in the target insect's midgut, has been proposed as one means of preventing the development of resistance. A factor which supports this particular strategy is the fact that the rate of discovery of new insecticidal proteins from various new *Bt* strains is increasing (Feitelson *et al.*, 1992). This increased rate of discovery will provide more active components which can be used for preventing the development of resistance. Further, as agricultural biotechnology continues to advance, new insecticidal constituents derived from sources other than *Bt* δ-endotoxins will be discovered and developed. For instance, cholesterol oxidase has been shown to possess insecticidal activity (Purcell *et al.*, 1993; see also Chapter 6). New insecticidal activities have been discovered in vegetative *Bacillus* which are based on proteins not related to the endotoxins (see also Chapter 7). Such new insecticidal proteins will furnish additional options for insect control and for resistance management. Sensible use of transgenic plants combined with careful planning and other available pest control alternatives will ensure that resistance does not become a significant issue.

Summary

Maize plants expressing a synthetic maize optimized version of the *cryIA(b)* δ-endotoxin gene are protected from severe ECB infestations in the field. Various promoters, including four different maize promoters, have proven effective in expressing insecticidal levels of CryIA(b) in ECB target tissues. Plants producing a δ-endotoxin from *Bt*, like those derived from Event 176, are the first to be commercialized as insect tolerant transgenic crops, offering growers a new source of protection from damage caused by insect pests. These plants provide protection even from insects which burrow into the plant and are not accessible by chemical sprays. In addition to higher profits the customer has greater peace of mind because, although ECB is widespread, it is sporadic, requiring scouting and spraying, the costs of which can exceed losses from the insect. Treatment efficacy is highly variable, often ineffective. Timing of pesticide treatments must coincide with flights of ECB moths and multiple applications are often necessary. ECB, if left untreated, can cause 5–25% yield reduction. Even with treatment, US maize growers suffer about $800 million annual revenue loss from ECB infestations. There are several environmental benefits of transgenic *Bt* maize. There is reduced use of chemical pesticides and the resulting exposure of workers and environment to pesticide pollution. Because the *Bt* protein is produced and contained within the maize plant there is no need for repeat applications; the *Bt* protein is rapidly degraded in the environment and there are no adverse effects on beneficial insect populations. Transgenic ECB resistant maize hybrids provide a safe and efficacious solution to a pest problem that causes huge losses in grain yield globally each year.

As additional sources of insecticidal proteins or compounds are identified and characterized, new alternatives will also become available. These new alternatives will allow the control of insect pests not susceptible to known *Bt* δ-endotoxins and will further assist the prevention or management of resistance to the new traits and chemical insecticides if such a resistance should arise.

References

Armstrong, C. L., Parker, G. B., Pershing, J. C., Brown, S. M., Sanders, P. R., Duncan, D. R., Stone, T., Dean, D. A., DeBoer, D. L., Hart, J., *et al.* (1995) Field evaluation of European corn borer control in progeny of 173 transgenic corn events expressing an insecticidal protein from *Bacillus thuringiensis*, *Crop Sci.* **35**, 550–557.

Christensen, D., Beland, G. and Meghji, M. (1993) Yield loss due to European corn borer in normal and transgenic hybrids, *Proceedings of the 48th Annual Corn and Sorghum Research Conference*, American Seed Trade Association, pp. 43–52.

de Frammond, A. J. (1991) A metallothionein-like gene from maize (*Zea mays*), *FEBS* **290**, 103–106.

Estruch, J. J., Kadwell, S., Merlin, E. and Crossland, L. (1994) Cloning and characterization of a maize pollen-specific calcium-dependent calmodulin-independent protein kinase, *Proc. Natl. Acad. Sci. USA* **91**, 8837–8841.

Evola, S. V. (1996) Transgenic European corn borer resistant maize, *Gesellschaft für Pflanzenzüchtung, Vorträge für Pflanzenzüchtung* **33**, 22–34.

Feitelson, J. S., Payne, J. and Kim, L. (1992) *Bacillus thuringiensis*: insects and beyond, *Bio/Technology* **10**, 271–275.

Fromm, M. E., Armstrong, C., Williams, R., Thomas, J. and Klein, T. M. (1990) Inheritance and expression of chimeric genes in the progeny of transgenic maize plants, *Bio/Technology* **8**, 833–839.

Gordon-Kamm, W. J., Spencer, T. M., Mangano, M. L., Adams, T. R., Daines, R. J., Start, W. G., O'Brien, J. V., Chambers, S. A., Adams, W. R., Willetts, N. G. *et al.* (1990) Transformation of maize cells and regeneration of fertile transgenic plants, *Plant Cell* **2**, 603–618.

Guthrie, W. D., Dicke, F. F. and Neiswander, C. R. (1960) Leaf and sheath feeding resistance to the European corn borer in eight inbred lines of dent corn, *Ohio Agricultural Experiment Station Research Bulletin* No. 860.

Hudspeth, R. L. and Grula, J. W. (1989) Structure and expression of the maize gene encoding the phosphoenolpyruvate carboxylase isozyme involved in C4 photosynthesis, *Plant Molec. Biol.* **12**, 579–589.

Koziel, M. G., Beland, G. L., Bowman, C., Carozzi, N. B., Crenshaw, R., Crossland, L., Dawson, J., Desai, N., Hill, M., Kadwell, S. *et al.* (1993a) Field performance of elite transgenic maize plants expressing an insecticidal protein derived from *Bacillus thuringiensis*, *Bio/Technology* **11**, 194–200.

Koziel, M. G., Carozzi, N. B., Currier, T. C., Warren, G. W. and Evola, S. V. (1993b) The insecticidal crystal proteins of *Bacillus thuringiensis*: past, present and future uses, In: Tombs, M. P. (Ed.), *Biotechnology and Genetic Engineering Reviews*, Andover: Intercept, pp. 171–228.

Kramer, V. C. and Koziel, M. G. (1995) Structure of a maize tryptophan synthase alpha subunit gene with pith enhanced expression, *Plant Mol. Biol.* **27**, 1183–1188.

Lynch, R. E. (1980) European corn borer: yield losses in relation to hybrid and stage of corn development, *J. Econ. Entomol.* **73**, 159–164.

McGaughey, W. H. and Whalon, M. E. (1992) Managing insect resistance to *Bacillus thuringiensis* toxins, *Science* **258**, 1451–1455.

Murray, E., Lotzer, J. and Eberle, M. (1989) Codon usage in plants, *Nucl. Acid Res.* **17**, 477–498.

Perlak, F. J., Fuchs, R. L., Dean, D. A., McPherson, S. L. and Fischhoff, D. A. (1991) Modification of the coding sequence enhances plant expression of insect control protein genes, *Proc. Natl. Acad. Sci. USA* **88**, 3324–3328.

Purcell, J. P., Greenplate, J. T., Jennings, M. G., Ryerse, J. S., Pershing, J. C., Sims, S. R., Prinsen, M. J., Corbin, D. R., Tran, M., Sammons, R. D. and Stonard, R. J. (1993) Cholesterol oxidase: a potent insecticidal protein active against boll weevil larvae, *Biochem. Biophys. Res. Comm.* **196**, 1406–1413.

Showers, W. B., Witkowski, J. F., Mason, C. E., Calvin, D. D., Higgins, R. A. and Dively, G. P. (1989) *European Corn Borer: Development and Management*, North Central Region Ext. Pub. No. 327, Ames, Iowa: Iowa State University.

Thompson, C. J., Movva, N. R., Tizard, R., Crameri, R., Davies, J. E., Lauwereys, M. and Botterman, J. (1987) Characterization of the herbicide-resistance gene *bar* from *Streptomyces hygroscopicus*, *EMBO J.* **6**, 2519–2523.

5

Enhancing Insect Resistance in Rice Through Biotechnology

JOHN BENNETT, MICHAEL B. COHEN, SANJAY K. KATIYAR, BEHZAD GHAREYAZIE
and GURDEV S. KHUSH

Introduction

Multiple insect resistance is an essential attribute of modern rice varieties and a cornerstone of integrated pest management (IPM) for rice production. Biotechnology is beginning to contribute to the enhancement of insect resistance in several ways:

(1) DNA marker technology increases the efficiency with which breeders move known resistance genes into new cultivars.

(2) Wide hybridization and transformation extend the gene pool available to rice breeders.

(3) DNA fingerprinting of insects provides insights into genetic change in insect populations.

We present case studies on two groups of insects: the Asian rice gall midge (*Orseolia oryzae*) and rice stem borers (*Chilo suppressalis* and *Scirpophaga incertulas*). These insects exert their economic impact by preventing the development of whole panicles at a time when the host is unable to compensate for the loss. The study of resistance to gall midge, the major dipteran pest of rice, illustrates the application of DNA markers to rice breeding and the use of DNA fingerprinting to understand insect population structure and diversity. The study of stem borers, the major lepidopteran pests of rice, illustrates the importance of transformation in rice breeding.

The combination of improved host plant resistance with new insights into biocontrol and deployment may increase the durability of resistance mechanisms, clarify the agenda for rice pest research, and address concerns in the areas of biosafety and biodiversity.

Insects and Rice Production

Rice is attacked by more than 100 insect species but relatively few cause significant economic loss. Past successes in conventional breeding for resistance, the ability of

75

plants to compensate for many types of damage and biological control by natural enemies combine to limit many pest problems to a tolerable level (Khush, 1995; Way and Heong, 1994). This situation enables rice scientists to focus on the few insects that cause significant yield losses on a chronic or sporadic basis.

It is particularly important to protect modern high-yielding cultivars from insect attack in a sustainable manner. The population of rice consumers is increasing at a rate of 2% annually, and yet the rate of growth of rice production has slowed to only 1.2% (Khush, 1995). If the present trend continues, the demand for rice will exceed production by the end of the century (Pinstrup-Andersen, 1994). Rice production will have to increase by 70% by the year 2020 to keep pace with demand. With little scope for expanding the irrigated area, this requirement can be met only by increasing production from existing rice lands. The future challenge of rice improvement is to develop varieties with higher yield potential and greater yield stability for both the favorable irrigated environment and the favorable and unfavorable rainfed environments.

New cultivars must possess multiple insect and disease resistance to enable them to reach their yield potential. The challenges facing insect resistance breeding in rice are therefore:

— to increase the efficiency with which known resistance genes are incorporated into new breeding lines;

— to introduce new resistance genes as insects and diseases inevitably adapt to existing defense mechanisms; and

— to find ways of prolonging the useful life of resistance genes through novel deployment strategies linked with a greater understanding of insect ecology.

In this paper we examine the growing contribution of biotechnology to insect resistance breeding.

Rice Biotechnology and Insect Resistance

Plant breeding occurs in two phases. Phase 1 is the production of novel genetic variation, and Phase 2 is the selection of improved variants. In the context of insect resistance in rice, the two most important components of Phase 1 have been the screening of rice germplasm to identify novel donors of resistance (Jackson, 1995), and the use of these donors in sexual hybridization with elite cultivars to create novel combinations of genes (Panda and Khush, 1995). Selection of improved variants from among the progeny of sexual hybrids has followed standardized phenotypic test procedures in multilocational trials in greenhouse and field. This approach has been extremely successful in the past in enhancing insect resistance (Khush, 1995) and will continue to be important in the future.

Nevertheless, plant breeders are looking to biotechnology to provide new approaches to both phases. Four biotechnological approaches are contributing to the enhancement of insect resistance in rice (Table 5.1). Wide hybridization and transformation contribute to the generation of novel genetic variation by increasing the gene pool available to breeders (Phase 1), while DNA markers and DNA fingerprinting of insects help to make selection and phenotypic analysis more efficient and powerful in certain situations (Phase 2). We shall now briefly discuss each approach.

Table 5.1 Biotechnological approaches to enhancing insect resistance in rice

Approach	Contribution	Examples
Wide hybridization	Gives access to genes in non-AA genome wild rices	Whitebacked planthopper
DNA markers	Accelerates breeding, permits pyramiding of genes of similar effect	Brown planthopper, gall midge
Transformation	Gives access to genes from any organism and synthetic genes, allows novel modifications of rice genes	Stem borers
DNA fingerprinting	Reveals biodiversity of insect at all levels from field to region, improves deployment strategies	Gall midge

Wide Hybridization/Embryo Rescue

Brar and Khush (1995) provide an account of wide hybridization in rice. The genus *Oryza* contains a total of 23 species. Cultivated rice, *O. sativa*, and seven other diploid species contain the AA genome, while 15 diploid and tetraploid species belong to six different genome groups (BB, CC, BBCC, CCDD, EE, FF). Most rice breeding is done within *O. sativa* (the primary gene pool). The AA genome wild species (the secondary gene pool) and the non-AA genome wild species (the tertiary gene pool) contain many useful resistance genes for biotic and abiotic stresses, including insects. Transfer of genes from the secondary gene pool to *O. sativa* may be accomplished by conventional sexual hybridization but the tertiary gene pool is not normally accessible by this method because of hybrid embryo abortion.

Rescue of hybrid embryos through repeated cycles of tissue culture and back-crossing to *O. sativa* allows recovery of new lines containing segments of the wild genome introgressed into the *O. sativa* background. Screening identifies lines into which insect resistance has been introgressed. Table 5.2 summarizes the status of the International Rice Research Institute's (IRRI) wide hybridization program with

Table 5.2 Progress in the transfer of genes for insect resistance from wild species to *Oryza sativa* by wide hybridization and embryo rescue (from Brar and Khush, 1995)

Resistance transferred	Donor wild species (genome)	Status
Brown planthopper	*O. officinalis* (CC)	Completed
	O. minuta (BBCC)	Completed
	O. latifolia (CCDD)	Completed
	O. australiensis (EE)	Completed
	O. granulata (unknown)	Material under test
Whitebacked planthopper	*O. officinalis* (CC)	Completed
Yellow stem borer	*O. brachyantha* (FF)	Material under test
	O. ridleyi (unknown)	Material under test

respect to insect resistance. Several of these lines are now grown in farmers' fields. One of the ongoing projects is the transfer of yellow stem borer resistance from *O. brachyantha* to *O. sativa*.

The Need to Map Genes for Insect Resistance

One of IRRI's breeding priorities over the last 25 years has been to incorporate genes for insect resistance into elite inbred indica lines (Khush, 1995). Although highly successful, the efficiency of this program is limited by a number of factors. First, for some insects such as the stem borers, the resistance obtainable from known donors is only moderate at best, despite the screening of over 15 000 accessions from IRRI's rice germplasm collection (Jackson, 1995). Second, hybridization between a susceptible elite line and a resistant landrace commits the breeder to 6–8 cycles of backcrossing and selection to recover insect resistance in the elite background while minimizing the contribution of donor genome. Third, screening for insect resistance, although simple in concept, can encounter severe difficulties. It is quite common for even simple breeding programs to become quite protracted, with selection possible only once a year and perhaps even less frequently because of adverse weather conditions. Fourth, populations of a given insect species often show differences in adaptation to host genotypes, indicating genetic diversity among populations. The term biotype is frequently used to refer to populations differing in host adaptation. The existence of biotypes increases the load on the breeder by requiring different resistance genes to be introduced to cope with different biotypes.

The above problems offer scope for DNA marker-aided selection. Once a gene for insect resistance has been tagged, its inheritance may be followed at any season, in any weather and at any location and genes of similar effect may be combined ('pyramided') for durability against several biotypes. DNA markers may also accelerate a breeding program by distinguishing between plants that are heterozygous and homozygous for a dominant resistance gene, and for identifying resistant lines in a backcross population when the resistance gene is recessive.

Progress to Date in Mapping Genes for Insect Resistance

Genes have been mapped for resistance to several homopteran insects including green leafhopper (*Nephotettix virescens*), brown planthopper (*Nilaparvata lugens*) and whitebacked planthopper (*Sogatella furcifera*), and one dipteran insect (gall midge) (Table 5.3). A gene for brown planthopper resistance gene was mapped after introgression into *O. sativa* from *O. australiensis*. The near-isogenic lines produced after wide hybridization, embryo rescue and repeated backcrossing are ideal for mapping. Two gall midge resistance genes (*Gm2* and *Gm6t*) have been mapped, and a third (*Gm4*) has been tagged and is currently being mapped.

In spite of their importance, stem borers do not appear in Table 5.3, because the moderate level of resistance obtained with stem borer resistance genes has discouraged mapping. Recently, however, a program to map stem borer resistance genes was initiated at IRRI. The combination of these genes with insecticidal genes introduced by transformation (see next section) will provide a level and mechanistic diversity of resistance that may prove effective and durable.

Table 5.3 Progress in mapping genes for insect resistance in rice (from Zheng *et al.*, 1996)

Gene	Insect	Donor	Chromosome	Linked RFLP marker (and distance, cM)	Reference
Bph10(t)	Brown planthopper	*Oryza australiensis*	12	RG457 (3.68)	Ishii *et al.* (1994)
Gm2	Gall midge	Phalguna	4	RG329 (1.3) RG476 (3.4)	Mohan *et al.* (1994)
Gm6(t)	Gall midge	Duokang #1	4	RG214 (1.2)	Katiyar *et al.* (1995)
Gm4	Gall midge	Abhaya	tagged	RAPD? (?)	Katiyar (unpublished)
Glh	Green leafhopper	ARC 11554	4	RZ262 (2.1)	Sebastian *et al.* (1996)
Wph1	Whitebacked planthopper	N-22	7	?	McCouch (1990)

The cost of DNA marker-aided selection is currently a factor limiting its application in rice breeding but the cost is expected to drop as scale-up and refinements in the technology occur. However, DNA markers are competitive in certain situations:

(1) When the real cost of traditional methods is also high because of labor costs, transport costs, need for replications, etc.

(2) When DNA markers permit novel outputs such as pyramids of genes of similar effect.

(3) When speed outweighs cost.

Transgenic Approach to Host Plant Resistance

The transgenic approach to host plant resistance is attractive for three reasons:

(1) It provides access to non-rice genes that are otherwise unavailable to rice breeders.

(2) It allows purified rice genes to be returned to rice after modifications that give enhanced performance not attainable through mutation and recombination *in vivo*.

(3) It allows the addition of specific characters into rice without the linkage drag and the requirement for backcrossing that accompany sexual hybridization.

The third point is especially important in a crop like rice in which many quality and adaptive traits are under complex and easily disrupted genetic control. Grain quality, yield and adaptation to adverse environments are some of the key polygenic traits in rice that are invariably disrupted by sexual hybridization but are unlikely to be disrupted by transformation.

Fertile transgenic rice may be produced by protoplast methods (Datta *et al.*, 1990; Shimamoto *et al.*, 1989), by microprojectile bombardment (Christou *et al.*,

Table 5.4 Transgenic mechanisms reported to enhance insect resistance in rice

Gene	Promoter	Insect	Reference
Synthetic *cryIA(b)*	Chlorophyll a/b protein	Leaffolder, striped stem borer	Fujimoto *et al.* (1993)
	CaMV35S	Striped stem borer, yellow stem borer	Wünn *et al.* (1996)
	Maize PEP carboxylase	Striped stem borer, yellow stem borer	This chapter
Synthetic *cryIA(c)*	Maize ubiquitin	Leaffolder	Riazuddin *et al.* (1996)
Potato proteinase inhibitor 2	Potato proteinase inhibitor 2	Striped stem borer, leaffolder	Duan *et al.* (1996)
Cowpea trypsin inhibitor	Rice actin-1	Striped stem borer, leaffolder	Xue *et al.* (1996)

1991) and by use of *Agrobacterium tumefaciens* (Hiei *et al.*, 1994). Of the six groups of rice plants within *O. sativa* (Glaszmann, 1987), the true indicas (Group 1) are usually the most difficult to regenerate from tissue culture after transformation, but the number of successful reports continues to grow, especially since the introduction of microprojectile bombardment.

Several insecticidal genes have been inserted into rice by transformation and have proved effective in enhancing resistance (Table 5.4). The genes include synthetic *Bacillus thuringiensis cryIA(b)* and *cryIA(c)* genes and several proteinase inhibitor genes under the control of various promoters. The insects shown to be controlled by these genes include: the leaffolder, *Cnaphalocrocis medinalis* (Fujimoto *et al.*, 1993; Wünn *et al.*, 1996), the yellow stem borer, *Scirpophaga incertulas* (Nayak *et al.*, 1996; Riazuddin *et al.*, 1996; Wünn *et al.*, 1996) and striped stem borer, *Chilo suppressalis* (Duan *et al.*, 1996; Fujimoto *et al.*, 1993; Wünn *et al.*, 1996; Xue *et al.*, 1996).

DNA Fingerprinting of Insects

Genetic markers have been used extensively to study insect population structure, distinguish biotypes and species and monitor the spread of insecticide resistance, among many other topics (Hoy, 1994). Studies with allozyme markers began in the 1960s. More recently a diversity of DNA techniques have been applied, including RFLP (restriction fragment length polymorphism), RAPD (polymerase chain reaction (PCR) based random amplified polymorphic DNA), and microsatellites. A frequent objective of research on host plant resistance to insects has been the development of genetic markers to distinguish biotypes collected from different regions and/or adapted to different crop resistance genes (Black *et al.*, 1992; Puterka *et al.*, 1993; Shufran and Whalon, 1996). In the case of rice gall midge, where different biotypes occur in different regions (Katiyar *et al.*, 1995), DNA fingerprinting is proving useful in characterizing population diversity, guiding breeding programs and improving strategies for the varietal deployment (see below).

Biotechnology and Integrated Pest Management

The adoption of IPM has been one of the most significant recent developments in rice production (Matteson *et al.*, 1994), and it is important that the products of rice biotechnology be compatible with existing IPM components. A central tenet of IPM for tropical rice is that insecticide use should be kept to a minimum, and that insecticide use is rarely justified (Matteson *et al.*, 1994; Way and Heong, 1994). Rice fields support a diverse and highly effective community of natural enemies, such as spiders and parasitic wasps. The fundamental importance of naturally occurring biological control in rice was not recognized until it was severely disrupted by calendar-based insecticide use, which was introduced to rice farmers in the 1960s along with the improved semi-dwarf varieties of the green revolution. Insecticide overuse resulted in massive outbreaks of the brown planthopper, previously a minor pest, and accelerated the breakdown of a series of brown planthopper-resistant varieties (Gallagher *et al.*, 1994; Way and Heong, 1994).

More recently, the tremendous ability of rice plants to compensate for insect feeding damage has also come to be appreciated (Way and Heong, 1994). The highly visible damage to rice leaves caused by foliage-feeding insects such as leaffolders and the rice whorl maggot was long assumed to result in yield losses. However, it has been demonstrated that large amounts of rice foliage can be removed without reducing yield; amounts exceed those damaged by typical population levels of leaffolders occurring in the field (Graf *et al.*, 1992; Hu *et al.*, 1993). In addition, improved rice varieties have a high tillering capacity and readily replace tillers damaged by stem borers at the vegetative stage of growth, although compensation for reproductive tillers damaged at later stages of crop growth is more limited (Rubia *et al.*, 1989).

Priority Setting

Given this current view of insect pest management in rice, the potential role of genetically engineered insect resistance must be carefully considered. Investments in genetic engineering should not divert attention from the importance of biological control, plant compensation, and the training of farmers to appreciate these processes. New strategies for the deployment of insect resistant cultivars are needed, to avoid cycles of resistance breakdown. A process of priority setting is necessary, so that genetic engineering is targeted at pests not adequately managed by natural enemies and existing forms of host plant resistance. As one component of this process, a project is underway at IRRI to quantify the impact of insects, diseases, and weeds through an empirical approach, based on field surveys of pest populations and experimental determination of the yield reducing effects of such populations (Savary *et al.*, 1995).

Because of the ability of rice to compensate for damaged foliage, the targeting of leaf-feeding insects for control by insecticides, conventional breeding, or genetic engineering cannot be justified in most cases. Less clear is the role of genetic engineering for control of homopteran pests, such as brown planthopper and whitebacked planthopper. In areas where insecticide use has diminished, outbreaks of rice planthoppers have become rare (Gallagher *et al.*, 1994; Way and Heong, 1994). The combination of biological control undisrupted by insecticides and host plant resistance produced by conventional breeding has proven to be very stable. Biotechnology is being used to introduce conventional brown planthopper resistance efficiently into new cultivars, such as IRRI's high-yielding new plant type for the irrigated ecosystem, using DNA markers to select *for* a resistance gene (Zheng *et al.*, 1996) and *against* the rest of the donor genome (Openshaw *et al.*, 1994).

Cultivars genetically engineered for resistance to planthoppers may be useful only in areas where outbreaks persist despite reduced insecticide use, or in certain temperate areas where waves of brown planthopper immigration from tropical regions can overwhelm populations of natural enemies. Genetic engineering may also contribute to control of rice tungro disease, the most important viral disease of rice, through control of the vectors, green leafhoppers (*Nephotettix* spp.). Progress has been reported in the transformation of rice with a lectin from the snowdrop plant, *Galanthus nivalis*, for control of brown planthopper and green leafhopper (Gatehouse *et al.*, 1996).

Stem borers appear to be an appropriate target for genetically engineered resistance because many rice production regions experience chronic yield loss from these insects, and because sources of high levels of stem borer resistance have not been identified in rice germplasm. In many areas, stem borers occur at low levels, causing yield losses less than 5%. In such areas, while individual rice farmers would not gain much yield increase from rice with enhanced stem borer resistance, the aggregate effect of a few per cent yield increase over large areas would be substantial. Pests characterized by outbreaks not linked to insecticide overuse, such as gall midges, are also important targets for biotechnological approaches. We present below two case studies on the use of biotechnology to enhance resistance to stem borers and gall midge, both of which reduce yield by preventing proper development of whole panicles at times when the plant is unable to compensate for the loss.

Case Study I: Rice Gall Midge

Occurrence and Life Cycle

The Asian rice gall midge (*Orseolia oryzae* Wood-Mason) is the most important dipteran pest of rice. It is found principally in southern China, Thailand, India and Sri Lanka, with frequent outbreaks also in Laos, Vietnam, Cambodia, Indonesia and Myanmar. A related species *O. oryzivora* is found in western Africa. As a Cecidomyid, rice gall midge is a relative of the Hessian fly of wheat (*Mayetiola destructor*) and the sorghum shootfly (*Contarinia sorghicola*).

The female gall midge lays eggs on leaf blades. After hatching, the neonate larvae migrate to the shoot apex where they feed on the meristem and disturb vegetative development. The last leaf primordium to be initiated prior to infestation ceases to develop as a blade and instead develops symmetrically as a 10–30 cm cylindrical gall, within which the larva feeds. After about ten days, the adult fly emerges at night from near the tip of the gall to mate and die within about 24 hours. No panicle is formed by the infested tiller, and the plant is unable to compensate for this loss.

A relative humidity of more than 90% is essential for growth of gall midge (Panda and Khush, 1995). The life cycle is adversely affected by even brief periods of low humidity. In most countries it is found only in the wet season, and even then it may not appear in adequate numbers for screening purposes. When conditions are favorable, however, it can be locally devastating because of the loss of entire panicles to gall formation.

Resistance Genes

Host plant resistance is the most effective means of controlling gall midge, and several excellent sources of resistance are known. Resistance can take the form of a hypersensitive reaction within the shoot meristem. The all-or-nothing nature of gall midge resistance in rice means that resistant varieties are vulnerable to the emergence of new forms of the insect, biotypes with differing host range. There are four biotypes of gall midge in southern China, at least five in India, and others in Sri Lanka, Thailand and Indonesia (Katiyar *et al.*, 1995).

The first gall midge resistance gene to be mapped was *Gm2* (Mohan *et al.*, 1994). This gene provides resistance against biotypes 1 and 2 in India. The second gene to be mapped was *Gm6(t)*, providing resistance against biotypes 1–4 in southern China (Katiyar *et al.*, 1995). Both genes map to similar positions on chromosome 4 of rice but it is not yet clear whether they are allelic or clustered. Mapping is in progress for other genes providing resistance to other biotypes in India and Sri Lanka. *Gm4*, which provides resistance against biotypes 1, 4 and 5 in India, has been tagged with three RAPD markers (Katiyar, unpublished results).

When phenotypic testing is straightforward, the use of DNA markers for single gene traits is normally unnecessary. DNA markers are helpful in the case of gall midge because:

— they enable breeding to proceed even in seasons and years when low humidity prevents the development of adequate selection pressures;
— they permit breeding to proceed in locations and countries where gall midge is absent (e.g., at IRRI); and
— they allow pyramiding of gall midge resistance genes for more durable resistance.

DNA Marker-Aided Selection

During 1996–97, DNA marker-aided selection will be used in breeding for gall midge resistance in India and China as part of a collaborative program on gall midge under the Asian Rice Biotechnology Network. This program is coordinated by IRRI and involves: the Indira Gandhi Agricultural University, Raipur; the Directorate of Rice Research, Hyderabad, India; the Guangdong Academy of Agricultural Sciences, Guangzhou, China; Kasetsart University, Kamphaengsaen; and the Department of Agriculture, Bangkok, Thailand. The objective in India will be to transfer *Gm2* to new germplasm developed for the rainfed lowlands, while in China the aim is to transfer *Gm6(t)* to the parental lines used in hybrid rice production.

Wherever possible, DNA markers for gall midge resistance genes will be PCR-based (Robeniol *et al.*, 1996; Zheng *et al.*, 1996). To facilitate PCR-based marker-

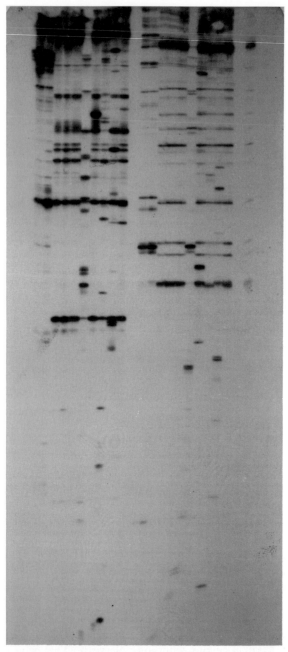

Figure 5.1 DNA fingerprinting of Asian rice gall midge (*Orseolia oryzae*) larvae collected in four countries. AFLP analysis (Vos *et al.*, 1995) was performed with a *Pst*I primer in combination with two different *Mse*I primers (left and right panels). The ten lanes in each panel correspond to the following gall midge biotypes collected from the indicated areas (left to right): 1, Chinese biotype 1 (Guangdong); 2, Chinese biotype 4 (Guangdong); 3, 4, Indian biotype 1 (Madhya Pradesh); 5, Indian biotype 2 (Orissa); 6, new Indian biotype from Manipur; 7, Indian biotype 5 (Kerala); 8, Nepal biotype/Indian biotype 3 (Nepal); 9, Sri Lanka biotype (Bombuwela); 10, Sri Lanka biotype (Batalagoda)

aided selection, we recently converted 350 cloned RFLP markers of the rice genetic map to sequence-tagged sites by sequencing of both ends of each clone (Robeniol *et al.*, 1996). The total of about 140 000 bases of sequence provides information for identifying polymorphic PCR-based markers within 5 cM of agronomically important genes. Polymorphisms have already been identified between the donors of the *Gm2* and *Gm6(t)* genes and planned recipients.

DNA Fingerprinting of Gall Midge across Asia

One of the major threats to the durability of gall midge resistance genes is the emergence of new forms or biotypes of the insect, presumably through selection, mutation or migration. DNA fingerprinting is proving to be a powerful tool to look at genetic diversity of gall midge within an area as small as a single field or as large as Asia itself. RFLP (Ehtesham *et al.*, 1995), RAPD and AFLP (amplification fragment linked polymorphism) markers are used in such studies.

The AFLP technique (Vos *et al.*, 1995) is probably the most powerful method in that it gives a high degree of reproducible variation between different gall midge isolates and is sensitive enough to be applied to individual insects in a population. During 1995, gall midge was sampled in south China (Guangdong), Nepal, India (Manipur, Orissa, Madhya Pradesh, Kerala) and Sri Lanka. A total of eight fingerprints were accomplished with these DNA samples using different AFLP primers. Figure 5.1 shows two AFLP fingerprints for these samples.

Several conclusions emerge from AFLP analysis of gall midge isolated across Asia:

— Gall midge isolates so far examined belong to two groups that are readily distinguishable with all AFLP primers utilized.

— One group is represented in our sample by the two isolates from Guangdong and the isolate from Manipur in eastern India.

— The two Guangdong isolates (biotypes 1 and 4 of China) are very closely related to one another but less closely related to the isolate from Manipur.

— The other Indian isolates (biotypes 1, 2 and 5 from Madhya Pradesh, Orissa and Kerala, respectively), the Nepalese isolate (biotype 3) and the two isolates from Sri Lanka are related to one another but are distinct.

— The Sri Lankan isolates are sufficiently different from one another to be likely to correspond to different biotypes with different host ranges.

The current Manipur biotype was first detected only 2–3 years ago. It replaced the previous Manipur biotype, which showed the same host range as the Nepal biotype and was classified as biotype 3. The marked divergence in fingerprints between the current Manipur and Nepal biotypes is consistent with the emergence of a new biotype in Manipur and would explain the sudden susceptibility of previously resistant varieties. Further gall midge collections will be made in 1996 in Myanmar, Thailand, Laos, Cambodia, Vietnam and Indonesia to make a complete survey of the Asian rice gall midge. However, the most interesting regions at present are in eastern India where the two major groups of gall midge appear to meet and in Sri Lanka, where gall midge diversification seems to be underway.

Case Study II: Transgenic Resistance to Stem Borers

Occurrence and Life Cycle

More than 40 species of stem boring caterpillars are known to attack rice, the most important of which are the yellow stem borer, *Scirpophaga incertulas*, in tropical Asia, and the striped stem borer, *Chilo suppressalis*, in temperate areas (Pathak and Khan, 1994). Stem borer larvae bore into the plant shortly after hatching. Early instars may feed on various tissues of the plant, including leaf sheaths, the developing panicle, and spikelets, but the older larvae feed principally on the nodes and internodes of the stem. The larvae may kill several tillers of the rice plant while completing development. Pupation usually occurs in the stem, and older larvae may also diapause in the stem between rice crops. Adult emergence, mating, and oviposition generally occur at night. Adults can be collected on rice vegetation during the day.

Non-Transgenic Host Plant Resistance

Although only partial levels of stem borer resistance have been achieved by conventional breeding, most modern rice varieties are more resistant to stem borers than the traditional varieties that they replaced. IRRI began evaluating germplasm for resistance to stem borers in 1962 and began breeding for resistance in 1972 (Heinrichs, 1994). Varieties such as TKM6 and IR-1820-52-5 were used as sources of resistance and provided the moderate resistance now available in many advanced lines (Khan *et al.*, 1991). These donors possess morphological traits such as narrower stems, short internodes, tighter leaf sheaths, reduced tillering and ovipositional deterrents such as oryzanone.

One possible route to high-level stem borer resistance is the use of wide hybridization and embryo rescue to transfer resistance genes from wild species of rice such as *O. brachyantha*. These experiments are in progress at IRRI and their outcome is awaited (Table 5.2). A second approach, transfer of novel mechanisms of insect resistance into rice by genetic engineering, is discussed in the following sections.

The δ-Endotoxins of **Bacillus thuringiensis**

The first report of transformation of rice with a *Bt* toxin gene and regeneration of fertile plants was published by Fujimoto *et al.* (1993). These authors transformed protoplasts of the japonica variety Nipponbare with a codon-optimized, truncated *crylA(b)* gene under control of the CaMV 35S promoter. Two independent lines, B2-9 and F2-11, carried a single, stably inherited and functional *crylA(b)* gene. Levels of *CrylA(b)* in leaves were estimated to be up to 0.5 μg/mg of total soluble protein. Preliminary bioassays with the striped stem borer and a species of rice leaffolder, *Cnaphlacrocis medinalis*, indicated that the transgenic lines reduced larval growth rate and survival.

Attention is now focused on the transformation of indica rice varieties with *Bt* genes. Indica varieties are grown far more widely than are japonica varieties, but it has proven more difficult to regenerate transformed indica plants. Wünn *et al.*

(1996), Riazuddin *et al.* (1996) and Nayak *et al.* (1996) have reported successful regeneration of indica plants transformed with δ-endotoxin genes. Wünn *et al.* (1996) used a codon-optimized, truncated *cryIA(b)* gene (Koziel *et al.,,* 1993) with a CaMV 35S promoter. Several lines were derived from a single transformation event of immature embryos of the variety IR58, and showed stable integration at a single locus. Levels of up to 84 ng/mg soluble leaf protein were detected by ELISA. Reduced survival and growth of larvae of the striped stem borer and yellow stem borer were detected in bioassays.

Riazuddin *et al.* (1996) reported transformation of the variety Basmati 370 with a codon-optimized, truncated *cryIA(c)* gene under control of a ubiqutin promoter. The construct is described by Altosaar *et al.* (1996). Seven independent lines are under evaluation for stable inheritance of *cryIA(c)*, three of which were reported to have repellent or toxic effects on *C. medinalis*. Nayak *et al.* (1996) reported success in obtaining fertile transformants of IR64 and IR72 containing codon-optimized, truncated forms of *cryIA(b)* and *cryIA(c)* with CaMV 35S and ubiquitin promoters, respectively.

At IRRI, transformation of rice with a synthetic *cryIA(b)* gene under the control of the maize PEP carboxylase promoter produced a line with high expression of the *cryIA(b)* gene and enhanced resistance to striped stem borer and yellow stem borer (Ghareyazie, 1996). The PEP carboxylase-*cryIA(b)* plasmid (pCIB4421) was transferred to rice along with a second plasmid carrying the hygromycin phosphotransferase (*hpt*) gene for selection on hygromycin B. Microprojectile bombardment was performed on embryogenic callus derived from mature seeds of variety Tarom Molaii, a high quality aromatic rice from Iran. Inclusion of 50 mg/l hygromycin B in culture media from bombardment through to rooting of plantlets resulted in a very low frequency of escapes and a high proportion of transformants. Line 827 contained the *cryIA(b)* and *hpt* genes and produced truncated (67 kDa) *Bt* toxin (about 0.1% of total leaf protein as judged by immunoblotting). The maize PEP carboxylase promoter is expected to give expression in leaf blades and sheaths but not in the endosperm. Protein immunoblots confirm that the *cryIA(b)* protein accumulates in leaves but not in immature spikelets or mature grain. Line 827 contained a cluster of 2–3 copies of the *cryIA(b)* gene which segregated $3:1::$ PCR-positive : PCR-negative in the second generation (Table 5.5). The gene co-segregated with extreme resistance to larvae of striped stem borer. In the third generation, plants homozygous for the *cryIA(b)* gene were identified and were used as donors of stem borer resistance in the breeding program. T_2 progeny of heterozygous T_1 plants segregated $3:1::$ PCR-positive : PCR-negative for the *cryIA(b)* gene. T_2 plants showed resistance to yellow stem borer (Alinia *et al.*, unpublished data).

Table 5.5 Inheritance of enhanced stem borer resistance in rice line 827 expressing *cryIA(b)* gene under the control of the maize C4 PEP carboxylase promoter

Generation	Insect	Assay	Segregation
First (T_0)	Striped stem borer	Cut stem segment	—
Second (T_1)	Striped stem borer	Cut stem segment	$3:1$
		Whole plant	$3:1$
Third (T_2)	Yellow stem borer	Cut stem segment	$3:1$ in segregating lines

Proteinase Inhibitors

The major proteinases in the midguts of lepidopteran insects belong to group I of the serine proteinases (Applebaum, 1985; Ryan, 1989). In the case of the yellow stem borer, midgut extracts contain trypsin and chymotrypsin activities (Mazumdar, 1995). Enhanced resistance to striped stem borer has been achieved in rice transformed with serine proteinase inhibitor genes from cowpea and potato (Duan *et al.*, 1996; Xue *et al.*, 1996) (Table 5.4).

Environmental and Biosafety Issues

At the present time, China, India, Japan and Thailand permit field testing of transgenic plants. The Philippines allows the production, importation and greenhouse testing of transgenic plants, but does not yet allow field testing of transgenic plants that could outcross with plants already grown in the country.

The availability of transgenic rice with enhanced insect resistance gives urgency to the discussion of related environmental and biosafety issues. Among the key issues are: the durability of resistance mechanisms; the effects of transgenes on non-target insects, especially natural enemies of rice pests; and the spread of transgenes to wild relatives of rice. We comment briefly on these issues below.

Resistance Management

The breakdown of single-gene resistance to pests such as brown planthopper, green leafhopper and gall midge has long been a problem in rice pest management (Heinrichs, 1994). There is a high level of awareness among rice scientists that single-gene, high-level resistance conferred by genetic engineering will not in itself provide sustainable host plant resistance. Pyramids of toxins, especially unrelated toxins differing in mode of action, can be used to enhance durability. However, the use of toxin-free refuges is likely to be the most successful approach to sustaining the effectiveness of transgenic, insect resistant plants (Tabashnik, 1994).

The optimal spatial scale of a refuge, e.g., toxin-free tissues within plants, or mixtures of toxic and non-toxic plants within fields or among fields, is highly dependent on the biology of the target pest. Studies on relevant aspects of stem borer biology have been initiated (Bottrell *et al.*, 1992; Cohen *et al.*, 1996); larval movement and adult dispersal are of particular importance (Gould, 1994; Mallet and Porter, 1992). Extensive movement of stem borer larvae both among tissues within plants (Magalit and Bottrell, 1996) and among plants within fields (Romena *et al.*, unpublished data) indicates that use of tissue specific promoters and seed mixtures may not be effective resistance management strategies for these species. If extensive premating adult dispersal takes place, then field-to-field mixtures of transgenic and non-transgenic plants may be more effective. Such mixtures may become naturally established in areas where farmers traditionally plant a diversity of rice varieties. However, they will be more difficult to maintain in areas where a single variety tends to become dominant, such as is currently the case with IR64 in Central Luzon, Philippines. The combination of refuges with plants expressing high doses of toxins is a promising resistance management strategy that is under study in several crops (Gould, 1994). We do not yet know if the available *Bt* rice lines express satisfactory levels of toxin for the high-dose strategy.

Effects on Non-Target Arthropods

There are two potential effects of genetically engineered insect resistant plants on spiders and beneficial insects: toxicity of foreign proteins following ingestion; and ecological effects caused by changes in the food web of the crop ecosystem. The toxicity of foreign proteins to non-target arthropods can be evaluated by standard toxicological tests, as are required prior to the release of transgenic plants in the United States. Food web effects are of particular interest in rice because of the hundreds of interacting species that underlie the powerful biological control active in tropical rice ecosystems (Schoenly *et al.*, 1996). As an example, cultivation of *Bt* rice plants expressing toxins in the stem and leaves will likely result in substantial reductions in the populations of most rice field Lepidoptera. More than 20 species of stem-boring and leaf-feeding Lepidoptera are common in tropical rice fields. The great majority of these species are not pests, and in fact may serve a beneficial role by providing alternate food sources to polyphagous predators and parasitoids that contribute to the control of other pests such as planthoppers. Studies using *Bt* sprays to simulate the effects of *Bt* rice on the rice food web are in progress at IRRI, and may be continued with *Bt* rice if approval for field testing of these plants is granted. The use of tissue specific promoters to minimize the exposure of non-target insects to the products of transgenes is also receiving close attention.

Spread of Transgenes to Wild Species

Cultivated rice, *O. sativa*, is sexually compatible with other rice species sharing the AA genome. In Asia, hybridization between *O. sativa* and two AA-genome wild species, *O. rufipogon* and *O. nivara*, occurs in nature (Oka, 1991). Given the large area planted to rice, it is reasonable to assume that if outcrossing can occur, it will occur, and so the focus of the debate should be on the consequences, rather than the probability, of transfer. Will the spread to *O. rufipogon* and *O. nivara* of genes conferring high levels of insect resistance alter the ecology of these species? Observations of these species suggest that insects do not have an important impact on their distribution or abundance, but this question has not been rigorously studied. Among Asian countries, India, Indonesia and Thailand have large populations of *O. nivara* and/or *O. rufipogon* and will undoubtedly devote resources to appropriate studies. The Philippines has only one small stand of *O. rufipogon* located around the edge of a lake on the country's highest mountain, Mt Apo, and that is not near a rice growing area.

Conclusion

Agricultural biotechnology is potentially a major contributor to national development in many rice growing countries. It is part of IRRI's mandate to evaluate the new approaches arising from developments in plant tissue culture and molecular biology. Hence, IRRI has active programs in wide hybridization, DNA markers, rice transformation and DNA fingerprinting and is evaluating them in relation to the wider questions of IPM, biosafety and biodiversity to ensure that rice producers and consumers reap a sustainable benefit from the new technology. The current availability of rice varieties with insect resistance enhanced through biotechnology means that the evaluation phase can now begin.

References

Altosaar, I., Sardana, R., Kaplan, H. (1996) Synthetic *Bt* toxin genes for insect resistance and their rapid bioassay using maize endosperm suspension cultures, In: *Rice Genetics III. Proceedings of the Third International Rice Genetics Symposium, 16–20 October 1995,* Manila, Philippines: International Rice Research Institute, pp. 737–42.

Applebaum, S. W. (1985) Biochemistry of digestion, In: Kerkut, G. A. and Gilbert, L. I. (Eds), *Comprehensive Physiology, Biochemistry and Pharmacology of Insects,* Vol. 4, New York: Pergamon Press, pp. 279–312.

Black, W. C. IV, Du Teau, N. M., Puterka, G. J., Nechols, J. R. and Pettorini, J. M. (1992) Use of random amplified polymorphic DNA polymerase chain reaction (RAPD-PCR) to detect DNA polymorphisms in aphids (Homoptera: Aphididae), *Bull. Entomol. Res.* **82,** 151–159.

Bottrell, D. G., Aguda, R. M., Gould, F. L., Theunis, W., Demayo, C. G. and Magalit, V. F. (1992) Potential strategies for prolonging the usefulness of *Bacillus thuringiensis* in engineered rice, *Korean J. Appl. Entomol.* **31**(2), 247–255.

Brar, D. S. and Khush, G. S. (1995) Wide hybridization for enhancing resistance to biotic and abiotic stresses in rainfed lowland rice, *Proceedings of the International Rice Research Conference, 13–17 February 1995,* Laguna, Philippines: International Rice Research Institute, pp. 901–910.

Christou, P., Ford, T. L. and Kofron, M. (1991) Production of transgenic rice (*Oryza sativa* L.) plants from agronomically important indica and japonica varieties via electric discharge particle acceleration of exogenous DNA into immature zygotic embryos, *Bio/ Technology* **9**, 957–962.

Cohen, M. B., Aguda, R. M. Romena, A. M., Roderick, G. K. and Gould, F. L. (1996) Resistance management strategies for *Bt* rice: what have we learned so far? In: *Rice Genetics III. Proceedings of the Third International Rice Genetics Symposium, 16–20 October 1995,* Manila, Philippines: International Rice Research Institute, pp. 749–52.

Datta, S. K., Peterhans, A., Datta, K. and Potrykus, I. (1990) Genetically engineered fertile indica-rice recovered from protoplasts, *Bio/Technology* **8**, 736–740.

Duan, X., Li, X., Wu, T., Abo-El-Saad, M., Chen, P., Xu, D. and Wu, R. (1996) Production and analysis of transgenic rice plants transformed with the potato proteinase inhibitor II gene, In: *Rice Genetics III. Proceedings of the Third International Rice Genetics Symposium, 16–20 October 1995,* Manila, Philippines: International Rice Research Institute, pp. 743–8.

Ehtesham, N. Z., Bentur, J. S. and Bennett, J. (1995) Characterization of a DNA sequence that detects repetitive DNA elements in the Asian rice gall midge (*Orseolia oryzae*) genome: potential use in DNA fingerprinting of biotypes, *Gene* **153**, 179–183.

Fujimoto, H., Itoh, K., Yamamoto, M., Kyozuka, J. and Shimamoto, K. (1993) Insect resistant rice generated by introduction of a modified δ-endotoxin gene of *Bacillus thuringiensis, Bio/Technology* **11**, 1151–1155.

Gallagher, K. D., Kenmore, P. E. and Sogawa, K. (1994) Judicial use of insecticides deter planthopper outbreaks and extend the role of resistant varieties in Southeast Asian rice, In: Denno, R. F. and Perfect, T. J. (Eds), *Planthoppers, their Ecology and Management,* London: Chapman & Hall, pp. 599–614.

Gatehouse, J. A., Powell, K. and Edmonds, H. (1996) Genetic engineering of rice for resistance to homopteran insect pests. In: *Rice Genetics III. Proceedings of the Third International Rice Genetics Symposium, 16–20 October 1995,* Manila, Philippines: International Rice Research Institute, pp. 189–200.

Ghareyazie, B. (1996) Transformation of indica and other rices (*Oryza sativa* L.): genetic integration, expression, inheritance and enhanced insect resistance. PhD thesis, University of the Philippines at Los Banos, Los Banos, Laguna, Philippines.

Glaszmann, J. C. (1987) Isozymes and classification of Asian rice varieties, *Theor. Appl. Genet.* **74**, 21–30.

Gould, F. (1994) Potential and problems with high-dose strategies for pesticidal engineered crops, *Biocontrol Sci. Tech.* **4**(4), 451–462.

Graf, B., Lamb, R., Heong, K. L. and Fabellar, L. (1992) A simulation model for the population dynamics of rice leaffolders (Lepidoptera: Pyralidae) and their interactions with rice, *J. Appl. Ecol.* **29**, 558–570.

Heinrichs, E. A. (1994) Host plant resistance, In: Heinrichs, E. A. (Ed.), *Biology and Management of Rice Insects*, New Delhi: Wiley Eastern, pp. 517–548.

Hiei, Y., Ohta, S., Komari, T. and Kumashiro, T. (1994) Efficient transformation of rice (*Oryza sativa* L.) mediated by Agrobacterium and sequence analysis of the boundaries of the T-DNA, *Plant J.* **6**, 271–282.

Hoy, M. A. (1994) *Insect Molecular Genetics: An Introduction to Principles and Applications*, San Diego: Academic Press.

Hu, G. W., Chen, Z. X., Liu, G., Ma, J. F. and Pan, Q. W. (1993) Dynamic modelling of compensation of late season hybrid to rice leaf-folder (*Cnaphalocrocis medinalis*) defoliation, *Sci. Agric. Sinica* **26**, 24–29 (in Chinese with English summary).

Ishii, T., Brar, D. S., Multani, D. S. and Khush, G. S. (1994) Molecular tagging of genes for brown planthopper resistance and earliness introgressed from *Oryza australiensis* into cultivated rice, *O. sativa*, *Genome* **37**, 217–221.

Jackson, M. T. (1995) Protecting the heritage of rice biodiversity, *GeoJournal* **35**(3), 267–274.

Katiyar, S. K., Tan, Y., Zhang, Y., Huang, B., Xu, Y., Zhao, L., Huang, N., Khush, G. S. and Bennett, J. (1995) Molecular tagging of gall midge resistance genes in rice, *Proceedings of the International Rice Research Conference, 13–17 February 1995*, Laguna, Philippines: International Rice Research Institute, pp. 934–948.

Khan, Z. R., Litsinger, J. A., Barrion, A. T., Villanueva, F. F. D., Fernandez, N. J. and Taylor, L. D. (1991) *World Bibliography of Rice Stem Borers 1794–1990*, IRRI/ICIPE publication 19.

Khush, G. S. (1995) Modern varieties – their real contribution to food supply and equity, *GeoJournal* **35**(3), 275–284.

Koziel, M. G., Beland, G. L., Bowman, C., Carozzi, N. B., Crenshaw, R., Crossland, L., Dawson, J., Desai, N., Hill, M., Kadwell, S. *et al.* (1993) Field performance of elite transgenic maize plants expressing an insecticidal protein derived from *Bacillus thuringiensis*, *Bio/Technology* **11**, 194–200.

Magalit, V. F. and Bottrell, D. G. (1996) Feeding behavior of *Scirpophaga incertulas* (Lepidoptera: Pyralidae): implications in expressing *Bacillus thuringiensis* toxins in engineered rice (submitted to *J. Econ. Entomol.*).

Mallet, J. and Porter, P. (1992) Preventing insect adaptation to insect-resistant crops: are seed mixtures or refugia the best strategy?, *Proc. R. Soc. Lond. B.* **250**, 165–169.

Matteson, P. C., Gallagher, K. D. and Kenmore, P. E. (1994) Extension of integrated pest management for planthoppers in Asia irrigated rice: empowering the user, In: Denno, R. F. and Perfect, T. J. (Eds), *Planthoppers, their Ecology and Management*, London: Chapman & Hall, pp. 656–685.

Mazumdar, S. (1995) Molecular characterization of trypsin genes in two economically important insect pests (*Scirpophaga incertulas* Wk. and *Helicoverpa armigera* Hb.) and recombinant soybean Kunitz trypsin inhibitor gene expressed in *Escherichia coli*, PhD thesis, submitted to the Department of Botany, University of Delhi, India.

McCouch, S. R. (1990) Construction and applications of a molecular linkage map of rice based on restriction fragment length polymorphism (RFLP), PhD thesis, Cornell University, Ithaca, NY.

Mohan, M., Nair, S., Benur, J. S., Rao, U. P. and Bennett, J. (1994) RFLP and RAPD mapping of the rice Gm2 gene that confers resistance to biotype 1 of gall midge (*Orseolia oryzae*), *Theor. Appl. Genet.* **87**, 782–788.

Nayak, P., Basu, D., Da, S., Ghosh, D., Nandi, A., Basu, A. and Sen, K. (1996) Transfer of insect resistance genes against the damage caused by yellow stem borer to indica rice cultivars, In: *Rice Genetics III. Proceedings of the Third International Rice Genetics Symposium, 16–20 October 1995*, Manila, Philippines: International Rice Research Institute, pp. 735–6.

Oka, H. I. (1991) Genetic diversity of wild and cultivated rice, In: Khush, G. S. and Toenniessen, G. H. (Eds), *Rice Biotechnology*, Manila, Philippines: International Rice Research Institute, pp. 55–81.

Openshaw, S. J., Jarboe, S. G. and Beavis, W. D. (1994) Marker-assisted selection in backcross breeding, *Analysis of Molecular Marker Data. Joint Plant Breeding Symposia Series, 5–6 August 1994, Corvallis, Oregon*, pp. 41–43.

Panda, N. and Khush, G. S. (1995) *Host plant resistance to insects*, Wallingford, UK: CAB International.

Pathak, M. D. and Khan, Z. R. (1994) *Insect Pests of Rice*, Manila, Philippines: International Rice Research Institute.

Pinstrup-Andersen, P. (1994) *World Food Trends and Future Food Security, Food Policy Report*, Washington, DC: International Food Policy Research Institute.

Puterka, G. J., Black, W. C. IV, Steiner, W. M. and Burton, R. L. (1993) Genetic variation and phylogenetic relationships among worldwide collections of the Russian wheat aphid, *Diuraphis noxia* (Mordvilko), inferred from allozyme and RAPD-PCR markers, *Heredity* **70**, 604–618.

Riazuddin, S., Husnain, T., Khan, E., Karim, S., Khanum, F., Makhdoom, R. and Altosaar, I. (1996) Transformation of indica rice with Bt pesticidal genes, In: *Rice Genetics III. Proceedings of the Third International Rice Genetics Symposium, 16–20 October, 1995*, Manila, Philippines: International Rice Research Institute, pp. 730–4.

Robeniol, J. A., Constantino, S. V., Resurreccion, A. P., Villareal, C. P., Ghareyazie, B., Lu, B-R., Katiyar, S. K., Menguito, C. A., Angeles, E. R., Fu, H-Y. *et al.* (1996) Sequence-tagged sites and low-cost DNA markers for rice, In: *Rice Genetics III. Proceedings of the Third International Rice Genetics Symposium, 16–20 October 1995*, Manila, Philippines: International Rice Research Institute, pp. 293–306.

Rubia, E. G., Shepard, B. M., Yambao, E. B., Ingram, K. T., Arida, G. S. and Penning de Vries, F. (1989) Stem borer damage and grain yield of flooded rice, *J. Plant Prot. Trop.* **6**, 205–211.

Ryan, C. A. (1989) Proteinase inhibitor gene families: strategies for transformation to improve plant defenses against herbivores, *BioEssays* **10**, 20–24.

Savary, S., Elazegui, F. A., Pinnschmidt, H. O. and Teng, P. S. (1995) Characterization of rice pest constraints in Asia: an empirical approach, In: Teng, P. S. and Kropff, M. (Eds), *Systems Approaches for Agricultural Development*, Amsterdam: Kluwer.

Schoenly, K. G., Cohen, J. E., Heong, K. L., Arida, G. S., Barrion, A. T. and Litsinger, J. A. (1996) Quantifying the impact of insecticides on food web structure of rice-arthropod populations in a Philippine farmer's field: a case study, In: Polis, G. A. and Wisemiller, K. (Eds), *Food Webs: Integration of Patterns and Dynamics*, London: Chapman & Hall, pp. 343–351.

Sebastian, L. S., Ikeda, R., Huang, N., Imbe, T., Coffman, W. R. and McCouch, S. R. (1996) Molecular mapping of resistance to rice tungro spherical virus and green leafhopper in rice, *Phytopathology* **86**, 25–30.

Shimamoto, K., Terada, R., Izawa, T. and Fujimoto, H. (1989) Fertile transgenic rice plants regenerated from transformed protoplasts, *Nature* **388**, 274–276.

Shufran, K. A. and Whalon, M. E. (1996) Genetic analysis of brown planthopper biotypes using random amplified polymorphic DNA-polymerase chain reaction (RAPD-PCR), *Insect Sci. Applic.* (in press).

Tabashnik, B. E. (1994) Evolution of resistance to *Bacillus thuringiensis, Annu. Rev. Entomol.* **39**, 47–79.

Vos, P., Hoges, R., Bleeka, M., Reijans, M., van de Lee, T., Homes, M., Freijters, A., Pot, J., Peleman, T., Kaiper, M. and Zabeau, M. (1995) AFLP: a new concept for DNA finger-printing, *Nucl. Acid Res.* **23**, 4407–4414.

Way, M. J. and Heong, K. L. (1994) The role of biodiversity in the dynamics and management of insect pests of tropical irrigated rice – a review, *Bull. Entomol. Res.* **84**, 567–587.

Wünn, J., Kloti, A., Burkhardt, P. K., Ghosh Biswas, G. C., Launis, K., Iglesias, V. A. and Potrykus, I. (1996) Transgenic indica rice breeding line IR58 expressing a synthetic *cryIA(b)* gene from *Bacillus thuringiensis* provides effective insect pest control, *Bio/Technology* **14**, 171–176.

Xue, Q., Duan, X., Xu, D. and Wu, R. (1996) Field tests of herbicide- and insect-resistant transgenic rice plants, In: *Rice Genetics III. Proceedings of the Third International Rice Genetics Symposium, 16–20 October 1995*, Manila, Philippines: International Rice Research Institute, pp. 239–46.

Zheng, K., Huang, N., Bennett, J. and Khush, G. S. (1996) PCR-based marker-assisted selection in rice breeding, In: *IRRI Discussion Paper series*, no. 12, Manila, Philippines: International Rice Research Institute.

Acknowledgments

We gratefully acknowledge financial support from the Rockefeller Foundation (JB, MBC, SKK), the German Federal Ministry of Economic Cooperation (BMZ/GTZ) and the Asian Development Bank through the Asian Rice Biotechnology Network (JB, BG, GSK, SKK), and the International Network for the Genetic Evaluation of Rice (SKK). We are grateful to R. C. Chaudhary (INGER coordinator) for his encouragement of this work.

6

Cholesterol Oxidase for the Control of Boll Weevil

JOHN P. PURCELL

New Genes for Insect Control

The feasibility of using genetically modified plants for insect control has been convincingly demonstrated by the expression of genes for *Bacillus thuringiensis* (*Bt*) proteins. These proteins have been demonstrated to provide control of several insect pests in different crops (Barton *et al.*, 1987; Fischoff *et al.*, 1987; Koziel *et al.*, 1993; Perlak *et al.*, 1990, 1993; Vaeck *et al.*, 1987). There are however many other economically important insect pests which are not effectively controlled by known *Bt* proteins. Several strategies are available to discover new proteins for insect control. These strategies include directed screening of sources that were previously shown to yield insect control proteins and random screening of novel sources.

Directed Screening

One strategy for finding new insecticidal proteins is to search sources that have been previously shown to contain such proteins. *Bt* strains have obviously provided a rich supply of proteins for the control of insects. Many directed screening efforts have focused on *Bt* collections and have had very impressive results. The first transgenic crop products that will hit the marketplace for insect control will have *Bt* proteins as their active components. Plants are another source rich in proteins which may have utility in controlling insects. Much effort has been focused on directed screening of known classes of insect control proteins commonly found in plants such as proteinase inhibitors, lectins and amylase inhibitors (for review, see Boulter, 1993; see also Chapters 8–10). These proteins can retard insect growth and development but do not generally provide the acute mortality seen with *Bt* proteins.

Random Screening

Another useful strategy for finding novel insect control proteins is to screen protein samples with no bias as to their source. This type of random screening may involve plant samples, microbial fermentation broths (Purcell *et al.*, 1994b) or even proteins purchased through commercial suppliers (Purcell *et al.*, 1994a; Rahbe and Febvay, 1993). In most cases, the protein samples are added to an artificial diet which is then fed to insects to identify samples which have deleterious effects on the insect. Once an active sample is identified, several characteristics may be determined to ascertain whether the active component is proteinaceous. Commonly used parameters include heat lability, molecular sizing, proteinase digestibility and ability to precipitate with ammonium sulfate. An activity which is heat labile, high molecular weight (> 3000), proteinase digestible, and ammonium sulfate precipitable, is almost certainly proteinaceous. Of course, proteins which are relatively stable to heat and proteinase inactivation are found, so care must be taken in interpreting characterization results. Once a sample with an insecticidally active protein is found, bioassay guided purification can then be used to identify the active component. Once a protein band is identified as the protein responsible for the insecticidal activity that is seen, then N-terminal amino acid sequencing of the band is pursued. Subsequent searches of protein sequence databases can often be used to determine the identity of the bioactive protein. If the identity of the bioactive protein is determined, then subsequent purifications can be guided by enzymatic or other *in vitro* assays, thus avoiding the more time- and material-consuming bioassay guided purifications.

Parameters to Assess Utility of New Insect Control Proteins

Once a new insect control protein is identified, three parameters are useful in assessing the value of such a protein: the identity of the protein; its potency against target insects; and its mode of action. Criteria can be established for each of these areas which are helpful in deciding whether or not to advance a lead to gene cloning and subsequent expression in a plant. The identity of a protein may rule out proteins with broad spectrum toxicology issues. The potency of a protein is determined by evaluating EC_{50} and LC_{50} values using purified proteins in diet incorporation insect bioassays. A good measure of the potency of a protein is to compare the activity of a newly discovered agent with the activity of *Bt* endotoxins in similar diet incorporation bioassays. In these assays, *Bt* endotoxins are generally active in the low ppm ($\mu g \ ml^{-1}$ of diet) concentration range (MacIntosh *et al.*, 1990). It is important to have proteins of sufficient potency to ensure that enough of the active component can be expressed in a transgenic plant in order to control the insect pest. Potency determinations against several insect pests will allow the spectrum of insects controlled by the insect control protein to be assessed. We use the potency of a new protein identified in our insect screens as an important prioritization tool in deciding on the advancement of a lead into gene cloning and plant expression. If a protein is active at a low ppm concentration, then the decision to advance the lead is an obvious one. It is when the proteins are active only at higher concentrations that the decision becomes more difficult and may depend on resource considerations or the potency of other leads that are available. An example of proteins from our screening program that are active only at higher concentrations is the insecticidal proteinases. We found a number of proteinases which were active against a broad spectrum of insects but only at concentrations in the hundreds to thousands of $\mu g \ ml^{-1}$ (Purcell *et al.*, 1994a).

The third parameter useful in judging the value of a new insect control protein is the mode of action. The precise molecular mechanism may not be known but the general mechanism by which the protein is active in an insect bioassay can be determined. Three general modes of action can be envisioned for insect control proteins: direct post-ingestion; toxic molecule production; and dietary depletion. In the direct post-ingestion mode of action, the protein itself is toxic to the insect when it is ingested. The *Bt* endotoxins are the best examples of insect control proteins active by this mechanism (English and Slatin, 1992; Gill *et al.*, 1992). This type of activity is the easiest to deal with conceptually. The assumption is that these proteins will be toxic to the insect when produced in the plant and the insect ingests such a plant. The key with such proteins is to make sure that they are present at high enough levels in the part of the plant fed upon by the pest which needs to be controlled.

The toxic molecule production mechanism is where the protein reacts with some constituent of the plant (or diet) and produces a product which is toxic to the insect. In this type of mechanism of action, both the insect control protein and the substrate need to be present at a high enough level in the insect's diet to ensure that enough product is made to bring about the desired deleterious effects on the insect pest. When this type of activity is found in a screening program using artificial insect diets for a bioassay arena, it is important to consider whether the substrate that is required will be found in the plant that is the target for insect protection. An insect control protein which utilizes this mechanism is lipoxygenase, which appears to be active by producing lipid oxidation products from substrates such as linoleic acid (Mohri *et al.*, 1990; Shukle and Murdock, 1983).

A third general mechanism is dietary depletion, where the protein alters the plant (or insect diet) in such a way that some dietary requirement is altered or depleted and thus the insect cannot survive on the altered or depleted plant (or diet). An insect control protein with a mode of action of this type is polyphenol oxidase. This enzyme can significantly reduce the protein quality of an insect diet, thus reducing the growth of larvae feeding upon such a diet (Felton *et al.*, 1992). When proteins active by this mechanism are discovered, the ramifications of putting such a protein in a target crop must be considered before advancing such a lead protein. Our screening program identified two classes of proteins that were active in green peach aphid bioassays by depleting the aphid diet of sucrose (Purcell *et al.*, 1994b). Mittler *et al.* (1970) had previously established the dietary sucrose requirements of aphids. Even though the invertase and hexosyl transferase proteins were active in the aphid bioassays at low ppm concentrations (Purcell *et al.*, 1994b), these leads were not advanced further because of the deleterious effects of expressing such proteins in a target crop (von Schaewen *et al.*, 1990). This work demonstrates the importance of determining the identity and general mechanism of action of an insect control protein in order to assess the value of a newly discovered protein.

Cholesterol Oxidase – A New Protein for Insect Control

Discovery and Identification

We tested over 10 000 filtrates from microbial fermentations for insecticidal activity against major pests (Purcell *et al.*, 1993, 1994b). Two *Streptomyces* culture filtrates (Monsanto isolates A19241 and A19249) killed boll weevil larvae in feeding studies and the characterization studies described above indicated that the active components were proteins (Purcell *et al.*, 1993). Subsequent purification identified the presence of a major protein of M_r 52 500 in chromatography fractions which correlated with the boll weevil mortality (Purcell *et al.*, 1993) (see Figure 6.1) and amino acid sequencing of the purified protein showed a high degree of homology with cholesterol oxidase from *Streptomyces* species (Ishizaki *et al.*, 1989). The identity of the protein was confirmed by its ability to oxidize cholesterol in an *in vitro* enzymatic assay (Gallo, 1981).

Potency

A series of LC_{50} bioassays against boll weevil neonate larvae determined that the cholesterol oxidase LC_{50} for neonate larvae fell from 6.0 μg ml^{-1} in six-day bioassays to 1.5 μg ml^{-1} in 16-day bioassays (Greenplate *et al.*, 1995). Second instar larvae were equally susceptible to the enzyme in 16-day bioassays (Greenplate *et al.*, 1995). *Bt* proteins are generally active against lepidopteran larvae at concentrations in the low μg ml^{-1} range (MacIntosh *et al.*, 1990) in diet incorporation bioassays similar to the ones used to determine the potency of cholesterol oxidase against boll

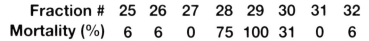

Fraction #	25	26	27	28	29	30	31	32
Mortality (%)	6	6	0	75	100	31	0	6

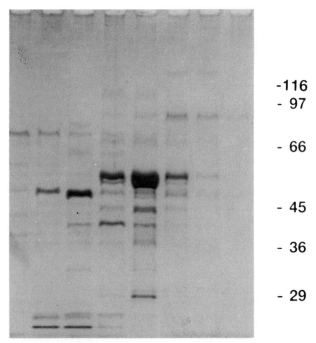

Figure 6.1 SDS-PAGE of boll weevil active chromatography fractions. Aliquots of the insecticidally active and adjoining chromatography fractions were electrophoresed on SDS-PAGE. The presence of a major protein of M_r 52 500 in fractions 28–30 correlates with the boll weevil mortality found in these fractions (reproduced from Purcell *et al.* (1993) with permission from Academic Press)

Table 6.1 LC$_{50}$ values of *Streptomyces* cholesterol oxidase and *Bt* proteins

Protein	Insect	LC$_{50}$ (μg ml^{-1})	Reference
B. thuringiensis CryIA(c)	Tobacco budworm	1	MacIntosh *et al.* (1990)
B. thuringiensis CryIA(b)	European corn borer	3.6	MachIntosh *et al.* (1990)
Cholesterol oxidase	Boll weevil	6.0	Greenplate *et al.* (1995)

weevil larvae. The *Bt* endotoxin LC$_{50}$ for tobacco budworm (*Heliothis virescens*), for example, is 1 μg ml^{-1} (MacIntosh *et al.*, 1990). The potency is an important factor to consider when assessing the value of a new protein in a transgenic crop strategy for insect control. There are limits to the amount of protein which can be produced in a transgenic crop plant. The protein must be potent enough so that the level of expression can meet the level of protein required for activity against the insect pest and protection of the plant. Proteins that are active in the same concentration range as the *Bt* proteins would be assumed to be potent enough to work in a transgenic crop. Cholesterol oxidase is active against the boll weevil in the same concentration range that *Bt* proteins are active against lepidopterans which are effectively controlled in transgenic crops expressing *Bt* proteins (Table 6.1).

Effects on Boll Weevil Adults

Cholesterol oxidase is not acutely toxic to newly emerged adult boll weevil but oviposition is severely reduced (Greenplate *et al.*, 1995). The effects of cholesterol oxidase on fecundity are entirely due to effects on the female boll weevil and females fed on cholesterol oxidase have poorly developed ovaries (Greenplate *et al.*, 1995). The decrease in oviposition and reduction in size of ovaries is similar to the effects seen with *Bt* CryIIIA protein on adult Colorado potato beetle (Perlak *et al.*, 1993). Reductions in fecundity have also been noted when *Bt* CryI proteins were fed to adult *Heliothis virescens* (Ali and Watson, 1982). These effects on fecundity may be due to a perturbation of the ability of the female adults properly and efficiently to take up the nutrients required for oogenesis. Effects on the adult midgut epithelia may be responsible for the adult effects seen with *Bt* proteins (Perlak *et al.*, 1993) and cholesterol oxidase (Greenplate *et al.*, 1995). Further work is required to test this hypothesis.

Spectrum of Insecticidal Activity

We have tested a number of insects for susceptibility to cholesterol oxidase. The enzyme was purified from *Streptomyces* organisms and used in artificial diet bioassays. Insects tested were from the orders Coleoptera, Lepidoptera, Diptera, Dictyoptera and Hemiptera (Purcell *et al.*, 1993). In most cases, only acute effects on insect growth and survival were determined in bioassays generally lasting from four to seven days. Long-term assays which would measure deleterious effects at all stages of development are needed to make a more comprehensive assessment of the utility of cholesterol oxidase for insect control. In a practical sense, however, it is short-term bioassays which are routinely used to look at the spectrum of bioactivity of a protein. Our bioassays demonstrated that boll weevil is the most susceptible to cholesterol oxidase but that a number of lepidopterans are also affected by the protein (Purcell *et al.*, 1993). Lepidopteran larvae were tested on artificial agar-based diet (Marrone *et al.*, 1985) containing 100 μg ml^{-1} cholesterol oxidase (*Streptomyces* – Monsanto A19249). The assays were carried out for six days; results are shown in Table 6.2. Lepidopteran larvae were stunted by cholesterol oxidase. Percent stunting is defined as:

$$\frac{[\text{Larval weights (control)} - \text{Larval weights (cholesterol oxidase)}]}{\text{Larval weights (control)}} \times 100\%.$$

Further work is being done to determine how effectively lepidopteran larvae are controlled by cholesterol oxidase. It is interesting that boll weevil larvae, unlike some other coleopterans, have major serine proteinase activity in their gut, as do lepidopteran larvae (Purcell *et al.*, 1992). Perhaps proteinase composition is one factor in determining the insects which are susceptible to the enzyme.

Alternate Bioassays

Alternate assays which introduce an insect control protein to the insect in a plant tissue based bioassay can offer important data on whether or not a protein retains its insecticidal activity in the presence of plant material. This type of data is important in assessing the value of a new insect control protein as these bioassays may provide insight into the mode of action of the insect control protein. For example, if a protein lead is identified in an artificial diet bioassay but has no activity in a plant tissue based bioassay, one explanation is that the protein is altering the artificial diet in some way to cause a deleterious effect on the insect being assayed. This could indicate that the protein is acting on a substrate found in the diet but not in the plant tissue, either producing a toxic product or depleting the artificial diet of some essential nutrient. An alternative explanation is that the physiology of the insect is inherently different when assayed on an artificial diet and that the insect control protein is only active under these conditions.

It is critical that cholesterol oxidase is insecticidally active when it is ingested with cotton tissue. Cholesterol oxidase bioactivity against boll weevil larvae was measured by addition of the enzyme to three different cotton derived assay media

Table 6.2 Stunting of lepidopteran larvae by *Streptomyces* – Monsanto A19249 cholesterol oxidase. The protein was present in the diet at 100 μg ml^{-1}

Test organism	Stunting (%)
Tobacco budworm (*Heliothis virescens*)	86
European corn borer (*Ostrinia nubilalis*)	46
Pink bollworm (*Pectniphora gossypiela*)	30

and to cotton callus (Purcell *et al.*, 1993). In all cases, strong bioactivity was retained demonstrating that the enzyme kills boll weevil larvae in bioassays where all the nutritional material is comprised solely of cotton tissue (seed, embryos, callus or leaves) (Purcell *et al.*, 1993). The demonstration that the bioactivity is independent of the type of bioassay used suggests that the enzyme is acting directly on the insect.

We have also tested the possibility that the bioactivity of cholesterol oxidase was due to an artifact of the larvae being from laboratory reared colonies. It was demonstrated that both wild-type and laboratory reared boll weevil larvae were equally susceptible to cholesterol oxidase, with more than 90% mortality at a concentration of 30 μg ml^{-1} (Greenplate *et al.*, 1995). This is an important experiment in assessing the value of cholesterol oxidase for insect control in cotton.

Diet Effect Studies

The requirement for sterols in insect nutrition is well documented (Dadd, 1985; Earle *et al.*, 1967). Two general modes of action for insect control proteins are described above which are based on the active protein reacting with a diet, to either produce a toxic molecule from the diet or deplete the diet of an essential nutrient. We have examined whether cholesterol oxidase activity against boll weevil is due to effects of the enzyme on the diet. Pretreatment of the diet with the enzyme did not compromise boll weevil survival, thus demonstrating that the diet was not altered prior to ingestion in a manner which resulted in bioactivity (Purcell *et al.*, 1993). The two products of the oxidation of cholesterol, cholestenone and hydrogen peroxide, also showed no toxicity against boll weevil larvae, suggesting that toxic molecule production in the diet was not the mode of action of this protein (Purcell *et al.*, 1993). These data offer more evidence that the enzyme is acting directly on the insect.

Microscopy Studies

We used light microscopy to show that ingestion of cholesterol oxidase has dramatic effects on the midguts of boll weevil larvae (Purcell *et al.*, 1993). Neonate and second instar larvae were fed diets containing cholesterol oxidase and the midguts of these larvae were dissected out and analyzed microscopically. Disruption of the epithelial cell layer was observed in the midguts of larvae ingesting low doses and complete lysis of the cells was observed at high doses (Purcell *et al.*, 1993) (see Figure 6.2). The concentrations of cholesterol oxidase causing these effects result in mortality to the larvae in parallel bioassays, presumably due to lysis (Purcell *et al.*, 1993). This is further evidence that the enzyme is acting directly on the insect. *Bt* insecticidal crystal proteins also cause lysis of the midgut epithelium (English and Slatin, 1992; Gill *et al.*, 1992). The cholesterol oxidase effects on boll weevil midguts is another demonstration of how critical a target the midgut is for insect control proteins (Federici, 1993).

There are membrane specific differences in the oxidation of cholesterol in various biological membranes (Crocket and Hazel, 1995; Lange, 1992). The cholesterol in many biological membranes is not very accessible to oxidation by cholesterol oxidase (Lange *et al.*, 1980; Pal *et al.*, 1980; Patzer *et al.*, 1978; Thurnhofer *et al.*, 1986). The rates of oxidation of membrane cholesterol can be influenced by several factors (Gronberg and Slotte, 1990; Lange, 1992; Lange *et al.*, 1984). One such factor is the cholesterol content of the membrane (Lange *et al.*, 1980; Pal *et al.*, 1980). In some membranes, the cholesterol is not accessible to the enzyme unless the cholesterol content is increased above the native level and the oxidation rate increases with increasing cholesterol content above this threshold value (Lange *et al.*, 1980; Pal *et al.*, 1980). Another factor influencing cholesterol oxidase reactivity with a membrane is the phospholipid composition of the membrane. Differences in the phospholipid environment of a membrane can influence the rate at which cholesterol oxidase can oxidize the cholesterol in the membrane (Gronberg and Slotte, 1990; Pal *et al.*, 1980; Patzer *et al.*, 1978). Cholesterol oxidase treatment of some membranes can affect the structure and function of such membranes (El Yandozi *et al.*, 1993; Kutryk *et al.*, 1991; Wood *et al.*, 1995). Cholesterol oxidase can cause lysis of cells susceptible to oxidation (Linder *et al.*, 1989). We have demonstrated that ingestion of cholesterol oxidase leads to lysis of the midgut epithelia of boll weevil larvae (Purcell *et al.*, 1993) (see Figure 6.2). The molecular mechanism for this lysis and the precise biological effects on boll weevil midgut cells is not yet known. Studies are underway to understand better the insecticidal activity on a molecular and cellular level. Membrane structure and function studies may be useful in understanding why boll weevil are so susceptible to the enzyme. Comparative studies of the membrane architecture and composition of the midgut epithelia of insects susceptible and not susceptible to cholesterol oxidase should also be useful in under-

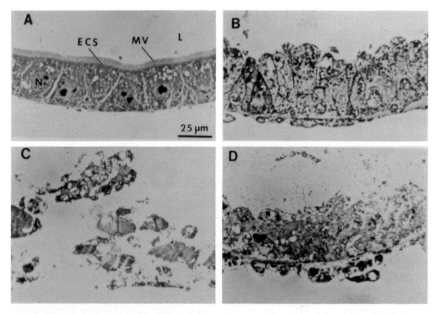

Figure 6.2 Effects of *Streptomyces* cholesterol oxidase on midgut histology in larval boll weevil. (A) Control larval midgut showing normal extracellular spaces (ECS) and nuclei (N), and well-defined microvilli (MV) on lumenal (L) surface. (B) Cholesterol oxidase at 10 μg ml^{-1} in the diet induces cell and nuclear swelling, loss of intercellular spaces, disruption of the apical microvilli and apical cell lysis. (C) Cholesterol oxidase at 30 μg ml^{-1} causes total epithelial disruption. (D) Total epithelial disruption of midguts of second instar larvae was seen with diets containing 100 μg ml^{-1} of cholesterol oxidase. Bar = 25 μm (reproduced from Purcell *et al.* (1993), with permission from Academic Press)

standing the spectrum of insect bioactivity and the mode of action of cholesterol oxidase. Insights gained from these lines of work will hopefully prove useful in discovering new insect control proteins.

Sources of Cholesterol Oxidase

The two *Streptomyces* isolates (designated A19241 and A19249) in our collection produced insecticidally active cholesterol oxidases and these enzymes were purified as reported previously (Purcell *et al.*, 1993). Cholesterol oxidases from a number of organisms are also available from commercial vendors. We purchased (Sigma) cholesterol oxidases from both *Streptomyces* sp. and *Pseudomonas fluorescens*. All four purified cholesterol oxidases were insecticidally active in boll weevil neonate larvae bioassays (Table 6.3). Western blot analyses were used to look for immunoreactivity of the purified cholesterol oxidases. Polyclonal antisera was generated against the Sigma *Streptomyces* enzyme. All three *Streptomyces* enzymes were recognized by the antisera (Figure 6.3). The *P. fluorescens* enzyme was not recognized by the antisera, demonstrating that it is not closely related immunologically to the *Streptomyces* enzymes. This demonstrates that immunologically distinct cholesterol oxidases from diverse species are insecticidally active.

Characterization of Cholesterol Oxidase Enzymatic Activity

The purified cholesterol oxidase from A19249 was determined to have a K_m for cholesterol of 100 μM and a pH optimum of 6–7. Cholesterol oxidases have been shown to oxidize sterols other than cholesterol (Smith and Brooks, 1976; Tomioka *et al.*, 1976). We tested the reactivity rates of the four insecticidally active cholesterol oxidases described above against a number of sterols. Sterols differing from cholesterol in both the side chain (stigmasterol, sitosterol, campesterol and fucosterol) and the ring structure (dihydrocholesterol, dehydrocholesterol and lathosterol) were used. Cholesterol was the preferred substrate for all four enzymes but other sterols were also oxidized by the enzymes (Table 6.4). This demonstrates why cholesterol oxidases are more correctly called 3-hydroxy sterol oxidases (Tomioka *et al.*, 1976). The ability to oxidize stigmasterol, sitosterol and campesterol is significant as these are the three major phytosterols found in cotton buds and anthers (Lusby *et al.*, 1987). Neither ecdysone nor 20-hydroxy ecdysone, two major insect hormones, were oxidized at a detectable rate.

Table 6.3 Boll weevil mortality caused by cholesterol oxidases from four different sources. Corrected mortalities were determined according to Abbott (1925) (some data reproduced from Purcell *et al.* (1993) with permission from Academic Press)

Source	Conc. (μg ml^{-1})	Mortality (%)
Streptomyces – Monsanto A19241	11	88
Streptomyces – Monsanto A19249	20	100
Streptomyces – Sigma	19	100
P. fluorescens – Sigma	44	100

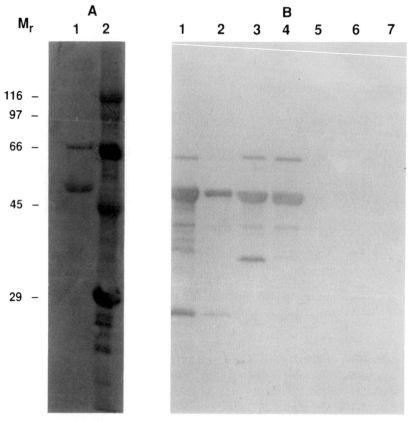

Figure 6.3 Western blot of cholesterol oxidases from various sources. (A) Amido black staining: lane 1 – 4.3 µg *Streptomyces* cholesterol oxidase purified from Sigma preparation; lane 2 – molecular weight standards. (B) Immunoblot staining: lane 1 – 2.9 µg *Streptomyces* cholesterol oxidase (Sigma); lane 2 – 0.6 µg *Streptomyces* cholesterol oxidase purified from Sigma preparation; lane 3 – 2.2 µg *Streptomyces* cholesterol oxidase purified from Monsanto A19241; lane 4 – 2.2 µg *Streptomyces* cholesterol oxidase purified from Monsanto A19249; lane 7 – 2.2 µg *P. fluorescens* cholesterol oxidase purified from Sigma preparation

Gene Cloning and Expression

The three parameters that we assess in deciding whether or not to advance a lead to gene cloning and expression are its identity, potency and mode of action. Cholesterol oxidase has potent bioactivity against boll weevil larvae, is active in cotton based bioassays and appears to be acting by a direct effect on the midgut (Purcell *et al.*, 1993; Greenplate *et al.*, 1995). Based on these data, cholesterol oxidase has been advanced into the next phase of our insect control discovery program. An important milestone in our cloning efforts is the demonstration that the heterologously expressed gene product is insecticidally active. This serves as confirmation that the gene that has been isolated and cloned is fully functional from an insect control standpoint. For cholesterol oxidase, the gene from *Streptomyces* sp. strain A19249 was cloned and expressed in *Escherichia coli* (Corbin *et al.*, 1994). The protein was purified from *E. coli* and shown to have enzymatic properties identical to the

Table 6.4 Substrate reaction rates with cholesterol oxidases from four different sources. Highest rate is set to 100% and all other rates are relative to the highest

Sterol substrate	*Streptomyces* (Mon A19241)	*Streptomyces* (Mon A19249)	*Streptomyces* (Sigma)	*P. fluorescens* (Sigma)
Cholesterol	100	100	100	100
Dihydrocholesterol	56	56	59	69
Dehydrocholesterol	13	12	7	47
Lathosterol	28	34	27	71
Stigmasterol	22	28	11	21
Sitosterol	88	65	49	50
Campesterol 65%/ dihydrobrassicasterol 35%	65	64	45	49
Fucosterol	22	20	12	68
Lanosterol	<1	<1	<1	1
Ecdysone	<1	<1	<1	<1
20-OH ecdysone	<1	<1	<1	<1

enzyme from the native organism (Corbin *et al.*, 1994). Most importantly, the purified protein from *E. coli* had potent insecticidal activity against boll weevil larvae and retained its stunting activity against tobacco budworm larvae (Corbin *et al.*, 1994). The demonstration that the heterologously expressed gene product was insecticidally active gave us the data we needed to advance this lead into the plant expression phase of the project. The next milestone we look for is the ability to express the genes in a transient plant expression system. Cholesterol oxidase has been successfully expressed in a plant cell expression system, retaining enzymatic activity and immunological reactivity (Corbin *et al.*, 1994). This is an important step in the expression of cholesterol oxidase in plants. The generation of plants expressing cholesterol oxidase has been initiated. Reports on the expression of cholesterol oxidase in whole plants and the insect control capabilities of such plants are forthcoming.

Future Directions

The boll weevil entered the United States over one hundred years ago. It is a major pest in cotton and is currently the focus of a federally operated eradication program (Perkins, 1980; Pencoe and Phillips, 1987; Simmonds and Brosten, 1991). Boll weevil eggs are oviposited within the squares (flower buds) and bolls of cotton and the larvae develop entirely within these structures. The life cycle is completed in 21 days so several generations per year are common. It is difficult to control the larvae with conventional insecticides since larval development is within the squares and bolls. A transgenic approach where cholesterol oxidase is expressed in the squares and bolls of cotton should lead to the death of the boll weevil larvae. This would prevent subsequent generations of boll weevil arising from these plants. If cholesterol oxidase ingestion by adult females is able to reduce fecundity, then the number of larvae resulting from any adults present in the field would also be reduced. The combined effects of cholesterol oxidase on boll weevil larvae and adults have the potential severely to reduce the population of this pest. The boll weevil represents an excellent target for control of a major insect pest using a transgenic crop. We are hopeful that expression of the cholesterol oxidase gene in cotton will provide an attractive and powerful tool to control the boll weevil.

References

Abbott, W. S. (1925) A method of computing the effectiveness of an insecticide, *J. Econ. Entomol.* **18**, 265–267.

Ali, A. A. and Watson, T. F. (1982) Effects of *Bacillus thuringiensis* var. *kurstaki* on tobacco budworm (Lepidoptera: Noctuidae) adult and egg stages, *J. Econ. Entomol.* **75**, 596–598.

Barton, K. A., Whitely, H. R. and Yang, K. S. (1987) *Bacillus thuringiensis* endotoxin expressed in transgenic *Nicotiana tabacum* provides resistance to lepidopteran insects, *Plant Physiol.* **85**, 1103–1109.

Boulter, D. (1993) Insect pest control by copying nature using genetically engineered crops, *Phytochemistry* **34**, 1453–1466.

Corbin, D. R., Greenplate, J. T., Wong, E. Y. and Purcell, J. P. (1994) Cloning of an insecticidal cholesterol oxidase gene and its expression in bacteria and in plant protoplasts, *Appl. Environ. Microbiol.* **60**, 4239–4244.

Crockett, E. L. and Hazel, J. R. (1995) Sensitive assay for cholesterol in biological membranes reveals membrane-specific differences in kinetics of cholesterol oxidase, *J. Exper. Zool.* **271**, 190–195.

Dadd, R. H. (1985) Nutrition: organisms, In: Kerkut, G. A. and Gilbert, L. I. (Eds), *Comprehensive Insect Physiology, Biochemistry and Pharmacology*, Vol. 4, New York: Pergamon Press, pp. 313–390.

Earle, N. W., Lambremont, E. N., Burks, M. L., Slatten, B. H. and Bennett, A. F. (1967) Conversion of β-sitosterol to cholesterol in the boll weevil and the inhibition of larval development by two aza sterols, *J. Econ. Entomol.* **60**, 291–293.

El Yandouzi, E. H. and Le Grimellac, C. (1993) Effect of cholesterol oxidase treatment on physical state of renal brush border membranes: evidence for a cholesterol pool interacting weakly with membrane lipids, *Biochemistry* **32**, 2047–2052.

English, L. and Slatin, S. L. (1992) Mode of action of delta-endotoxins from *Bacillus thuringiensis*: a comparison with other bacterial toxins, *Insect Biochem. Mol. Biol.* **22**, 1–7.

Federici, B. A. (1993) Insecticidal bacterial proteins identify the midgut epithelium as a source of novel target sites for insect control, *Arch. Insect Biochem. Physiol.* **22**, 357–371.

Felton, G. W., Donato, K. K., Broadway, R. M. and Duffey, S. S. (1992) Impact of oxidized plant phenolics on the nutritional quality of dietary protein to a noctuid herbivore, *Spodoptera exigua*, *J. Insect Physiol.* **38**, 277–285.

Fischhoff, D. A., Bowdish, K. S., Perlak, F. J., Marrone, P. G., McCormick, S. M., Niedermeyer, J. G., Dean, D. A., Kusano-Kretzmer, K., Mayer, E. J., Rochester, D. E., *et al.* (1987) Insect tolerant transgenic tomato plants, *Bio/Technology* **5**, 807–813.

Gallo, L. L. (1981) Pancreatic sterol ester hydrolase, *Methods Enzymol.* **71**, 665–667.

Gill, S. J., Cowles, E. A. and Pietrantonio, P. V. (1992) The mode of action of *Bacillus thuringiensis* endotoxins, *Annu. Rev. Entomol.* **37**, 615–636.

Greenplate, J. T., Duck, N. B., Pershing, J. C. and Purcell, J. P. (1995) Cholesterol oxidase: an oostatic and larvicidal agent active against the cotton boll weevil, *Anthonomus grandis*, *Entomol. Exper. Appl.* **74**, 253–258.

Gronberg, L. and Slotte, J. P. (1990) Cholesterol oxidase catalyzed oxidation of cholesterol in mixed lipid monolayers: effects of surface pressure and phospholipid composition on catalytic activity, *Biochemistry* **29**, 3173–3178.

Ishizaki, T., Hirayama, N., Shinkawa, H., Nimi, O. and Murooka, Y. (1989) Nucleotide sequence of the gene for cholesterol oxidase from a *Streptomyces* sp., *J. Bacteriol.* **171**, 596–601.

Koziel, M. G., Beland, G. L., Bowman, C., Carozzi, N. B., Crenshaw, R., Crossland, L., Dawson, J., Desai, N., Hill, M., Kadwell, S. *et al.* (1993) Field performance of elite transgenic maize plants expressing an insecticidal protein derived from *Bacillus thuringiensis*, *Biotechnology* **11**, 194–200.

Kutryk, M. J. B., Maddaford, T. J., Ramjawan, B. and Pierce, G. N. (1991) Oxidation of membrane cholesterol alters active and passive transsarcolemmal calcium movement, *Circ. Res.* **68**, 18–26.

Lange, Y. (1992) Tracking cell cholesterol with cholesterol oxidase, *J. Lipid Res.* **33**, 315–321.

Lange, Y., Cutler, H. R. and Steck, T. L. (1980) The effect of cholesterol and other intercalated amphipaths on the contour and stability of the isolated red cell membrane, *J. Biol. Chem.* **255**, 9331–9337.

Lange, Y., Matthies, H. and Steck, T. L. (1984) Cholesterol oxidase susceptibility of the red cell membrane, *Biochim. Biophys. Acta* **769**, 551–562.

Linder, R., Bernheimer, A. W., Cooper, N. S. and Pallias, J. D. (1989) Cytotoxicity of cholesterol oxidase to cells of hypercholesteremic guinea pigs, *Compar. Biochem. Physiol.* **94C**, 105–110.

Lusby, W. R., Oliver, J. E., McGibben, G. H. and Thompson, M. J. (1987) Free and esterified sterols of cotton buds and anthers, *Lipids* **22**, 80–83.

MacIntosh, S. C., Stone, T. B., Sims, S. R., Hunst, P. L., Greenplate, J. T., Marrone, P. G., Perlak, F. J., Fischhoff, D. A. and Fuchs, R. L. (1990) Specificity and efficacy of purified *Bacillus thuringiensis* proteins against agronomically important insects, *J. Invert. Pathol.* **56**, 258–266.

Marrone, P. G., Ferri, F. D., Mosley, T. R. and Meinke, L. J. (1985) Improvements in laboratory rearing of the southern corn rootworm, *Diabrotica undecimpunctata howardi* Barber (Coleoptera: Chrysomelidae) on artificial diet and corn, *J. Econ. Entomol.* **78**, 290–293.

Mittler, T. E., Dadd, R. H. and Daniels, S. C. (1970) Utilization of different sugars by the aphid *Myzus persicae*, *J. Insect Physiol.* **16**, 1873–1890.

Mohri, S., Endo, Y., Matsuda, K., Kitamura, K. and Fujimoto, K., (1990) Physiological effects of soybean seed lipoxygenases on insects, *Agric. Biol. Chem.* **54**, 2265–2270.

Pal, R., Barenholz, Y. and Wagner, R. R. (1980) Effect of cholesterol concentration on organization of viral and vesicle membranes, *J. Biol. Chem.* **255**, 5802–5806.

Patzer, E. J., Wagner, R. R. and Barenholz, Y. (1978) Cholesterol oxidase as a probe for studying membrane organisation, *Nature* **274**, 3994–3995.

Pencoe, N. L. and Phillips, J. R. (1987) Boll weevil: legend, myth and reality, *J. Entomol. Sci. Suppl.* **1**, 30–51.

Perkins, J. H., (1980) Boll weevil eradication, *Science* **207**, 1044–1050.

Perlak, F. J., Deaton, R. W., Armstrong, T. A., Fuchs, R. L., Sims, S. R., Greenplate J. T. and Fischhoff, D. A. (1990) Insect resistant cotton plants, *Biotechnology* **8**, 939–943.

Perlak, F. J., Stone, T. B., Muskopf, Y. M., Petersen, L. J., Parker, G. B., McPherson, S. A., Wyman, J., Love, S., Reed, G., Biever, D. and Fischhoff, D. A. (1993) Genetically improved potatoes: protection from damage by Colorado potato beetles, *Plant Mol. Biol.* **22**, 313–321.

Purcell, J. P., Greenplate, J. T. and Sammons, R. D. (1992) Examination of midgut luminal proteinase activities in six economically important insects, *Insect Biochem. Mol. Biol.* **22**, 41–47.

Purcell, J. P., Greenplate, J. T., Jennings, M. G., Ryerse, J. S., Pershing, J. C., Sims, S. R., Prinsen, M. J., Corbin, D. R., Tran, M., Sammons, R. D. and Stonard, R. J. (1993) Cholesterol oxidase: a potent insecticidal protein active against boll weevil larvae, *Biochem. Biophys. Res. Commun.* **196**, 1406–1413.

Purcell, J. P., Greenplate, J. T., Duck, N. B., Sammons, R. D. and Stonard, R. J. (1994a) Insecticidal activity of proteinases, *FASEB J.* **8**, A1372.

Purcell, J. P., Isaac, B. G., Tran, M., Sammons, R. D., Gillespie, J. E., Greenplate, J. T., Solsten, R. T., Prinsen, M. J., Pershing, J. C. and Stonard, R. J. (1994b) Two enzyme classes active in green peach aphid bioassays, *J. Econ. Entomol.* **87**, 15–19.

Rabhe, Y. and Febvay, G. (1993) Protein toxicity to aphids: an *in-vitro* test on *Acyrthosiphon pisum, Entomol. Exper. Appl.* **67**, 149–160.

Shukle, R. H. and Murdock, L. L. (1983) Lipoxygenase, trypsin inhibitor and lectin from soybeans: effects on larval growth of *Manduca sexta* (Lepidoptera: Sphingidae), *Environ. Entomol.* **12**, 787–791.

Simmonds, B. and Brosten, D. (1991) A process of elimination, *Agrichem. Age*, June, 6–7.

Smith, A. G. and Brooks, C. J. W. (1976) Cholesterol oxidases: properties and applications, *J. Steroid Biochem.* **7**, 705–713.

Thurnhofer, H., Gains, B., Mutsch, B. and Hauser, H. (1986) Cholesterol oxidase as a structural probe of biological membranes: its application to brush-border membranes, *Biochim. Biophys. Acta* **856**, 174–181.

Tomioka, H., Kagawa, M. and Nakamura, S. (1976) Some enzymatic properties of 3-hydroxysteroid oxidase produced by *Streptomyces violascens, J. Biochem.* **79**, 903–915.

Vaeck, M., Reynaerts, A., Hofte, H., Jansens, S., DeBeuckeleer, M., Dean, C., Zabeau, Van Montagu, M. and Leemans, J. (1987) Transgenic plants protected from insect attack, *Nature* **328**, 33–37.

von Schaewen, A., Stitt, M., Schmidt, R., Sonnewald U. and Willmitzer, L. (1990) Expression of a yeast-derived invertase in the cell wall of tobacco and *Arabidopsis* plants leads to accumulation of carbohydrate and inhibition of photosynthesis and strongly influences growth and phenotype of transgenic tobacco plants, *EMBO J.* **9**, 3033–3044.

Wood, W. G., Igbavboa, U., Rao, A. M., Schroeder, F. and Avdulov, N. A. (1995) Cholesterol oxidation reduces $Ca^{2+} + Mg^{2+}$-ATPase activity, interdigitation, and increases fluidity of brain synaptic plasma membranes, *Brain Res.* **683**, 36–42.

Acknowledgments

The author wishes to thank all members of the Monsanto Insect Control Team, Bioactive Sourcing Team, Protein Analysis Group, Crop Transformation Group and Dr Jan Ryerse (St Louis University) for their contributions to the work discussed in this chapter.

7

Vegetative Insecticidal Proteins: Novel Proteins for Control of Corn Pests

GREGORY W. WARREN

Introduction

Maize (corn) is the principle cash grain crop in the United States and is the third most important cereal food crop of the world. In the US, the upper Midwestern states, commonly referred to as the 'Corn Belt', contribute up to 80% of the annual corn production. However, the diversity of cultural zones within the Corn Belt and other areas of the US contribute to a varied and unique pest–insect complex.

Corn is subject to a combination of insect attacks from the time it is planted until it is harvested and used as food or feed. More than 50 insect species are known to cause economic losses to corn in the US. However, only about 34 insect pests, excluding stored grain pests, occur with enough frequency to warrant chemical applications for control (Luckmann, 1978). Luckmann (1982) categorized these 34 insect pests as major and consistent pests, major but sporadic pests, and pests of moderate to minor importance. Those pests which were considered as major problems to corn production are shown in Table 7.1.

Species of corn rootworm (CRW) are considered to be the most destructive corn pests (Metcalf, 1986). In the US the three important species are *Diabrotica virgifera virgifera* (LeConte), the western corn rootworm (WCR); *D. longicornis barberi* (Smith & Lawerence), the northern corn rootworm (NCR); and *D. undecimpunctata*

Table 7.1 Pest species considered to be a major problem in corn production

Major and consistent pests[a]
Western corn rootworm, *Diabrotica virgifera virgifera* (Leconte)
Northern corn rootworm, *Diabrotica longicornis barberi* (Smith & Lawerence)
European corn borer, *Ostrinia nubilalis* (Hübner)
Corn earworm, *Helicoverpa zea* (Boddie)
Fall armyworm, *Spodoptera frugiperda* (J. E. Smith)
Major but sporadic pests[a]
Black cutworm, *Agrotis ipsilon* (Hüfnagel)
Corn leaf aphid, *Rhopalosiphum maidis* (Fitch)
Southwestern corn borer, *Diatraea grandiosella* (Dyar)

[a] Pests categorized by Luckmann (1982)

howardi (Barber), the southern corn rootworm (SCR). Only WCR and NCR are considered primary pests of corn in the Corn Belt. CRW larvae cause the most substantial plant damage by feeding almost exclusively on corn roots (Branson and Ortman, 1970). This injury has been shown to increase plant lodging (Sutter *et al.*, 1990), to reduce grain yield (Sutter *et al.*, 1990; Turpin *et al.*, 1972) and vegetative yield (Spike and Tollefson, 1991) as well as alter the nutrient content of the grain (Kahler *et al.*, 1985). Larval feeding also causes indirect effects on maize by opening avenues through the roots for bacterial and fungal infections which allow root or stalk rots to develop (Palmer and Kommedahl, 1969). Adult CRW are active in corn fields in late summer where they feed on ears, silks and pollen, interfering with normal pollination (Maredia and Landis, 1993).

Another soil pest which causes significant damage to seedling corn is *Agrotis ipsilon* (Hüfnagel), the black cutworm (BCW). Although some young larvae feed on corn foliage, the majority feed on a number of weed species surrounding corn fields (Luckmann, 1982). As the larvae mature they move to the soil surrounding corn plants seeking protection under clods or in soil cracks during daylight hours. Injury to corn results from these large larvae cutting off plants at or near the soil surface. Damage is normally limited to plants less than one meter in height.

Use of insecticides is the principle tactic employed in corn insect control throughout most of the Corn Belt. Corn soil insecticides, applied at the time of planting, are the single largest insecticide market in the US today. These applications are primarily targeted against three of the species described above, WCR, NCR and BCW.

WCR and NCR account for over $1 billion annual loss to US corn producers in yield reduction and insecticide treatment costs (Thurston and Yule, 1990). Soil insecticides targeted against CRW larvae are applied annually to approximately 50% (12 million hectares) of the US corn acreage (Thurston and Yule, 1990), representing an expenditure of approximately $500 million. Approximately $50 million is spent each year for control of BCW (Allemann, personal communication).

The application of soil insecticides at planting does not insure protection from damage by CRW or BCW. Insecticides applied at planting may degrade or move too low into the soil profile before CRW eggs hatch or BCW larvae move into a field. Soil microorganisms have been demonstrated to inactivate some insecticides before larval feeding activity begins. Post-emergence contact insecticides are only minimally effective against BCW because once the damage is noticed, it progresses rapidly and results from large, late instar larvae which are less susceptible to insecticides. Insecticide resistance to several compounds, including the carbamates, has developed in CRW during the last 10–15 years (Chio *et al.*, 1978; Luckmann, 1982). The potential negative environmental impact from using a large volume of insecticides as well as problems with adequately delivering the chemical to the most susceptible pest stage warrants exploring alternative strategies for controlling these three species.

Host plant resistance has long been viewed as an important component in insect control. Extensive plant breeding research was initiated in the early 1960s to identify genetic sources of resistance to CRW feeding. No useful sources of resistance have been identified either in adapted or exotic maize germplasms. Among the maize relatives of the tribe Maydeae, only *Tripsacum dactyloides* appears to be highly resistant to CRW larvae (Branson, 1971). Breeding programs using this species have determined that the chances of transferring a desired trait from *Tripsacum* to agronomically useful maize are remote (Galinat, 1977). Natural host plant resistance also appears not to be a viable option for control of BCW. Over 6000 genotypes of corn have been evaluated for resistance to first and second instar BCW (leaf-feeding stages) and none were found to be resistant (Jarvis *et al.*, 1981; Wilson *et al.*, 1983).

Plant genetic engineering offers a more rapid means of providing pest-resistant corn plants. Central to this end is the discovery of proteins which are efficacious against the target insect pests. Most efforts to engineer useful genes into corn and other crops have focused on genes from the bacterium, *Bacillus thuringiensis (Bt). Bt* is the *Bacillus* species most widely used as a source of biopesticides. The pesticidal properties of *Bt* are due to crystal protein inclusions, called δ-endotoxins, which are made during sporulation and have activity against a wide range of insect and non-insect pests (Fietelson, 1993). The first successful insertion of a *Bt* δ-endotoxin gene into corn for control of the European corn borer, *Ostrinia nubilalis*, and other lepidopteran species was reported by Koziel *et al.* (1993a).

Very few proteins have adequate oral insecticidal activity against CRW or BCW. In fact, these species have proven quite recalcitrant to most *Bt* δ-endotoxins. Thousands of *Bt* strains have been screened at sporulation for δ-endotoxins with CRW activity (see for example Donovan *et al.*, 1993a). δ-Endotoxins active against other coleopteran species, CryIIIA, CryIIIB, CryIIIB2, CryIIIC, CryIIIC(b) and CryIIID (Donovan *et al.*, 1992, 1993b; Lambert *et al.*, 1992a, b; McPherson *et al.*, 1988; Sick *et al.*, 1990) have not been shown to be efficacious against WCR and NCR. Some δ-endotoxin activity has been described for SCR when larvae are exposed to high concentrations (Abdelhameed and Landen, 1994; Donovan *et al.*, 1992; Rupar *et al.*, 1991). Donovan *et al.* (1993a, b) showed that the LC_{50} for CryIIIA, CryIIIB, CryIIIC(a) and CryIIIC(b) on SCR was >500, >457, 118 and 154 mg/cm^2 of artificial diet, respectively.

Several *Bt* strains have been reported to be active against other *Diabrotica* species. Bradfisch *et al.* (1993) described two *Bt* strains, PS86A1 and PS86Q3, which produced growth inhibition in *D. undecimpunctata undecimpunctata*, the western spotted cucumber beetle, when fermentation broths containing spores and crystals were tested at 25 ml/cm^2 of artificial diet. No mortality was reported at any concentration tested.

Reports of *Bt* strains with WCR activity have only recently been published. Lambert *et al.* (1995) reported that two *Bt* strains, BTS02584B and BTS02584C, were active against WCR. WCR mortality was 86% and 91% for BTS02584B and BTS02584C, respectively, when larvae were exposed to treated maize leaf pieces dipped into suspensions containing 1×10^9 spores-crystals/ml. It was not clear from this study how active pure δ-endotoxin was against WCR. Ely and Tippett (1995) discovered a *Bt* strain, JHCC 5767, which produced 88% mortality in WCR when larvae were exposed to a spore-crystal preparation at a concentration of 4800 ppm. *Bt* strains or δ-endotoxins with activity against NCR have not been reported.

The high concentrations of δ-endotoxin necessary for CRW and BCW activity make the use of *Bt* δ-endotoxins impractical for CRW and BCW field control, particularly through expression in transgenic corn. The fact that δ-endotoxins with high specific activity toward WCR and NCR appear to be quite rare prompted our laboratory to break from the traditional search for insecticidal activities in *Bt* strains at sporulation and search for active proteins during the vegetative growth stage (log phase) of *B. cereus*, *B. thuringiensis* and other *Bacillus* species.

B. cereus (Bc) is a soil bacterium that is very closely related to *Bt* and *B. anthrasis* (Gordon *et al.*, 1973; Kaneko *et al.*, 1978; Shinagawa *et al.*, 1991). *Bc* strains are difficult to distinguish from acrystalliferous *Bt* strains and some authors have suggested that *Bt* is a subspecies of *Bc* (Baumann *et al.*, 1984; Kaneko *et al.*, 1978; Logan and Bereley, 1984; Lysenko, 1983; Priest *et al.*, 1988). However, other reports support the separate species status for *Bc* and *Bt* (Krieg, 1970; Nakamura, 1994; O'Donnell *et al.*, 1980; Rogoff and Yousten, 1969; Somerville and Jones, 1972). Today, essentially all researchers use the taxonomic scheme where *Bc* and *Bt* are separate species.

In this chapter, the isolation and characterization of a *Bc* and a *Bt* strain which produce two new classes of insecticidal proteins during vegetative growth (log phase) with activity against the CRW complex and BCW, respectively, are described. These new classes of insecticidal proteins have been named VIPs (vegetative insecticidal proteins).

Discovery and Characterization of CRW-Active VIPs

Bioassay Development

A reliable and efficient artificial diet bioassay is a primary tool to aid in the discovery of insecticidal proteins. This is particularly true for CRW larvae which typically feed below the soil surface, making feeding behavior or toxicity studies difficult. A number of artificial diets have been developed for SCR (Marrone *et al.*, 1985; Rose and McCabe, 1973; Sutter *et al.*, 1971) but we found these diets to be inadequate for WCR and NCR bioassays. To screen efficiently for insecticidal proteins with activity against WCR larvae we developed an artificial diet bioassay system based on modifications to the SCR diet described by Marrone *et al.* (1985).

Screen for CRW-Active Bacillus

Culture supernatants of more than 400 *Bacillus* isolates, obtained from diverse environmental samples, were evaluated for production of insecticidally active principles during vegetative growth against a variety of insect species, including CRW and BCW. A culture supernatant from an isolate designated AB78 produced 100% mortality in WCR and NCR in the primary and subsequent bioassays. AB78 was identified as *B. cereus* on the basis of standard bacteriological criteria (Parry *et al.*, 1983), the lack of δ-endotoxin production after sporulation and molecular characterization carried out by polymerase chain reaction analysis (Carozzi *et al.*, 1991).

As shown in Table 7.2, only WCR and NCR, out of 13 insect species tested, were highly susceptible to the AB78 culture supernatant indicating a narrow spectrum of activity of the active principle. All WCR and NCR neonates died within 36–48 h of being exposed to 50 ml of culture supernatant per ml diet. Figure 7.1 shows AB78 treated and untreated WCR larvae after a 48 h exposure.

Initial characterization showed that the CRW active principle was proteinaceous by its heat lability, susceptibility to proteases, its size (>30 kDa) as well as its precipitation with ammonium sulfate. Proteins from AB78 culture supernatant were separated using a Mono-Q anion exchange column. SDS-PAGE analysis identified the presence of two major proteins of sizes 80 kDa and 45 kDa in column fractions which correlated with WCR mortality. A degenerate oligonucleotide probe encoding

Table 7.2 Biological activity spectrum of *B. cereus* strain AB78 culture supernatant

Insect species tested	Order	Activity
Western corn rootworm (*Diabrotica virgifera virgifera*)	Coleoptera	+ + +
Northern corn rootworm (*D. longicornis barberi*)	Coleoptera	+ + +
Southern corn rootworm (*D. undecimpunctata howardi*)	Coleoptera	+
Colorado potato beetle (*Leptinotarsa decemlineata*)	Coleoptera	—
Yellow mealworm (*Tenebrio molitor*)	Coleoptera	—
European corn borer (*Ostrinia nubilalis*)	Lepidoptera	—
Black cutworm (*Agrotis ipsilon*)	Lepidoptera	—
Fall armyworm (*Spodoptera frugiperda*)	Lepidoptera	—
Beet armyworm (*S. exigua*)	Lepidoptera	—
Tobacco budworm (*Heliothis virescens*)	Lepidoptera	—
Tobacco hornworm (*Manduca sexta*)	Lepidoptera	—
Corn earworm (*Helicoverpa zea*)	Lepidoptera	—
Northern house mosquito (*Culex pipiens*)	Diptera	—

amino acids 3–9 of the 80 kDa protein N-terminal sequence was used for isolation of the CRW active genes.

Vip1A(a) and Vip2A(a) Define a New Insecticidal Family

The genes encoding the CRW active proteins were identified using cosmid cloning with a functional screen of all cosmid clones against WCR. An AB78 genomic DNA library was constructed using the Stratagene SuperCos 1 cosmid vector system (Stratagene Inc.) in *E. coli* strain HB101. Approximately 600 clones were tested for insecticidal activity against WCR. All clones which were active against WCR also hybridized to the oligonucleotide probe specific for the 80 kDa WCR active protein.

One active cosmid clone, P3-12, was chosen for further characterization. Fragments from a Sau3A partial restriction digestion of the P3-12 DNA were subcloned into an *E. coli* pBluescript vector. One of the WCR active clones, designated pCIB6022, contained a 6 kb insert which was sufficient to confer insecticidal activity. The 6 kb insert encodes two open reading frames designated *vip2A(a)* and *vip1A(a)*, respectively (Figure 7.2). The *vip2A(a)* and *vip1A(a)* genes encode proteins with predicted sizes of 52 kDa and 100 kDa, respectively.

To confirm that the Vip1A(a) open reading frame was necessary for insecticidal activity a translational frameshift mutation was created in the *vip1A(a)* gene. The resulting plasmid, pCIB6203, contains a four nucleotide insertion in the coding region of Vip1A(a) and is not active against WCR, confirming that Vip1A(a) is an essential component of WCR activity. To define further the role of the Vip1A(a) and Vip2A(a) proteins in WCR activity, two further subclones of the 6 kb fragment were made. One, pCIB6206, contains the entirety of Vip1A(a) but lacks most of the Vip2A(a) coding region. The other pCIB6033 contains the entirety of Vip2A(a), but lacks most of the Vip1A(a) coding region (Figure 7.2). Cells containing pCIB6206 have only minimal activity and cells containing pCIB6033 exhibit no activity against WCR. However, a mixture of cells containing pCIB6206 and cells containing pCIB6033 produces 100% mortality in WCR. Thus, pCIB6206 produces a functional *vip1A(a)* gene product, while pCIB6033 produces a functional *vip2A(a)* gene

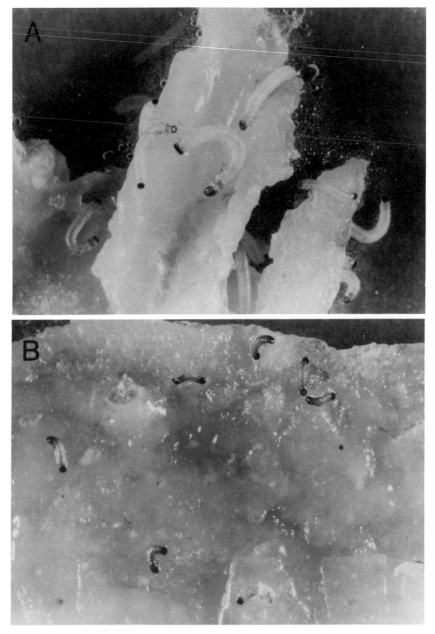

Figure 7.1 Comparison of healthy diet fed western corn rootworm larvae (A) with Vip1A(a) and Vip2A(a) treated larvae (B). Photograph taken 48 h after treatment

product. These results suggest a requirement for the gene product from the *vip*2A(a) region, in combination with *vip*1A(a), to confer maximal WCR activity.

Purified Vip1A(a) and Vip2A(a) protein were tested for toxicity against WCR separately and in combination. Full toxicity was observed only when Vip1A(a) was combined with Vip2A(a). The Vip1A(a) plus Vip2A(a) LC_{50} for neonate WCR was

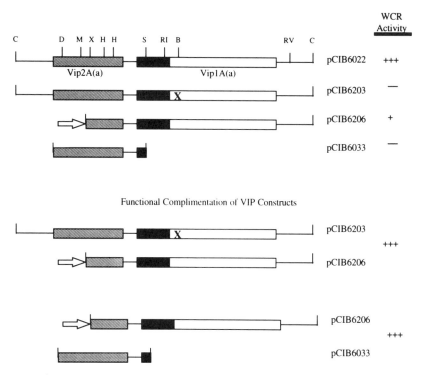

Figure 7.2 Characterization of Vip1A(a) and Vip2A(a) clones. Large 'X' in pCIB6203 represents the frameshift mutation introduced into Vip1. Arrow in pCIB6206 represents transcription by the β-galactosidase promoter. Restriction sites: C-Cla; X-Xba; S-Sca; RI-Eco RI; B-Bgl II; RV-Eco RV

40 ng Vip1A(a): 24 ng Vip2A(a)/g diet in four-day bioassays. These results confirm the binary mode of action of Vip1A(a) and Vip2A(a).

Discovery of Vip1A(a) and Vip2A(a) Homologues

As part of our screening program, we are interested in identifying *Bacillus* strains containing genes related to *vip*1A(a) and *vip*2A(a). Of 463 *Bacillus* strains screened by colony hybridization using *vip*1A(a) and *vip*2A(a)-derived probes, 55 strains showed strong hybridization signals. Twenty-two of these strains were grouped into eight classes according to Southern analysis of total DNA. Each of the 22 strains were bioassayed for activity against WCR and only three strains were found to be active. Among the strains identified which hybridized the Vip probes but were not active against WCR was *B. thuringiensis* var. *tenebrionis* (*Btt*). Although Western blot analysis indicated that *Btt* produces Vip1 and Vip2-like proteins, *Btt* supernatants did not exhibit any insecticidal activity against WCR.

The *vip*1A(a) and *vip*2A(a) homologues from *Btt*, designated *vip*1A(b) and *vip*2A(b), were cloned and sequenced. The Vip2A(a) and (b) proteins are similar throughout the protein, sharing a 95% amino acid identity (Figure 7.3). Vip1A(a) and (b) show a higher amino acid identity in the N-terminal (up to 97%) than in the C-terminal region (31%) with an overall amino acid identity of 78% (Figure 7.3). In

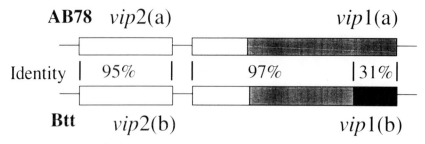

Figure 7.3   Comparison of *vip2* and *vip1* from *B. cereus* strain AB78 and *B. thuringiensis* var. *tenebrionis*. Dark gray area in Vip1 is the 80 kDa mature protein. Black box in *Btt* Vip1 is area of least homology to AB78 Vip1

complementation studies, the Vip2A(b) protein can substitute for Vip2A(a) and exhibits insecticidal activity against WCR when bioassayed in combination with Vip1A(a). However, no WCR insecticidal activity was observed when Vip1A(b) was used in combination with Vip2A(a). Considering that the degree of homology exhibited by the Vip1A proteins is strongly reduced in the C-terminal domain, it is likely that this domain is involved in WCR and NCR specificity.

Discovery and Characterization of BCW Active VIPs

Screen for BCW Active Bacillus

During the initial screen of vegetative *Bacillus* culture supernatants a *Bt* isolate, AB88, was identified with specific activity against many lepidopteran species (Table 7.3). Because some *Bt* strains are known to produce a non-proteinaceous thermostable toxin, designated β-exotoxin, during vegetative growth (Heimpel, 1967) the

Table 7.3   Biological activity spectrum of *B. thuringiensis* strain AB88 culture supernatant

Insect species tested	Order	Supernatant activity	
		Non-autoclaved	Autoclaved
Eurpean corn borer (*Ostrinia nubilalis*)	Lepidoptera	+ + +	—
Black cutworm (*Agrotis ipsilon*)	Lepidoptera	+ + +	—
Fall armyworm (*Spodoptera frugiperda*)	Lepidoptera	+ + +	—
Beet armyworm (*Spodoptera exigua*)	Lepidoptera	+ + +	—
Tobacco budworm (*Heliothis virescens*)	Lepidoptera	+ + +	—
Corn earworm (*Helicoverpa zea*)	Lepidoptera	+ + +	—
Western corn rootworm (*Diabrotica virgifera virgifera*)	Coleoptera	—	—
Colorado potato beetle (*Leptinotarsa decemlineata*)	Coleoptera	—	—
Yellow mealworm (*Tenebrio molitor*)	Coleoptera	—	—
Fruit fly (*Drosophila melanogaster*)	Diptera	—	—
Northern house mosquito (*Culex pipiens*)	Diptera	—	—

Table 7.4 Effect of Vip3A(a) protein on *Agrotis ipsilon* (BCW) and *Spodoptera fru-giperda* (FAW) larvae

Treatment	Vip3A concentration (ng/ml)	Percent mortality	
		BCW	FAW
Control		6	8
Bt AB88 supernatant		100	100
E. coli (pCIB7105)	28	82	44
	70	96	78
	140	100	100
	280	100	100

AB88 active principle was tested for heat stability. The reduction of insecticidal activity of the AB88 culture supernatant to insignificant levels by autoclaving indicated that the active principle was not β-exotoxin (Table 7.3). Further analysis and separation of the major proteins in the AB88 culture supernatant demonstrated that different proteins were responsible for the mortality of the different lepidopteran species. Mortality of BCW and fall armyworm (FAW) correlated with the presence of a protein with an approximate molecular weight of 80 kDa.

Cloning the Vip3A(a) Insecticidal Protein Gene

Size selected fragments of *Bt* strain AB88 genomic DNA were ligated into pBluescript and transformed into *E. coli* strain DH5a. Two recombinants out of 400 colonies hybridized to an oligonucleotide probe specific for the 88 kDa BCW active protein. Further subcloning produced a plasmid, pCIB7105, which contained a 3.5 Kb fragment of AB88 DNA. Sequence analysis of the 3.5 Kb fragment revealed an open reading frame which encodes a protein of 791 amino acids corresponding to a molecular weight of 88.5 kDa. This protein, named Vip3A(a), has no homology to any known insecticidal proteins including the *Bt* δ-endotoxins (Estruch *et al.*, 1996).

E. coli containing pCIB7105 exhibited selective toxicity to BCW and FAW larvae comparable to that of the AB88 culture supernatant (Table 7.4). MacIntosh *et al.* (1990) reported that the *Bt* δ-endotoxins CryIA(b) and CryIA(c) were active against BCW with LC_{50} values of more than 80 mg and 18 mg per ml of diet, respectively. The Vip3A(a) protein produced 96–100% mortality in BCW when added at 70–140 ng/ml of diet. This amount of protein is at least 128-fold lower than the amount of CryIA proteins needed to achieve just 50% mortality and it is similar to the LC_{50} of δ-endotoxins to susceptible insects (Koziel *et al.*, 1993b).

Summary

Secreted proteins expressed during log-phase growth of *Bacillus* species appear to be a rich source for insecticidal activities. Two new classes of insecticidal proteins were identified with activity against four of the major insect pests of corn after evaluating approximately 400 *Bacillus* isolates. All four insect species are recalcitrant to the *Bt* δ-endotoxins. Activity of the binary Vip1A(a)–Vip2A(a) toxin appears to be very highly specific to WCR and NCR. Interestingly, very little activity was apparent on SCR, a closely related *Diabrotica* species. Vip concentrations necessary for 100% mortality in WCR and NCR caused only 10–20% mortality in SCR. This may be due to differences in general gut physiology between WCR/NCR and SCR. These two groups differ considerably in their biology but most notably in food source utilization. WCR and NCR are oligophagous, feeding on a few closely related grasses, whereas SCR is polyphagous, feeding on a wide variety of host plants, including maize, peanuts, small grains and many wild grasses (Maredia and Landis, 1993). The midgut surface of SCR may be quite different from WCR and NCR since SCR has to cope with a wide variety of food sources. Thus, SCR may lack or have a limited number of binding receptors for the Vip1A(a) protein. Research exploring the Vip1A(a)–Vip2A(a) mode of action is in progress.

The Vip1A(a)–Vip2A(a) complex is unique in that it represents the first described binary toxin active against coleopteran species. Another binary toxin has been described from *B. sphaericus* (*Bs*) which is active against certain dipteran species. This toxin however is expressed during sporulation and is contained in parasporal crystal inclusions similar to the *Bt* δ-endotoxins (Berry *et al.*, 1991). There is no sequence homology between the Vip1A(a)–Vip2A(a) and *Bs* binary toxins.

The Vip1A(a)–Vip2A(a) and Vip3A(a) proteins, active against CRW and BCW respectively, appear to be more potent than any reported activity of other proteins including the *Bt* δ-endotoxins. Both classes are active at ng/g diet levels against their respective insect targets. Activities reported for the δ-endotoxins appear to be rather weak, suggesting that expression of these proteins in maize would not be a viable option for CRW and BCW control. Therefore, the Vip1A(a), Vip2A(a) and Vip3A(a) insecticidal proteins offer great potential as a viable means of controlling major corn pests when delivered through transgenic corn.

References

Abdelhameed, A. and Landen, R. (1994) Studies on *Bacillus thuringiensis* strains isolated from Swedish soils – insect toxicity and production of *B. cereus* diarrheal-type enterotoxin, *World J. Microbiol. Biotechnol.* **10**, 406–409.

Baumann, L., Okamoto, K., Unterman, B. M., Lynch, M. J. and Bauman, P. (1984) Phenotypic characterization of *Bacillus thuringiensis* and *Bacillus cereus*, *J. Invert. Pathol.* **44**, 329–341.

Berry, C., Hindley, J. and Oei, C. (1991) The *Bacillus sphaericus* toxins and their potential for biotechnological development, In: Maramorosch, K. (Ed.), *Biotechnology for Biological Control of Pests and Vectors*, Boca Raton: CRC Press, pp. 35–51.

Bradfisch, G. A., Michaels, T. and Payne, J. M. (1993) Biologically active *Bacillus thuringiensis* isolates, United States Patent No. 5 208 017.

Branson, T. F. (1971) Resistance in the grass tribe Maydae to larvae of the western corn rootworm, *Ann. Entomol. Soc. Am.* **64**, 861–863.

Branson, T. F. and Ortman, E. E. (1970) The host range of larvae of the western corn rootworm: further studies, *J. Econ. Entomol.* **63**, 800–803.

Carozzi, N. B., Kramer, V. C., Warren, G. W., Evola, S. and Koziel, M. G. (1991) Prediction of insecticidal activity of *Bacillus thuringiensis* strains by polymerase chain reaction product profiles, *Appl. Environ. Microbiol.* **57**, 3057–3061.

Chio, H., Chang, C. S., Metcalf, R. L. and Shaw, J. (1978) Susceptibility of four species of *Diabrotica* (corn rootworm) to insecticides, *J. Econ. Entomol.* **71**, 389–393.

Donovan, W. P., Rupar, M. J., Slaney, A. C. and Johnson, T. B. (1993a) *Bacillus thuringiensis* CryIIIC gene encoding toxic to coleopteran insects, US Patent No. 5 187 091.

Donovan, W. P., Rupar, M. J. and Slaney, A. C. (1993b) *Bacillus thuringiensis* CryIIIC(b) toxin gene and protein toxic to coleopteran insects, United States Patent No. 5 264 364.

Donovan, W. P., Rupar, M. J., Slaney, A. C., Malvar, T., Gawron-Burke, M. C. and Johnson, T. (1992) Characterization of two genes encoding *Bacillus thuringiensis* insecticidal crystal proteins toxic to Coleoptera species, *Appl. Environ. Microbiol.* **58**, 3921–3927.

Ely, S. and Tippett, J. M. (1995) *Bacillus thuringiensis* toxin with activity against *Diabrotica* species, United States Patent No. 5 446 019.

Estruch, J. J., Warren, G. W., Mullins, M. A., Nye, G. J. and Koziel, M. G. (1996) Vip3A, a novel *Bacillus thuringiensis* vegetative insecticidal protein with a wide spectrum of activities against lepidopteran insects, *Proc. Nat. Acad. Sci.* **93**, 5389–5394.

Fietelson, J. S. (1993) The *Bacillus thuringiensis* family tree, In: Kim, L. (Ed.), *Advanced Engineered Pesticides*, New York: Mercel Dekker, pp. 63–72.

Galinat, W. C. (1977) The origin of corn, In: Sprague, G. F. (Ed.), *Corn and Corn Improvement*, Madison: American Society of Agronomy, pp. 1–47.

Gordon, R. E., Haynes, W. C. and Pang, C. H. (1973) The genus *Bacillus, Agricultural Handbook No. 427*, Washington: US Department of Agriculture.

Heimpel, A. M. (1967) A critical review of *Bacillus thuringiensis* var. *thuringiensis* Berliner and other crystalliferous bacteria, *Ann. Rev. Entomol.* **12**, 287–322.

Jarvis, J. L., Guthrie, W. D. and Robbins, J. C. (1981) Evaluation of maize plant introductions for resistance to black cutworm larvae, *Maydica* **26**, 219–225.

Kahler, A. L., Olness, A. E., Sutter, G. R., Dybing, C. D. and Devine, O. J. (1985) Root damage by western corn rootworm and nutrient content in maize, *Agron. J.* **77**, 769–774.

Kaneko, T., Nozaki, R. and Aizawa, K. (1978) Deoxyribonucleic acid relatedness between *Bacillus anthrasis, Bacillus cereus* and *Bacillus thuringiensis, Microbiol. Immunol.* **22**, 639–641.

Koziel, M. G., Beland, G. L., Bowman, C., Carozzi, N. B., Crenshaw, R., Crossland, L., Dawson, J., Desai, N., Hill, M., Kadwell, S. *et al.* (1993a) Field performance of elite transgenic maize plants expressing an insecticidal protein derived from *Bacillus thuringiensis, Biotechnology* **11**, 194–200.

Koziel, M. G., Carozzi, N. B., Currier, T. C., Warren, G. W. and Evola, S. V. (1993b) The insecticidal crystal proteins of *Bacillus thuringiensis*: past, present and future uses, *Biotechnol. Gene. Eng. Rev.* **11**, 171–228.

Krieg, A. (1970) In vitro determination of *Bacillus thuringiensis, Bacillus cereus*, and related Bacilli, *J. Invert. Pathol.* **15**, 313–320.

Lambert, B., Hofte, H., Annys, K., Jansens, J., Soetaert, P. and Peferoen, M. (1992a) Novel *Bacillus thuringiensis* insecticidal crystal protein with a silent activity against coleopteran larvae, *Appl. Environ. Microbiol.* **58**, 2536–2542.

Lambert, B., Theunis, W., Aguda, R., van Audenhove, K., Decock, C., Jansens, S., Seurinck, J. and Peferoen, M. (1992b) Nucleotide sequence of gene cryIIID encoding a novel coleopteran-active crystal protein from strain BT109P of *Bacillus thuringiensis* subsp. *kurstaki, Gene* **110**, 131–132.

Lambert, B., Jansens, S. and Peferoen, M. (1995) Method of controlling Coleoptera using *Bacillus thuringiensis* strains MG P-14025 and LMG P-14026, United States Patent No. 5 422 106.

Logan, N. A. and Bereley, R. C. W. (1984) Identification of *Bacillus* strains using the API system. *J. Gen. Microbiol.* **130**, 1871–1882.

Luckmann, W. H. (1978) Insect control in corn-practices and perspectives, In: Smith, E. H. and Pimental, D. (Eds), *Pest Control Strategies*, New York: Academic Press, pp. 137–155.

Luckmann, W. H. (1982) Integrating the cropping system for corn insect pest management, In: Metcalf, R. L. and Luckmann, W. H. (Eds), *Introduction to Insect Pest Management*, New York: John Wiley & Sons, pp. 499–519.

Lysenko, O. (1983) *Bacillus thuringiensis*: evolution of a taxonomic conception, *J. Invert. Pathol.* **42**, 295–298.

MacIntosh, S. C., Stone, T. B., Sims, S. R., Hunst, P. L., Greenplate, J. T., Marrone, P. G., Perlak, F. J., Fischhoff, D. A. and Fuchs, R. L. (1990) Specificity and efficacy of purified *Bacillus thuringiensis* proteins against agronomically important insects, *J. Invert. Pathol.* **56**, 258–266.

Maredia, K. and Landis, D. (1993) *Corn Rootworms: Biology, Ecology and Management*, Extension Bulletin E-2438, Cooperative Extension Service, Michigan State University.

Marrone, P. G., Ferre, F. D., Mosley, T. R. and Meinke, L. J. (1985) Improvements in laboratory rearing of the southern corn rootworm, *Diabrotica undecimpunctata howardi* Barber (Coleoptera: Chrysomelidae), on an artificial diet and corn, *J. Econ. Entomol.* **78**, 290–293.

McPherson, S. A., Perlak, F. J., Fuchs, R. L., Marrone, P. G., Lavrik, P. B. and Fischhoff, D. A. (1988) Characterization of the coleopteran-specific protein gene of *Bacillus thuringiensis* var. *tenebrionis*, *Bio/Technology* **6**, 61–66.

Metcalf, R. L. (1986) Forward, In: Krysan, J. L. and Miller, T. A. (Eds), *Methods for the Study of Pest Diabrotica*, New York: Springer-Verlag.

Nakamura, L. K. (1994) DNA relatedness among *Bacillus thuringiensis* serovars, *Int. J. System. Bacteriol.* **44**, 125–129.

O'Donnell, A. G., MacFie, H. J. H. and Norris, J. R. (1980) An investigation of the relationship between *Bacillus cereus*, *Bacillus thuringiensis* and *Bacillus mycoides* using pyrolysis gas-liquid chromatography, *J. Gen. Microbiol.* **119**, 189–194.

Palmer, L. T. and Kommedahl, T. (1969) Root-infesting *Fusarium* species in relation to rootworm infestations in corn, *Phytopathology* **59**, 1613–1617.

Parry, J. M., Turnbill, P. C. B. and Gibson, J. R. (1983) *A Colour Atlas of* Bacillus *Species*, London: Wolfe Medical Publications.

Priest, F. G., Goodfellow, M. and Todd, C. (1988) A numerical classification of the genus *Bacillus*, *J. Gen. Microbiol.* **134**, 1847–1882.

Rogoff, M. H. and Yousten, A. A. (1969) *Bacillus thuringiensis*: microbiological considerations, *Annual Rev. of Microbiol.* **23**, 357–389.

Rose, R. L. and McCabe, J. M. (1973) Laboratory rearing techniques for the southern corn rootworm, *J. Econ. Entomol.* **66**, 398–400.

Rupar, M. J., Donovan, W. P., Groat, R. G., Slaney, A. C., Mattison, J. W., Johnson, T. B., Charles, J. F., Dumanoir, V. C. and Barjac, H. de (1991) Two novel strains of *Bacillus thuringiensis* toxic to coleopterans, *Appl. Environ. Microbiol.* **57**, 3337–3344.

Seurinck, J. and Peferoen, M. (1992b) Nucleotide sequence of gene cryIIID encoding a novel coleopteran-active crystal protein from strain BT109P of *Bacillus thuringiensis* subsp. *kurstaki*, *Gene* **110**, 131–132.

Shinagawa, K., Sugiyama, J., Terada, T., Matsusaka, N. and Sugii, S. (1991) Improved methods for purification of an enterotoxin produced by *Bacillus cereus*, *FEMS Microbiol. Lett.* **64**, 1–5.

Sick, A., Gaertner, F. and Wong, A. (1990) Nucleotide sequence of a coleopteran-active toxin gene from a new isolate of *Bacillus thuringiensis* subsp. *tolworthi*, *Nucl. Acid Res.* **18**, 1305.

Somerville, H. J. and Jones, M. L. (1972) DNA competition studies within the *Bacillus cereus* group of bacilli, *J. Gen. Microbiol.* **73**, 257–265.

Spike, B. P. and Tollefson, J. J. (1991) Response of western corn rootworm-infested corn to nitrogen fertilization and plant density, *Crop Sci.* **31**, 776–785.

Sutter, G. R., Krysan, J. L. and Guss, P. L. (1971) Rearing the southern corn rootworm on artificial diet, *J. Econ. Entomol.* **64**, 65–67.

Sutter, G. R., Fisher, J. R., Elliot, N. C. and Branson, T. F. (1990) Effects of insecticide treatments on root lodging and yields of maize in controlled infestations of western corn rootworm (Coleoptera: Chrysomelidae), *J. Econ. Entomol.* **83**, 2414–2420.

Thurston, G. S. and Yule, W. N. (1990) Control of larval northern corn rootworm (*Diabrotica barberi*) with two Steinernematid nematode species, *J. Nematol.* **22**, 127–131.

Turpin, F. T., Dumenil, L. C. and Peters, D. C. (1972) Edaphic and agronomic characteristics that affect potential for rootworm damage to corn in Iowa, *J. Econ. Entomol.* **65**, 1615–1619.

van Frankenhuyzen, K. (1993) The challenge of *Bacillus thuringiensis*, In: Entwistle, P. F., Cory, J. S., Bailey, M. J. and Higgs, S. (Eds), *Bacillus thuringiensis, An Environmental Biopesticide: Theory and Practice*, New York: John Wiley & Sons.

Wilson, R. L., Jarvis, J. L. and Guthrie, W. D. (1983) Evaluation of maize for resistance to black cutworm larvae, *Maydica* **28**, 449–453.

Acknowledgments

The author wishes to thank all members of the Insect Control–Seeds Group for their contribution to the work discussed in this chapter.

8

Plant Lectins as Insect Control Proteins in Transgenic Plants

THOMAS H. CZAPLA

Introduction

The protection of food plants from insect pests is a constant challenge to growers and scientists worldwide. The increasing demands placed on agriculture to produce high yielding crops with less input is a daunting challenge that will require the use of several technologies in the future. A new technology for crop protection has been the development of several different transformation systems that can incorporate foreign genes into the plant's genome. Plants such as cotton, potato and maize have been transformed to contain a gene that produces an insecticidal endotoxin protein from *Bacillus thuringiensis* (*Bt*). This unique approach has several advantages over current pesticide technology. The stage of insect targeted is now the most vulnerable, the neonate, rather than older instars. A second advantage is that the plant will continue to produce the agent as required throughout the season, thus avoiding the loss of control due to the breakdown of the insecticidal protein and the need for repeated insecticidal applications. However, this approach will not prevent the development of insect resistance to control agents such as *Bt*. This concern is greatly increased due to the lack of alternative genes currently available to control these insects and provide resistance management options. There are also other serious pest insects that are not susceptible to the current array of *Bt*s. This lack of genes has led to investigation of other potential proteins for insect control. One potential source of insect resistance genes may be plant lectins.

Lectins are carbohydrate-binding proteins of non-immune origin that can agglutinate cells and bind glycans of glycoproteins, glycolipids or polysaccharides (Goldstein *et al.*, 1980). Lectins have been isolated from various plant tissues, with seeds being the richest source (Etzler, 1986). This is especially true among plants in the Leguminosae family (Strosberg *et al.*, 1986). Lectins are widely used as biological tools due to their interaction with carbohydrates in solution and on cell surfaces (Liener *et al.*, 1986). Although the specific carbohydrate binding affinity of lectins has been intensely studied, there is much speculation as to the role lectins play in plants. Biological roles in defense (Chrispeels and Raikel, 1991; Peumans and Van Damme, 1995a,b), cell–cell recognition (Knox *et al.*, 1976) or as catalytic enzymes (Hankins *et al.*, 1980) have been proposed. The lectin from *Phaseolus vulgaris* was

123

demonstrated to have a lethal effect on larvae of the bruchid beetle *Callosobruchus maculatus* when added to an artificial diet (Janzen *et al.*, 1976). Ironically, it has now been determined that this lectin is not detrimental to the beetle, but the active material was an alpha-amylase inhibitor (Huesing *et al.*, 1991a; see also Chapter 9). However, other lectins have been shown to have a detrimental effect on insects (Czapla and Lang 1990; Murdock *et al.*, 1990).

This chapter will focus on the *in vitro* assays conducted to date on several different insect species and transgenic plant bioassays from both greenhouse and field trials. The potential mode of action regarding lectins as well as the insect's ability to develop resistance to lectins will also be discussed. The chapter will conclude with the outlook for lectins as a means of insect control in transgenic plants.

Artificial Diet Bioassays

The ability to rear insects and test potentially insecticidal proteins via an artificial diet is a critical component in the development of transgenic insect resistant plants. The time needed to clone a gene, make the appropriate vectors, transform the plants and then grow the plants to be infested with the insect is too long and expensive to screen new proteins efficiently in this manner. A more methodical approach is needed in which proteins can be tested using artificial diets and then only the best genes are cloned and tested in transgenic plants. There are two major problems with this approach. The first is that some insects are unable to complete their development or even survive on an artificial diet. The second problem is how well the artificial diet mimics the host plant. The ability to predict which proteins will work in a transgenic plant depends upon how well the artificial diet bioassay handles these two problems. Proteins that work very effectively in an artificial diet may have no effect at all on the insect infesting the transgenic plant.

Most proteins are tested using either an overlay assay in which the material is transferred to the diet surface or an incorporation test in which the protein is mixed into the diet itself. Several insect species have been screened via artificial diets for their susceptibility to lectins. These data have been grouped according to the insect screened.

Lectins as Insect Control Proteins

Callosobruchus maculatus (*Cowpea Weevil*)

The effect of a lectin on the normal development of insects was first investigated utilizing an artificial seed diet and the cowpea weevil (Janzen *et al.*, 1976). The phytohemagglutinin (PHA) from black bean, *Phaseolus vulgaris*, was detrimental to the survival of cowpea weevil larvae. Female weevils oviposited on the seeds and all but 10 eggs were then subsequently removed. Approximately 4.5 larvae per artificial seed survived to adulthood when reared on the control seeds. Two small adult beetles emerged from 14 artificial seeds containing 1% PHA (0.1 beetles per seed) and no adult beetles emerged on diets containing 5% PHA. These results led to the hypothesis that the major function of PHA in black beans and other legumes was as a protective agent against seed insects.

Similar feeding trial results were obtained that supported the hypothesis that PHA was lethal to the cowpea weevil larvae (Gatehouse *et al.*, 1984). However, it was observed that activity was reduced in the more highly purified preparations compared to the commercial preparation of the lectin. Interactions with the different PHA subunits or impurities in the commercial preparations were cited as potential reasons for this discrepancy.

PHA is a tetrameric protein containing two subunits, E and L, designated by their ability to bind erythrocytes or lymphocytes respectively (Goldstein and Poretz, 1986). Purified preparations of PHA containing only E or L subunits were fed to cowpea weevil larvae (Boulter *et al.*, 1986). Neither the E4 nor the L4 preparations displayed any detrimental effect on the cowpea weevil larvae. The toxic effects of PHA were due to a synergism of the subunits as concluded by the author for the above observation. Insect activity from the *Pisum sativum* (pea) lectin was also identified. Cowpea weevil larvae survival to adulthood was only 33% when reared on diets containing 2% pea lectin (Boulter *et al.*, 1986; Hepher *et al.*, 1989).

Seventeen plant lectins were bioassayed to determine if the biological effects of lectins on cowpea weevil were widespread (Murdock *et al.*, 1990). Five of the 17 lectins significantly delayed the development time of the cowpea weevil larvae. Lectins, with N-acetylgalactosamine/galactose (GalNAc/Gal) carbohydrate binding specificity, from *Maclura pomifera* (osage orange) and *Arachis hypogaea* (peanut) delayed development by approximately 8.5 and 10 days respectively at the 1% level in the artificial seeds. However, other lectins with similar binding specificity from *Glycine max* (soybean), *Vicia villosa* (hairy vetch), *Dolichos biflorus* (horse gram) and *Sophora japonica* (pagoda tree) had no apparent effect on cowpea weevil development. The three lectins with N-acetylglucosamine (GlcNAc) carbohydrate binding specificity displayed significant effects in delaying the development time of the cowpea weevil larvae. Lectins from *Solanum tuberosum* (potato) and *Datura stramonium* (jimson weed) were of a semi-purified nature and contained other proteins that may have contributed to their activity. A delay of 3.6 and 4.1 days, respectively, was observed for these lectin preparations. The lectin from *Tritcum aestivum* (wheat), known as wheat germ agglutinin (WGA), significantly delayed development by approximately 6.5 days at the 0.2% level and 22.8 days at the 1% level; this was the most effective lectin tested. WGA was shown by electrophoresis to be essentially pure. PHA lectins were shown to have no activity against the cowpea weevil in this experiment. Three preparations were tested consisting of E only, L only and a naturally occurring isolectin mixture. No developmental delay was observed with any of the preparations or levels tested. Subsequent experimentation demonstrated that the lethal effect of *P. vulgaris* preparations was due to an alpha-amylase inhibitor in the lectin preparation (Huesing *et al.*, 1991a).

The effect of GlcNAc binding lectins was further investigated by Huesing *et al.* (1991b,c). All three WGA isolectins were determined to be equally effective against the cowpea weevil (Huesing *et al.*, 1991b). GlcNAc binding lectins from *Oryza sativa* (rice) and *Urtica dioica* (stinging nettle) also displayed increased mortality and increased development time when fed to the cowpea weevil (Huesing *et al.*, 1991c). Rice lectin activity was identical to WGA, while the stinging nettle was approximately 2–4 times less effective than WGA.

Lectins from *Allium sativum* (garlic) and *Galanthus nivalis* (snowdrop) were discovered to have excellent activity against the cowpea weevil larvae (Gatehouse *et al.*, 1992). Larvae reared on artificial seeds containing 2% lectin suffered 90% mortality compared to the control larvae. Larval mortality was greater than 50% at the 1% level. The lectin from tulip displayed no biological activity against the cowpea weevil.

Ostrinia nubilalis (European Corn Borer)

The European corn borer (ECB) is a major pest of maize in North America and has two or more generations per year in the United States corn belt region. The yearly estimated damage and control costs due to this insect are approximately 800 million dollars.

Neonate ECB larvae were reared on artificial diets containing lectins either topically applied to the diet surface or incorporated into the diet (Czapla and Lang, 1990). Twenty-six plant lectins from six different families were bioassayed. Only three lectins were discovered to be effective at a 2% topical level against ECB larvae. They were WGA, *Bauhinia purpurea* agglutinin (BPA) and the lectin from *Ricinus communis*. The activity seen in the *R. communis* preparation has been determined to be due to a contaminant and not the lectin. The LC_{50} for WGA and BPA was determined to be 0.59 and 0.73 mg/g diet, respectively, when incorporated into a diet. This is approximately 0.06% of the diet for WGA and 0.07% for BPA. WGA has a carbohydrate binding specificity for GlcNAc residues (Goldstein and Poretz, 1986); however, it was the only lectin with this generalized binding specificity that was effective against ECB. Lectins from *Lycopersicon esculentum* (tomato), potato, *Phytolacca americana* (pokeweed), *Datura stramonium* (jimson weed) and *Wistaria floribunda* (Japanese wisteria) had no effect. BPA carbohydrate binding is specific to GalNAc/Gal (Goldstein and Poretz, 1986). BPA was very similar in activity to WGA, inhibiting weight gain and causing mortality of the neonate ECB larvae. Sixteen other lectins with GalNAc/Gal specific binding were tested but only BPA was effective. Although lectins are classified by their specific binding to a single carbohydrate residue, their binding affinities to oligomers or complex carbohydrates may be dramatically different (Goldstein and Poretz, 1986). These differences may account for the variability in insect activity.

Two additional lectins showed ECB activity out of 127 lectin preparations tested via a topical bioassay (Cavalieri *et al.*, 1995). These were the rice lectin and a lectin from the leaves of another grass, *Agropyrum repens*. Both lectins are very similar in structure to WGA. The activity seen in the preparations from *Crocus vernus* was recently determined to be due to a contaminating material (Rao, personal communication).

Diabrotica Species (Corn Rootworm)

The corn rootworm complex in the United States consists of three species, *Diabrotica barberi* (northern), *D. undecimpunctata howardi* (southern) and *D. virgifera virgifera* (western). The easiest species to rear utilizing an artificial diet is the southern corn rootworm, but the western and northern cause the greatest economic damage to maize. The economic damage and control costs exceed 1 billion dollars a year (Metcalf, 1986).

Several lectins were identified as having a detrimental effect on the growth of southern corn rootworm (Czapla and Lang, 1990). Out of 26 lectins, eight inhibited weight gain by at least 40% when tested at a 2% topical level. Four of these lectins inhibited weight gain by 70% or more. These four lectins were from *Bandeiraea simplicifolia*, *Maclura pomifera* (MPL, osage orange), *Artocarpus integrifolia* (jackfruit), known as jacalin and *Codium fragile* (CFL, green marine algae). The activity of the lectin preparation from *R. communis* has been recently determined to be due to a contaminant (personal observation). WGA, BPA and the lectin from hairy vetch each inhibited weight gain by approximately 40%.

Topical bioassays identified over 60 plant lectins that inhibited weight gain of neonate CRW by at least 40% when tested at the 2% level (Cavalieri *et al.*, 1995). The most effective lectin identified was from *Eranthis hyemalis* (EHL). This was the only lectin to cause 100% mortality to neonate CRW larvae at the 2% level. Twenty-seven other lectin preparations inhibited weight gain by over 80%. The best of this group were lectins from *Urtica dioica* (stinging nettle), *Cytisus scoparius* (scotch broom), *Crocus vernus*, and *Cicer arientinum* (chick-pea). The activity seen in the preparations from *C. vernus* was recently determined to be due to a contaminating material (Rao, personal communication).

Dilution series bioassays have been conducted with the most promising lectins using a diet incorporation bioassay. WGA inhibits CRW growth by only 40% at the 0.3% level. The LC_{50} values for jacalin and CFL were determined to be 0.3% and 0.2% respectively, while the level needed to inhibit weight gain by 50% was approximately 0.1% for jacalin and 0.05% for CFL (personal observations).

The EHL lectin, although very active in a topical assay, displayed only 30% weight loss against CRW larvae at incorporation concentrations of 0.1% (personal observations). Surprisingly, EHL and a reduced and alkylated EHL were reported to have 90% and 100% mortality at the same level (Kumar *et al.*, 1993; Rao and Kumar, 1994) in a different set of bioassays. This discrepancy may be due to the heat labile nature of EHL. The early experiments were not conducted using low melt agar, hence those EHL preparations were subject to higher temperatures while being incorporated into the diet.

The latter experiments were conducted using a low melt agar. Unfortunately, efforts to duplicate this excellent activity have been unsuccessful (Meyer and Balasubramaniam, personal communication).

Empoasca fabae (Potato Leafhopper)

The potato leafhopper is a serious pest of alfalfa in the midwestern and eastern United States. Its feeding injury to alfalfa, know as hopperburn, results in severe stunting and reduced stand longevity (Kindler *et al.*, 1973).

Fourteen plant lectins were studied for activity against potato leafhopper using an artificial diet (Habbi *et al.*, 1993). Adult females (1–2 days old) were placed on acclimating diets for two days and then transferred to diets containing the lectin to be tested. Mortality was observed every 12 hours for 15 days and comparisons were made based upon mean survival time. Six lectins – PHA, WGA, jacalin, and lectins from *Lens culinaris* (lentil), pea and horse gram – were found to be effective in reducing mean survival time at a concentration of 1.5%. PHA (E + L) was the most effective lectin and significantly reduced mean survival time from an average of 10 days for the control treatment to 1.5 days at 0.5%.

Experiments using PHA-E and PHA-L at the 0.5 % level indicated that PHA-E is the effective subunit. Leafhoppers feeding on diet containing PHA-E had a mean survival of 3 days, while those that fed on diet containing PHA-L had a mean survival time of almost 10 days, which was slightly greater than the control (8.5 days). Quality control assurances from Vector (source of PHA lectins) and SDS-PAGE were cited by the authors as indicating that the samples were of a highly purified nature. However, based upon the findings by Huesing *et al.* (1991a), minor contaminants cannot be completely excluded as the source of the observed activity.

Nilaparvata lugens (Rice Brown Planthopper) and Nephotettix cinciteps (Rice Leafhopper)

Two major pests of rice are the rice brown planthopper and the rice leafhopper. Both pests can cause severe damage to the host plant and act as vectors of plant viruses (Saxena and Khan, 1989).

Four lectins were tested against first instar nymphs of the rice brown plant-hopper, while eight lectins were tested against third instar nymphs (Hilder *et al.*, 1993; Powell *et al.*, 1993). Both *Galanthus nivalis* lectin (GNA) and WGA were very effective against both instars at the 0.1% level. GNA and WGA caused greater than 75% mortality compared to the control treatment. PHA had only a minor affect on the first instar nymphs (27% mortality). Concanavalin A, horse gram lectin, and lentil lectin had only a minor effect on the third instar larvae. WGA and GNA were also tested against the rice leafhopper. WGA had a minor impact (15% mortality), but GNA was very effective (87% mortality) at the 0.1% level.

Rice brown planthopper adults were challenged with WGA, pea lectin, and GNA at the 0.1% level (Powell *et al.*, 1995). Both WGA and GNA appear to act as anti-feedants – honeydew excretion volume and droplet production are significantly reduced compared to controls. GNA was the most effective as it reduced honeydew droplet production by 96% during the first 24 hours and 64% at 36 hours. Honeydew volume was reduced by 66% at 36 hours. The pea lectin had no effect on the adult planthopper.

Aphids

Three species of aphid have been tested for susceptibility to lectins in limited bio-assays. GNA was effective against the green peach aphid, *Myzus perscae* (Hilder *et al.*, 1993, 1995). Aphids feeding on artificial diets containing 0.1% GNA suffered 25% mortality and survivors were significantly reduced in size compared to the control treatment. This was the only lectin bioassayed against the green peach aphid.

Two lectins were screened against the pea aphid, *Acyrthosiphon pisum* (Rahbé and Febvay, 1993). Concanavalin A displayed toxicity and growth inhibition in the bioassays. The LC_{50} for concanavalin A was determined to be approximately 134 μg/ml diet and the IC_{50} was 182 μg/ml diet. WGA had no effect on the pea aphid at the levels tested.

Three chitin binding lectins, WGA, stinging nettle lectin (UDL), and lectin puri-fied from *Brassica fruticulosa* and *B. Brassicae* (BL), were studied for their effect on the cabbage aphid, *Brevicornye brassicae*, utilizing a chemically defined synthetic diet (Cole, 1994). UDL and BL caused 100% mortality within four days when tested at 2.5 mg/ml diet. WGA caused 100% mortality within eight days.

Other Insects

Bioassays conducted with the soybean lectin showed that it significantly inhibited *Manduca sexta* larval growth at the 1% level (Shukle and Murdock, 1983). Larvae reared on diet containing the lectin weighed approximately 50% less than control larvae by day 12. GNA reduced weight gain and survival when bioassayed against *Spodoptera littoralis* at the 5% level (Gatehouse *et al.*, 1992). PHA was reported to reduce survivorship of *Callosobruchus assimilis* by 70% when reared on artificial seeds at the 2% level (Boulter *et al.*, 1986). However, it is likely that this PHA preparation contained an alpha-amylase inhibitor. PHA was also reported to have no detrimental effect on the weevil *Acanthoscelides obtectus* (Gatehouse *et al.*, 1989). The lectin from *Phaseolus acutifiolius*, tepary bean, was toxic to *A. obtectus*, when incorporated into the artifical seed diet at 1–5% levels (Pratt *et al.*, 1990). No deter-mination was made by the authors for the presence of an alpha-amylase inhibitor.

Jacalin, WGA, concanavalin A and the lectin from peanut, pea, and soybean were bioassayed against the tobacco budworm, *Heliothis virescens*. All lectins were tested at a level of 1% in a topical artificial diet bioassay and had no detrimental effect on weight or survival of neonate larvae. Neither WGA nor jacalin had a detrimental effect on neonate black cutworm, *Agrotis ipsilon*, at the 1% topical level (personal observation).

Artificial Bioassay Summary

The artificial bioassay data accumulated by several investigators prove that some lectins can produce a detrimental effect on the growth and survival of insects. Although several insect species have been tested for susceptibility to lectins, the actual number of lectins tested per insect is extremely small. Even the largest studies pale in comparison with the thousands of bacterial isolates that have been screened for insecticidal activity. The reasons for this large discrepancy include the avail-ability of source materials, lack of appropriate growth media, the required protein purification process, and the amount of purified protein needed for insect bioassays. Therefore, excellent commercial activity in transgenic plants will need to be demon-strated before large-scale screening efforts can probably be justified. Lectins have shown the capability to affect insect development in artificial diets. The next step is to prove their effectiveness when expressed in a transgenic plant.

Transgenic Plant Bioassays

Several lectins have been identified with excellent activity against serious insect pests. The genes encoding these lectin proteins have been cloned (Higgins *et al.*, 1983; Raikhel and Wilkins, 1987; Van Damme *et al.*, 1991; Wilkins and Raikhel,

1989; Yang and Czapla, 1993). These genes can be transformed into important plant species via *Agrobacterium* or microprojectile bombardment. Tobacco and maize have been the primary species transformed with the lectin genes. Tobacco has been successfully transformed, via *Agrobacterium*, with genes encoding for the pea lectin and GNA (Boulter *et al.*, 1986; Gatehouse *et al.*, 1992; Hepher *et al.*, 1989; Hilder *et al.*, 1993, 1995). Maize has been successfully transformed with genes encoding WGA, rice lectin, and jacalin (Maddock *et al.*, 1991; Maddock and Peterson, personal communication). The transgenic plants and their progeny have been subsequently screened for insect resistance in both the greenhouse and field environments.

Tobacco

Tobacco plants transformed with the pea lectin were infested with neonate tobacco budworm larvae. There were no differences observed in larval survival after seven days compared to the control, but a dramatic difference was observed in weights of the surviving larvae. Larvae from the control plants weighed approximately 9.9 mg, while larvae reared on the transgenic plants containing the pea lectin weighed only 1.7 mg. (Hepher *et al.*, 1989).

Tobacco plants transformed with GNA were bioassayed against two insects, tobacco budworm and green peach aphid. No differences in tobacco budworm larval mortality were observed between the leaf discs from the GNA plants and the control plants. However, larvae reared on the GNA leaf discs weighed approximately 60% less than the control larvae. GNA leaf disc area consumed was reduced to 50% compared to the control leaf disc (Gatehouse *et al.*, 1992).

Leaf disc and whole plant bioassays were used to screen transgenic tobacco against the green peach aphid. Aphid numbers present on the leaf disc after seven days were significantly reduced for the GNA containing material compared with the control. Approximately 40 aphids were found per control leaf disc, while only nine were found on the GNA expressing leaf disc (Hilder *et al.*, 1993). A whole plant bioassay produced a similar result. Tobacco plants were infested with eight late instar nymphs and sealed. Aphid numbers were then determined 27 days later. Approximately 600 aphids were found on the control plants, compared to 300 aphids on the GNA positive plants (Hilder *et al.*, 1993, 1995). Transgenic experiments utilizing potato and lettuce were stated to show similar results.

Maize

WGA was first transformed into Black Mexican sweet maize callus (BMS). BMS calli expressing WGA were then evaluated against neonate ECB larvae (Cavalieri *et al.*, 1995; Czapla *et al.*, 1990). Nine transformed lines of BMS callus were bioassayed for insecticidal activity. Larvae reared on control callus weighed 4.5 mg at seven days. Three lines of BMS callus producing WGA significantly reduced ECB larval growth. One line reduced weight gain by 30%, while the other two lines reduced weight gain by 95%. No differences in larval growth were observed for the other six lines of BMS callus. The three lines that inhibited weight gain also had the highest agglutination activity per mg protein compared to the control. There were no significant differences in agglutination activity among the three insect active BMS callus lines.

WGA was successfully transformed into a maize suspension cell culture (W23/T66) (Maddock *et al.*, 1991). Leaf material from regenerated plants was characterized by PCR, ELISA and insect bioassays for WGA activity. ELISA and subsequent Western blot analysis identified two transformation events positive for WGA. PCR analysis identified two additional transgenic events, but they did not express the WGA protein. Insect bioassays, conducted with neonate ECB larvae and leaf strips, indicated that the two lines expressing WGA reduced leaf damage from ECB feeding by approximately 60% compared to the non-WGA expressing plants. Although there were no differences observed regarding larvae mortality, larvae reared on the WGA expressing leaf material weighed approximately 50% less than the control larvae. Unfortunately, no progeny were obtained because the plants were sterile.

Three seeds were obtained from a single WGA transformed plant in 1991 (Maddock *et al.*, 1991). This material was increased in a greenhouse and a winter nursery for a 1992 field trial in Johnston, Iowa. Unfortunately, no significant resistance was observed in this material as determined by ECB first generation scoring and second generation stalk tunneling. There was however a trend seen in the winter nursery progeny line designated HW0047 (Campbell and Isenhour, personal communication). This 'increased' resistance was not enough to warrant further efforts for the development of a commercial product, but was exciting. Hemagglutination assays indicated that WGA could be produced up to a level of 100 μg/g of plant tissue (personal observation). This level was five times lower than the artificial diet LC_{50}. This probably explains the lack of insect resistance observed in this field trial. Different promoters, targeting of the protein and modifications to increase WGA activity have been proposed to alleviate this problem.

Genes that encode the rice lectin and jacalin have also been successfully transformed into maize to confer CRW resistance. Rice lectin and jacalin were both expressed in plants as determined by ELISA. Both successfully agglutinated red blood cells, indicating proper folding of the protein. Bioassays with the initial transgenic plants (T0) indicated that there was no CRW activity in the roots expressing the rice lectin.

Several T0 plants expressing jacalin recorded a higher bioassay root score compared to the control plants (Moellenbeck, personal communication). Several plants scored a 3, on a scale of 1–9, compared to the control score of 1. Unfortunately, the T1 progeny of these plants displayed no insecticidal activity. Jacalin was estimated to be expressed at approximately 100 μg/g root tissue in these T1 plants (personal observation). This value was 30 times less than the LC_{50} determined utilizing artificial diets.

Excellent progress has been made with tobacco plants expressing the lectin protein and subsequent correspondence of insect activity. This should improve with increased expression levels from new promoters and identification of lectins with increased activity. This will be critical for maize as the current levels of expression, or insecticidal activity of the lectins, need to be increased in order to develop a commercial product.

Mode of Action

The exact mechanism of action underlying the toxicity of lectins toward insects is not clear. It likely involves the specific binding of a lectin to glycoconjugates located somewhere in the midgut of the insect. There are several possible interactions that could occur:

(1)   binding of the lectins to the chitin matrix of the peritrophic membrane;

(2)   binding of the lectin to glycoconjugates within the peritrophic membrane;

(3)   binding of the lectin to glycoconjugates exposed on the surface of the midgut epithelial cells;

(4)   binding of lectins to glycosylated digestive enzymes; and

(5)   binding to glycosylated proteins from the host plant preventing the protein's digestion

(Chrispeels and Raikhel, 1991; Czapla and Lang, 1990; Peumans and Van Damme, 1995a,b).

Indirect immunofluorescence showed that the PHA lectin, when ingested by the cowpea weevil larvae, would bind strongly to the luminal side of the midgut epithelial cells. Lectin binding also occurred on the distal cell surface indicating the disruption of the cell. This dissipation would allow the passage of lectin across the cell which could then cause more disruption (Gatehouse *et al.*, 1984). PHA, which was not toxic to the bean weevil, *A. obtectus*, did not bind to the luminal surface of midgut epithelial cell of this insect (Gatehouse *et al.*, 1989). This proved that lectins would bind differentially when fed to different insect species. However, this did not prove that lectin binding alone was sufficient for insect toxicity as subsequent work proved that the activity in the PHA preparations was due to an alpha-amylase inhibitor (Huesing *et al.*, 1991a).

Structure–function experimentation with WGA and BPA indicated that the toxicity observed for ECB larvae was due to the specific binding properties of the lectin (Balasubramaniam *et al.*, 1991, 1993). Cyanogen bromide treated WGA was not toxic to neonate ECB larvae. The treated lectin did not dimerize properly and was unable to bind to GlcNAc or its polymers (Balasubramaniam *et al.*, 1991). The toxicity of WGA is therefore intrinsically related to its ability to dimerize and bind to sugar residues.

Succinylation of BPA dramatically changed the conformation and biological activity of this lectin. Succinylated BPA did not bind to GalNAc residues and had no effect on neonate ECB larvae. Other BPA modifications, which did not lose the ability to bind to N-GalNAc, retained their insect activity (Balasubramaniam *et al.*, 1993). Thus, BPA activity is related to its ability to bind GalNAc, but in some manner different to other GalNAc lectins tested.

The toxicity of WGA could be removed by titrating GlcNAc and its polymers in artificial diets (personal observations). Chitotriose reduced and eliminated the effect of WGA on neonate ECB larvae when incorporated into artificial diets. Results indicated that larvae reared on diet containing WGA at 500 μg/g diet were completely unaffected by WGA if chitotriose was present at 2.5 mg/g diet.

Newly ecdysed fifth instar ECB larvae were reared on diets containing 1 mg/ml of either WGA, BPA, jacalin or tomato lectin (Czapla *et al.*, 1992). WGA and tomato lectin are N-GlcNAc binding lectins and BPA and jacalin are N-GalNAc lectins (Goldstien and Poretz, 1986). There were no differences observed in weight gain between the larvae reared on the jacalin, tomato lectin or control diets. WGA and BPA significantly reduced weight gain compared to the control larvae. The larvae gained only 5 mg after four days on the WGA or BPA diets, compared to the control weight gain of 50 mg.

Light microscopy examination showed no observable differences in the midgut region of larvae reared on jacalin, tomato lectin, or control diets. However, within 24 hours, larvae reared on the WGA and BPA diets displayed a major reduction in the amount of fat body and the absence of food inside the peritrophic membrane. No fat body or peritrophic membrane were present at 72 hours. The diet surface showed very little feeding on the WGA and BPA diets, indicating a possible feeding deterrent with these lectins.

No observable differences were seen by scanning electron microscopy in the midgut epithelial cells of larvae from all treatments. There were also no differences observed in the peritrophic membrane for larvae reared on jacalin, tomato lectin and control diets. However, disruption of the peritrophic membrane (PM) was easily seen within 24 hours in larvae reared on WGA and BPA (Cavalieri *et al.*, 1995; Czapla *et al.*, 1992). Several large pores developed within the anterior region of the PM. These large pores could easily allow food particles or bacteria to pass through this protective barrier and attach to or disrupt the epithelial cells. Damage to the PM increased in both severity and in a posterior direction until no PM was observable by 72 hours. It is uncertain if WGA prevents the formation of the PM by preventing its formation or its secretion from the epithelial cells.

Biotinylated WGA, BPA, jacalin and tomato lectin were used to probe blots containing protein extracts from the midgut epithelial cells and peritrophic membrane of fifth instar ECB larvae. Although some lectin binding was seen in the midgut extracts, the majority of binding was associated with the PM preparations. Several proteins are bound by all four lectins. However, the ECB active lectins, WGA and BPA, each stained a 22 Kd protein very intensely. Neither jacalin nor tomato lectin bound to any proteins in this region. This 22 Kd protein may be an essential component in the formation of the PM. The binding of this protein was hypothesized to be the reason why WGA and BPA are toxic to ECB (Cavalieri *et al.*, 1995; Czapla *et al.*, 1992). The lack of binding and activity seen by the GlcNAc binding tomato lectin or the GalNAc binding jacalin also supported this hypothesis.

The impact of lectins binding to specific glycoconjugates on the cell surface of midgut cells could also play a major role in the observations reported above. Binding of WGA and BPA to a specific protein on the cell surface could easily prevent the secretion of the PM components. This would also produce the large pores observed. This glycoconjugate would not be bound by the tomato lectin or jacalin. It is also possible that several sites need to be bound both on the cell surface and the PM for the insecticidal activity to occur, again demonstrating that our knowledge of why certain lectins are effective and others are not is incomplete.

Lectin binding to brush border membranes was recently investigated using ECB and CRW larvae (Harper *et al.*, 1995). Thirty-eight lectins were used to probe brush border membrane extract protein blots. Lectins that caused significant mortality to ECB were also bound strongly to the brush border membrane proteins, but not all lectins that bound were insecticidal. There was even less correlation between brush border membrane protein binding and mortality in the CRW tests.

Insect Resistance to Lectins

The ability of insects to evolve resistance to several different types of crop protection methods is well documented. Transgenic plants with insect control proteins will face the same consequences as any insecticide if used indiscriminately. The discovery of new genes and the management of existing ones will be critical for the long-range success of transgenic plants.

Resistance potential to lectins was evaluated for ECB using WGA in artificial diets (Lang and Sabus, 1993). Three colonies of ECB were reared for 20 generations on artificial diets containing 0, 100 and 200 ppm WGA respectively. Efficacy bio-assays were conducted at 250, 500, 1000 and 2000 ppm. There were no differences observed between the three colonies at the 1000 and 2000 ppm levels. However, the colonies reared at 100 and 200 ppm WGA had significantly higher mean weights (twofold) after 20 generations than the control colony at the 500 ppm treatment. The 200 ppm chronic dose colony also had significantly higher weights (twofold) than the control colony at the 250 ppm treatment. These results indicated that ECB larvae continuously exposed to low levels of WGA would acquire a limited level of resistance to WGA over time.

Concerns Regarding the Utilization of Lectins in Transgenic Plants

The major concerns regarding the use of lectins in transgenic plants are their anti-nutrient properties in food for higher animals including humans and their impact on non-target insect species (Belzunces *et al.*, 1994; Pusztai, 1991). Lectins from kidney bean (PHA), soybean and others have displayed nutritionally toxic effects for most animals. The lectins bind to and can be endocytosed by the epithelial cells of the small intestines (Pusztai, 1991). WGA and other N-GlcNAc binding lectins were detrimental to rats when incorporated into diets at 7 g/kg (Pusztai *et al.*, 1993). WGA did not display acute toxicity toward *Apis mellifera*, but the feeding tests were conducted for only four days, so long-term effects were not determined (Belzunces *et al.*, 1994). GNA and the lectin from *Vicia faba*, with specific binding properties for mannose or mannose/glucose displayed only a slight effect on rats (Pusztai *et al.*, 1990; Rubio *et al.*, 1991). Thus, GNA may be a more suitable lectin for use in transgenic plants than WGA. Unfortunately GNA has no effect on ECB or CRW (Cavalieri *et al.*, 1995). However, other lectins may exist which are toxic to ECB and CRW and not other organisms.

The antinutritive effect of lectins becomes a major concern if the lectin is present at high levels in seed or other edible parts of the plant. Implications of long-term exposure to lectins at transgenic plant levels will need to be addressed for consumer and livestock safety as well as their impact on non-target organisms. The identification of insecticidal lectins to pest insects that have little or no effect on other organisms will be a difficult challenge for researchers in this area.

The high concentration of lectins needed for insecticidal activity, compared to that of the *Bt* endotoxins, is also a major concern. Although ECB growth was significantly affected when reared on WGA transgenic leaves, there were no observable differences seen during the field trials. Transgenic plants will need to have resistance levels that are of economic significance, not statistical significance.

A possible solution to reduce the high lectin levels needed for insect control would be the combination of different insect resistance proteins. These combinations, if synergistic, could dramatically reduce the levels needed for achieving economic control.

The effects of WGA combined with a serine protease inhibitor, aprotinin, were tested in combination against ECB larvae in an artificial bioassay. The LC_{50} for WGA and aprotinin were approximately 0.5 and 10 mg/ml, respectfully. The level needed for 50% inhibition in weight gain after seven days was 0.25 mg/ml for WGA and 2.5 mg/ml for aprotinin. When these proteins were combined, the LC_{50} was 0.15 mg/ml for WGA and 0.5 mg/ml for the inhibitor. A 50% weight loss was observed for ECB larvae at 0.1 mg/ml for both WGA and the inhibitor (personal observations). Other combinations of different inhibitors, lectins, or other proteins may provide an even more dramatic effect. These reduced levels may reduce some concerns regarding lectins, provided that the synergistic effect is limited only to the targeted insects.

Future Outlook

Lectins have demonstrated the capability of disrupting the development of key insect pest species, not only in artificial bioassays, but in transgenic plants as well. Thus, they have the potential to be a viable alternative or supplement to *Bt* transgenic plants. However, several hurdles remain regarding the economic effectiveness of lectins and their overall safety to non-target organisms. These issues will be the major challenge to researchers in this area (and other potential control agents) as crop protection moves into the twenty-first century.

References

Balasubramaniam, N. K., Czapla, T. H. and Rao, A. G. (1991) Structural and functional changes associated with cyanogen bromide treatment of wheat germ agglutinin, *Arch. Biochem. Biophys.* **288**(2), 374–379.

Balasubramaniam, N. K., Timm, D. E., Neet, K. E. and Rao, A. G. (1993) The lectin from *Bauhinia purpurea*: effect of eodification of lysine residues on conformation and biological properties, *J. Agric. Food Chem.* **41**(11), 1844–1850.

Belzunces, L. P., Lenfant, C., Di Pasquale, S. and Colin, M. E. (1994) *In vivo* and *in vitro* effects of wheat germ agglutinin and Bowman–Birk soybean trypsin inhibitor, two potential transgene products, on midgut esterase and protease activities from *Apis mellifera*, *Comp. Biochem. Physiol.* **109B**(1), 63–69.

Boulter, D. (1993) Insect pest control by copying nature using genetically engineered crops, *Phytochemistry*, **34**(6), 1453–1466.

Boulter, D., Croy, R. R. D., Ellis, R. J., Evans, I. M., Gatehouse A. M. R., Gatehouse J. A., Shhirsat, A. and Yarwood, J. H. (1986) Isolation of genes involved in pest and disease resistance, In: Nijhoff, M. (Ed.), *Report of the EEC Biomolecular Engineering Programme*, The Hague: Martinus Nijhoff, pp. 715–725.

Cavalieri, A., Czapla, T., Howard, J. and Rao G. (1995) Larvicidal lectins and plant insect resistance based thereon, United States Patent No. 5 407 454.

Chrispeels, M. J. and Raikhel, N. V. (1991) Lectins, lectin genes, and their role in plant defense, *The Plant Cell* **3**, 1–9.

Cole, R. A. (1994) Isolation of a chitin-binding lectin, with insecticidal activity in chemically-defined synthetic diets, from two wild brassica species with resistance to cabbage aphid *Brevicoryne brassicae*, *Entomol. Exper. Appl.* **72**, 181–187.

Czapla, T. H. and Lang, B. A. (1990) Effect of plant lectins on the larval development of European corn borer (Lepidoptera: Pyralidae) and southern corn rootworm (Coleoptera: Chrysomelidae), *J. Econ. Entomol.* **83**(6), 2480–2485.

Czapla, T. H., Rao, A. G. and Hopkins, T. L. (1992) Effect of lectins on the peritrophic membrane of the European corn borer, poster presented at The North Central Branch Entomological Society of America Annual Meeting, Kansas City, MO, March.

Czapla, T. H., Maddock, S. E., Huffman, G. and Raikhel, N. V. (1990) Expression and biological activity of a European corn borer resistant gene transformed into maize callus, poster presented at The Entomological Society of American Annual Meeting, New Orleans, LA, December.

Etzler, M. E. (1986) Distribution and function of plant lectins, In: Liener, I. E., Sharon, N. and Golstein, I. J. (Eds), *The Lectins: Properties, Functions, and Applications in Biology and Medicine*, New York: Academic Press, pp. 371–435.

Gatehouse, A. M. R., Dewey, F. M., Dove, J., Fenton, K. A. and Pusztai, A. (1984) Effect of seed lectins from *Phaseolus vulgaris* on the development of larvae of *Callosobruchus maculatus*: mechanism of toxicity, *J. Sci. Food Agric.* **55**, 63–74.

Gatehouse, A. M. R., Shackley, S. J., Fenton, K. A., Bryden, J. and Pusztai, A. (1989) Mechanism of seed lectin tolerance by a major insect storage pest of *Phaseolus vulgaris, Acanthoscelides obtectus, J. Sci. Food Agric.* **47**, 269–280.

Gatehouse, A., Hilder, V., van Damme, E., Peumans, W., Newell, C. and Hamilton, W. (1992) Insecticidal proteins, World Intellectual Patent Organization Application, No. WO 92/02139.

Gatehouse, A. M. R., Shi, Y., Powell, K. S., Brough, C., Hilder, V. A., Hamilton, W. D. O., Newell, C. A., Merryweather, A., Boulter, D. and Gatehouse, J. A. (1993) Approaches to insect resistance using transgenic plants, *Phil. Trans. R. Soc. Lond.* **342**, 279–286.

Goldstein, I. J. and Poretz, R. D. (1986) Isolation, physiochemical characterization, and carbohydrate-binding specificity of lectins, In: Liener, I. E., Sharon, N. and Golstein, I. J. (Eds), *The Lectins: Properties, Functions, and Applications in Biology and Medicine*, New York: Academic Press, pp. 33–247.

Goldstein, I. J., Hughes, R. C., Monsigny, M., Osawa, T. and Sharon, N. (1980) What should be called a lectin?, *Nature* **285**, 66.

Gould, A. R., Cowen, N. M., Merlo, D. J., Petolino, J. F., Thompson, S. A. and Walsh, T. A. (1993) Insect control via transgenic hybrid maize, *48th Annual Corn and Sorghum Research Conference*, **48**, 63–75.

Habibi, J., Bacus, E. A. and Czapla, T. H. (1993) Plant lectins affect survival of the potato leafhopper (Homoptera: Cicadellidae), *J. Econ. Entomol.* **86**(3), 945–951.

Hankins, C. N., Kindinger, J. I. and Shannon, L. M. (1980) Legume α-galactosidases which have hemagglutinin properties, *Plant Physiol.* **65**, 618–622.

Harper, S. M., Crenshaw, R. W., Mullins, M. A. and Privalle, L. S. (1995) Lectin binding to insect brush border membranes, *J. Econ. Entomol.* **88**(6), 1197–1202.

Hepher, A., Edwards, G. and Gatehouse, J. (1989) Improvements relating to transgenic plants, European Patent Application No. 89201899.5.

Higgins, T. J. V., Chandler, P. M., Zurawski, G., Button, S. C. and Spencer, D. (1983) The biosynthesis and primary structure of pea seed lectin, *J. Biol. Chem.* **258**, 9544–9549.

Hilder, V. A., Gatehouse, A. M. R., Powell, K. S. and Boulter, D. (1993) Proteins with insecticidal properties against homopteran insects and their use in plant protection, World Intellectual Patent Organization Application No. WO 93/04177.

Hilder, V. A., Powell, K. S., Gatehouse, A. M. R., Gatehouse, J. A., Gatehouse, L. N., Shi, Y., Hamilton, W. D. O., Merryweather, A., Newell, C. A., Timans, J. C. *et al.* (1995) Expression of snowdrop lectin in transgenic tobacco plants results in added protection against aphids, *Transgen. Res.* **4**, 18–25.

Huesing, J. E., Shade, R. E., Chrispeels, M. J. and Murdock, L. L. (1991a) α-Amylase inhibitor, not phytohemagglutinin, explains resistance of common bean seeds to cowpea weevil, *Plant Physiol.* **96**, 993–996.

Huesing, J. E., Murdock, L. L. and Shade, R. E. (1991b) Effect of wheat germ isolectins on development of cowpea weevil, *Phytochemistry* **30**, 785–788.

Huesing, J. E., Murdock, L. L. and Shade, R. E. (1991c) Rice and stinging nettle lectins: insecticidal activity similar to wheat germ agglutinin, *Phytochemistry* **30**, 3565–3568.

Janzen, D. H., Juster, H. B. and Liener, I. E. (1976) Insecticidal action of the phytohemagglutinin in black beans on a bruchid beetle, *Science* **192**, 795–796.

Kindler, S. D., Kehr, W. R., Ogden, R. L. and Schalk, J. M. (1973) Effect of potato leafhopper injury on yield and quality of resistant and susceptible alfalfa clones, *J. Econ. Entomol.* **66**, 1298–1302.

Knox, R. B., Clarke, A., Harrison, S., Smith, P. and Marchalonis, J. J. (1976) Cell recognition in plants: determinants of the stigma surface and pollen interactions, *Proc. Natl. Acad. Sci. USA* **73**, 2788–2792.

Kumar, M. A., Timm, D. E., Neet, K. E., Owen, W. G., Peumans, W. J. and Rao, G. A. (1993) Characterization of the lectin from the bulbs of *Eranthis hyemalis* (winter aconite) as an inhibitor of protein synthesis, *J. Biol. Chem.* **268**(33), 25176–25183.

Lang, B. A. and Sabus, B. T. (1993) Evaluating resistance to *Triticum vulgaris* lectin in European corn borer (Lepidoptera: Pyralidae) using artificial diet, poster presented at The Entomological Society of American Annual Meeting, Indianapolis, IN, December.

Liener, I. E., Sharon, N. and Goldstein, I. J. (1986) Preface, In: Liener, I. E., Sharon, N. and Golstein, I. J. (Eds), *The Lectins: Properties, Functions, and Applications in Biology and Medicine*, New York: Academic Press, pp. xi–xv.

Maddock, S. E., Hufman, G., Isenhour, D. J., Roth, B. A., Raikhel, N. V., Howard, J. A. and Czapla, T. H. (1991) Expression in maize plants of wheat germ agglutinin, a novel source of insect resistance, poster presented at The Third International Congress for Plant Molecular Biology, Tucson, AZ.

Metcalf, R. L. (1986) Foreword, In: Krysan, J. L. and Miller, T. A. (Eds), *Methods for the Study of Pest Diabrotica*, New York: Springer-Verlag, pp. vii–xv.

Murdock, L. L., Huesing, J. E., Nielsen, S. S., Pratt, R. C. and Shade, R. E. (1990) Biological effects of plant lectins on the cowpea weevil, *Phytochemistry* **29**(1), 85–89.

Peumans, W. J. and Van Damme, E. J. M. (1995a) Lectins as plant defense proteins, *Plant Physiol.* **109**, 347–352.

Peumans, W. J. and Van Damme, E. J. M. (1995b) The role of lectins in plant defence, *Histochem. J.* **27**, 253–271.

Powell, K. S., Gatehouse, A. M. R., Hilder, V. A. and Gatehouse, J. A. (1993) Antimetabolic effects of plant lectins and plant and fungal enzymes on the nymphal stages of two important rice pests, *Nilaparvata lugens* and *Nephotettix cinciteps*, *Entomol. Exper. Appl.* **66**, 119–126.

Powell, K. S., Gatehouse, A. M. R., Hilder, V. A. and Gatehouse, J. A. (1995) Antifeedant effects of plant lectins and an enzyme on the adult stage of the rice brown planthopper, *Nilaparvata lugens*, *Entomol. Exper. Appl.* **75**, 51–59.

Pratt, R. C., Singh, N. K., Shade, R. E., Murdock, L. L. and Bressan, R. A. (1990) Isolation and partial characterization of a seed lectin from tepary bean that delays bruchid beetle development, *Plant Physiol.* **93**, 1453–1459.

Pusztai, A. (1991) *Plant Lectins*, Cambridge: Cambridge University Press.

Pusztai, A., Ewen, S. W. B., Grant, G., Peumans, W. J., Van Damme, E. J. M., Rubio, L. and Bardocz, S. (1990) The relationship between survival and binding of plant lectins during small intestine passage and their effectiveness as growth factors, *Digestion* **46**, 308–316.

Pusztai, A., Ewen, S. W. B., Grant, G., Brown, S. D., Stewart, J. C., Peumans, W. J., Van Damme, E. J. M. and Bardocz, S. (1993) Antinutritive effects of wheat-germ agglutinin and other N-acetylglucosamine-specific lectins, *Br. J. Nutr.* **70**, 313–321.

Rahbé, Y. and Febvay, G. (1993) Protein toxicity to aphids: an *in vitro* test on *Acyrthosiphon pisum*, *Entomol. Exper. Appl.* **67**, 149–160.

Raikhel, N. V. and Wilkins, T. A. (1987) Isolation and characterization of cDNA clone encoding wheat germ agglutinin, *Proc. Natl. Acad. Sci. USA* **84**, 6745–6749.

Rao, A. G. and Kumar, M. A. (1994) Derivatives of *Eranthis hyemalis* lectin, World Intellectual Patent Organization Application, No. WO 94/02513.

Rubio, L. A., Grant, G., Bardocz, S., Dewey, P. and Pusztai, A. (1991) Nutritional response of growing rats to faba beans (*Vica faba* L., minor) and faba bean fractions, *Br. J. Nutr.* **66**, 533–542.

Saxena, T. and Khan, Z. R. (1989) Factors affecting resistance of rice varieties to planthopper and leafhopper pests, *Agric. Zool. Rev.* **3**, 97–132.

Shukle, R. H. and Murdock, L. L. (1983) Lipoxygenase, trypsin inhibitor, and lectin from soybeans: effects on larval growth of *Manduca sexta* (Lepidoptera: Sphingidae), *Environ. Entomol.* **12**, 787–791.

Strosberg, A. D., Buffard, D., Lauwereys, M. and Foriers, A. (1986) In: Liener, I. E., Sharon, N. and Golstein, I. J. (Eds), *The Lectins: Properties, Functions, and Applications in Biology and Medicine*, New York: Academic Press, pp. 251–263.

Van Damme, E. J. M., Goldstein, J. J. and Peumans, W. J. (1991) Comparative study of related mannose-binding lectins from Amaryllidaccae and Alliaceae species, *Phytochemistry* **30**, 509–514.

Wilkins, T. A. and Raikhel, N. V. (1989) Expression of rice lectin is governed by two temporally and spatially regulated mRNAs in developing embryos, *Plant Cell* **1**, 541–549.

Yang, H. and Czapla, T. H. (1993) Isolation and characterization of cDNA clones encoding jacalin isolectins, *J. Biol. Chem.* **268**(8), 5905–5910.

9

Transfer of Bruchid Resistance from the Common Bean to Other Starchy Grain Legumes by Genetic Engineering with the α-Amylase Inhibitor Gene

MAARTEN J. CHRISPEELS

Introduction

The seeds of starchy legumes are an important source of food for man, but as with other crops, insect pests cause considerable crop losses. Particularly devastating is the damage done by different species of bruchids, coleopteran insects that belong to the family *Bruchidae*. The damage is caused by the larvae of these insects when they burrow into the seeds. In some cases, as with the pea weevil, the damage is done entirely in the field, when the seeds are developing in the pods; in other cases, as with the cowpea weevil, the damage occurs almost entirely during the storage of the seeds. The losses are greatest among the poorest farmers who store their seeds in baskets or granaries where the bruchids freely multiply during the months of storage.

The Bruchid Life Cycle

Many bruchids are storage pests and to understand why they cause such extensive crop losses, it is necessary to consider their life cycle. The life cycle of the three *Callosobruchus* species can be summarized as follows (Talekar, 1987). Adults emerge from seeds and mate within an hour after emergence. Subsequently, they lay 80–100 eggs, depositing only one to three eggs per seed. An oily substance that attaches the eggs to the seeds also functions as an inhibitor of hatching, ensuring that only one or two eggs on each seed will hatch even if more are deposited. This mechanism assures that there is enough food for all developing larvae. Soon after hatching, the larva eats a hole in the seed coat (and the seed pod if the egg was laid on the pod) and burrows into the seed where it continues to feed and grow until it pupates. Just before the larva pupates, it eats a circular hole, leaving only a thin layer, or 'window', of the seed coat intact. Development from egg to adult takes 22–30 days

depending on the species and the conditions. With a 30-day life cycle and a reproductive rate of 80 eggs per female, it is obvious that the population of bruchids will
mushroom when seeds are stored for several months, even if the initial infestation is
moderate.

Seed Defenses

Bruchids are able to infest legume seeds in spite of the presence in those same seeds
of proteins such as lectins and inhibitors of digestive enzymes, as well as secondary
metabolites (alkaloids, saponins, cyanogenic glucosides) which are toxic to insects.
Why should this be so? If these are 'defense chemicals' why do they not defend the
seeds against predation by the insects? In many cases, these chemicals are present at
higher levels in the wild progenitors of our crop plants than in the crops themselves.
Indeed, plant domestication led to the elimination or dilution of defense chemicals
because they have an unpleasant taste, or are directly toxic to humans (as well as to
insects). More important is the fact that insects co-evolved with their food sources,
and in the course of evolution, mutant insect strains arose which were able to
'handle' the defense chemicals in the seeds. These insect strains quickly became
dominant in the population because they were able to use as their exclusive food
source the plant which was previously resistant. A good example of this phenomenon is provided by the α-amylase inhibitor of the common bean (*Phaseolus vulgaris*),
which is the focus of this chapter. Two bruchid species, the Mexican bean weevil
(*Zabrotes subfasciatus*) and the bean weevil (*Acanthoscelides obtectus*), which evolved
in the Americas, feast on all cultivated varieties of the common bean, also a New
World plant (Figure 9.1). Proteases and amylases in the midgut of these larvae
readily digest the seed proteins and starch that they ingest. The bean seeds contain
high levels (1–2%) of an α-amylase inhibitor, but this inhibitor either does not
inhibit the α-amylases in the larval gut of these bruchids, or the inhibitor is digested

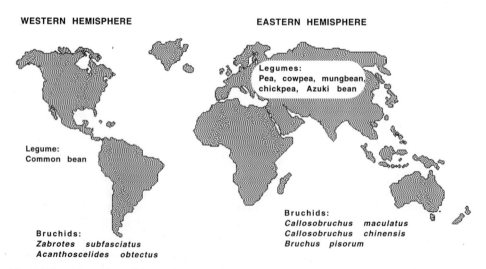

Figure 9.1 Origin of starchy seed legumes and bruchid species in the Western and Eastern
hemispheres

Table 9.1 Bruchid and starchy grain legume species

Bruchids	
Cowpea weevil	*Callosobruchus chinensis*
Azuki bean weevil	*Callosobruchus maculatus*
	Callosobruchus analis
Mexican bean weevil	*Zabrotes subfasciatus*
Bean weevil	*Acanthoscelides obtectus*
Starchy grain legumes	
Pea (garden pea, field pea)	*Pisum sativum*
Common bean	*Phaseolus vulgaris*
Cowpea	*Vigna unguiculata*
Mungbean	*Vigua radiata*
Azuki bean	*Vigna angustifolia*
Lentil	*Lens culinaris*
Chick-pea (pigeon pea)	*Cicer arietinum*
Broad-bean, horse bean	*Vicia faba*

by gut proteases (Ishimoto and Kitamura, 1992). Either way, the inhibitor is ineffective in defending the seeds against the bruchids. However, bruchid species that evolved in the Old World, such as the pea weevil (*Bruchus pisorum*) and the cowpea weevil (*Callosobruchus maculatus*), cannot handle this α-amylase inhibitor, and its presence is sufficient to starve larvae of these bruchids to death when the inhibitor is present in their diet.

In nature, resistance to a certain insect species may result from a combination of chemicals – secondary metabolites and proteins – rather than from a single component able to cause resistance by itself. Thus, a combination of factors, all present at levels that are too low to give resistance by themselves, may be responsible for resistance. Resistance to insects – as well as to pathogens – is often found in the wild ancestors of our crop plants. In some cases, it has been possible to transfer this resistance from the wild accession to a cultivar by backcrossing, indicating that resistance was encoded at a single genetic locus and may result from a single resistance factor. In other cases, backcrossing failed to transfer resistance, possibly because resistance was caused by a combination of seed factors encoded at different genetic loci. Resistance which is caused by a combination of seed factors, all present at suboptimal levels, may be more durable and may give the plant an evolutionary advantage.

These considerations suggest several approaches for transferring bruchid resistance by genetic engineering:

(1) A good place to look for resistance factors is in legume species that have not co-evolved with the bruchids that damage them and/or in wild accessions of the same species. This approach may necessitate expressing genes of one plant species in another plant species.

(2) By elevating the level of a single resistance factor that is already present in the seed, it may be possible to make the seeds resistant. Thus, switching genes between species may not always be necessary.

(3) It may be prudent to express several resistance factors at suboptimal levels simultaneously. This mimics the natural situation and may improve the durability of resistance.

141

It should be pointed out that crop protection does not require a level of inhibitor that results in complete mortality of the larvae, even when a single resistance factor is expressed. Lengthening the developmental time of the larvae by 50% or 100% (increasing it from 25 days to 37 or 50 days) will also provide considerable benefit because in the case of storage pests, much of the damage is done by the later generations when the numbers of bruchids increase dramatically.

Phytohemagglutinin, Arcelin and α-Amylase Inhibitor: Seed Defense Proteins of the Common Bean

The common bean has long been known to be toxic to mammals and to contain factors that protect it against certain species of bruchid. Of particular interest is the family of defense proteins which includes phytohemagglutinin (PHA), arcelin and α-amylase inhibitor (αAI) (Chrispeels and Raikhel, 1991). The genes for these three proteins are encoded at a single locus in the *P. vulgaris* genome (Nodari *et al.*, 1993) and it is likely that the homologous genes have arisen by duplication of an ancestral gene. The derived amino acid sequences obtained from cDNA clones have been determined for all three proteins and they show 45–85% amino acid sequence identity. The evolutionary relationship of a number of sequenced genes is shown in Figure 9.2. It is therefore predicted that the proteins will have similar structures as well (Rougé *et al.*, 1993), and that they will be similar to those of the other legume lectins, four of which have been crystallized and their three-dimensional structures determined. Legume lectins are β-barrel proteins in which each monomer consists of two antiparallel β-pleated sheets; one sheet has six strands and the other one seven, and the strands are connected by short loops. The structures of the three proteins differ primarily by the shortening or absence of specific loops. In particular, a long loop of phytohemagglutinin which has amino acids that are implicated in sugar binding and metal binding in legume lectins, is partially deleted from arcelin and almost completely missing from αAI.

PHA and αAI are found in most domesticated bean varieties as well as in wild accessions; arcelin however has been found only in certain wild accessions of *P. vulgaris* from Mexico. PHA consists of two homologous polypeptides, PHA-E and PHA-L, in the same seed, and comprises 4–8% of total seed protein, whereas αAI accounts for 1% of total seed protein. Arcelin however is the major seed storage protein in those wild beans where it is present and comprises 30% of total protein. The three proteins have quite different modes of action in protecting seeds from being eaten by mammals or infested by bruchids. PHA is a lectin that binds to the glycans on the glycoproteins of the intestinal mucosa of mammals, and acts as a mitogen. Its toxicity and mitogenicity result from this binding (see Pusztai, 1993 for review). Mutant PHA, which does not bind glycans, is not a mitogen (Mirkov and Chrispeels, 1993) and is therefore probably not toxic to mammals. PHA was first thought to be toxic to bruchids (Gatehouse *et al.*, 1984; Janzen *et al.*, 1976), but the observed effect was subsequently shown to be caused by the presence of αAI in the commercial PHA preparations used for the feeding studies (Huesing *et al.*, 1991).

The mechanism of action of arcelin, which occurs in six different electrophoretic forms, is still unknown. The wild bean accessions that contain arcelin are resistant to the Mexican bean weevil and there is both genetic evidence and evidence from feeding trials that arcelin is toxic to the Mexican bean weevil (Osborn *et al.*, 1988).

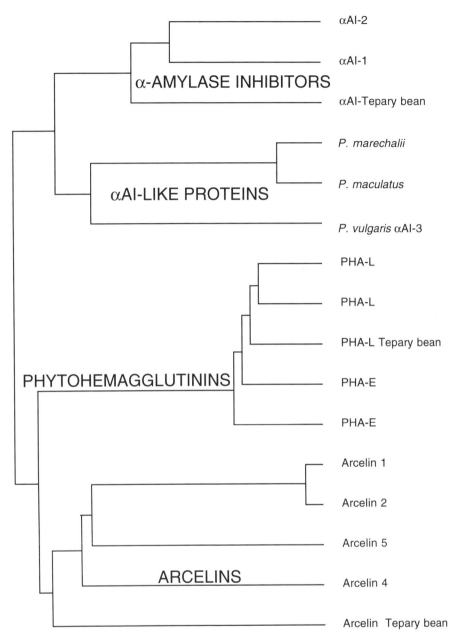

Figure 9.2 Evolutionary relationships between αAI, αAI-like proteins, phytohemagglutinins and arcelins (for details, see Mirkov *et al.*, 1995a)

Arcelin may bind to the peritrophic membrane of the insect gut and restrict nutrient uptake in some unknown manner and/or it may also be indigestible (Minney *et al.*, 1990). One of the four conserved amino acids that make up the sugar binding pocket in all legume lectins is missing from all sequenced arcelins (Mirkov *et al.*, 1995b) and for this reason, arcelins are probably not lectins (carbohydrate-binding proteins) or at best very weak lectins (Goossens *et al.*, 1994; Hartweck *et al.*, 1991).

143

Bean αAI inhibits the activity of certain mammalian and insect α-amylases but not that of plant enzymes (Jaffé *et al.*, 1972; Lajolo and Finardi-Filho, 1985; Marshall and Lauda, 1975; Pick and Wöber, 1978; Powers and Whitaker, 1977a,b) and αAI has therefore long been considered a plant defense protein rather than a metabolic protein. The effect of bean αAI on α-amylases of different origin is particularly well illustrated by the work of Ishimoto and Kitamura (1989), who measured the inhibition of α-amylase obtained from the porcine pancreas and from the guts of cowpea and Mexican bean weevil larvae (see Figure 9.3). It is clear from these data that there is a differential effect of the inhibitor on the activities of α-amylases of different origin. Least inhibited is the α-amylase of the Mexican bean weevil, and this weevil is a major pest of beans. Subsequent experiments (Ishimoto and Kitamura, 1992; Ishimoto and Chrispeels, 1996) showed that this lack of inhibition is caused by the digestion of the inhibitor by gut proteases. The α-amylase of the cowpea weevil is strongly inhibited by bean αAI and this bruchid does not attack beans. Such experiments led to the conclusion that αAI is a major factor, perhaps the major factor, in the resistance of beans to the cowpea weevil.

This conclusion is confirmed by artificial feeding experiments in which purified bean αAI was added to the flour of azuki beans (Ishimoto and Kitamura, 1989) or of cowpeas (Huesing *et al.*, 1991) and shown to inhibit larval development and the emergence of adult cowpea weevils and azuki bean weevils from the artificial seeds produced with this meal (Figure 9.4). This leaves unresolved the question as to whether the inhibition of larval development is caused by the inhibition of gut α-amylase. Although this is likely, it cannot be excluded that αAI may have another physiological effect on bruchid larvae and that this effect is responsible in part or entirely for the inhibition of larval development. Suzuki *et al.* (1993) showed that an electrophoretic variant of αAI, called αAI-2, is a poor inhibitor of azuki bean larval

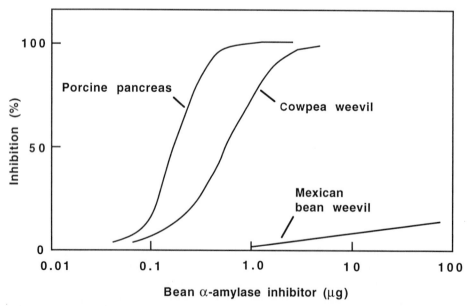

Figure 9.3 Inhibition of α-amylases from porcine pancreas, cowpea weevil and Mexican bean weevil (see Ishimoto and Kitamura, 1989 for details)

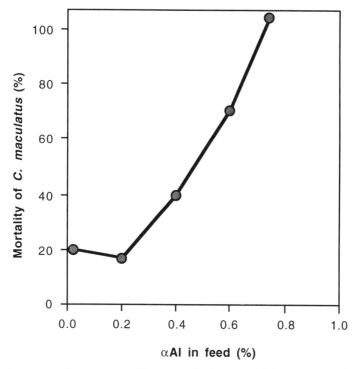

Figure 9.4 Responses of cowpea weevil larvae developing in artificial seeds made of cowpea meal to different levels of common bean αAI mixed with the meal (see Huesing *et al.*, 1991 for details)

α-amylase but a strong inhibitor of larval growth. This suggests that αAI may have effects on larval growth distinct from its effect on starch digestion. This question could be answered by utilizing mutants of αAI that do not have inhibitory activity (Mirkov *et al.*, 1995a).

Expression of PHA, Arcelin and αAI in Developing Bean Seeds and Seeds of Transgenic Tobacco Plants

The plant defense proteins PHA, arcelin and αAI are typical seed glycoproteins which accumulate in the mid-phase of seed development when the cotyledons expand in size and the seed reserves are being deposited. Like the 7S (vicilins) and 11S (legumins) storage proteins, they accumulate in protein storage vacuoles. These small protein-filled vacuoles are formed *de novo* during seed maturation (Hoh *et al.*, 1995). The transport of these seed proteins to the vacuoles is mediated by the secretory system, a system of interconnected cisternae and vesicles which includes the endoplasmic reticulum (ER) and the Golgi apparatus (Golgi). The primary translation products of the proteins are characterized by the presence of an amino-terminal signal peptide which directs the nascent protein into the lumen of the ER. As the polypeptide is being synthesized, glycans are transferred from dolichol pyrophosphate carriers in the ER membrane to specific Asn residues, converting the nascent polypeptide chain in the ER lumen into a glycoprotein with typical high mannose glycans. These glycoproteins are then transported to the Golgi where some of these glycans are modified by Golgi enzymes – glycosidases and glycosyltransferases – to yield glycoproteins with high mannose and complex (Golgi-modified) glycans. Because of a certain amount of heterogeneity in the glycans, a single gene product may give rise to several glycopolypeptides when the analysis is done by SDS-PAGE.

The transport of these proteins to protein storage vacuoles is mediated by Golgi derived vesicles and depends on the presence within the proteins of specific vacuolar targeting signals. In some vacuolar proteins, vacuolar targeting information is present in an aminoterminal or a carboxyterminal domain which is proteolytically removed after the proteins arrive in the vacuoles (Chrispeels and Raikhel, 1992). However, in these bean proteins targeting information appears to be contained in the mature protein itself. PHA and arcelin are not proteolytically processed after arrival in the vacuole, except perhaps by the removal of a few carboxyterminal residues by vacuolar carboxypeptidase. Bean αAI however is processed by a vacuolar endopeptidase which is specific for Asn residues, but this processing is not related to the vacuolar targeting of the protein. The endopeptidase cleaves αAI at the carboxyl-side of Asn[77], producing two different glycopolypeptides that remain bound together as an αβ complex. The calculated molecular weights of the α and β polypeptides are 8700 and 16 500 Da respectively, suggesting a molecular weight of approximately 30 000 Da for the glycoprotein.

Changes in the mRNA levels of αAI and arcelin have not been described in detail, but it is likely that they follow the same course as mRNAs for PHA and storage proteins in bean and other legumes. The mRNAs all dramatically increase in abundance during the expansion phase of seed development and decrease and disappear again during seed maturation. The promoters of these genes all confer seed-specific expression and there is little or no RNA in other plant organs. When these promoters are used in transgenic plants, they also confer seed-specific expression.

When vacuolar proteins are synthesized in transgenic tobacco seeds, they undergo the same co-translational and post-translational changes – removal of the signal peptide, attachment of glycans, modification of glycans in the Golgi, proteolytic processing and accumulation in vacuoles – as in beans (Sturm *et al.*, 1988). This is an important consideration for genetic engineering because many of these steps are necessary to obtain active protein which accumulates in stable form. However, proteolytic processing may be more extensive in tobacco than in bean and result in partial breakdown of the protein, indicating that the proteolytic complement of the vacuoles is an important consideration when expressing vacuolar proteins in transgenic plants (Altabella and Chrispeels, 1990). This is especially true for αAI which must be processed to be active (see below), but should not be broken down subsequently. We failed to obtain more than trace levels of αAI in leaves of potato when the expression of the αAI gene was driven by the CaMV 35S promoter. The likely cause was the breakdown of the protein in the protease-rich vacuoles of the leaves.

αAI Forms a Small Gene Family in *P. vulgaris*, its Wild Accessions and Closely Related Species in the Genus *Phaseolus*

The first αAI sequence was identified (Moreno and Chrispeels, 1989) on the basis of amino acid sequence identity of a purified heat stable αAI and an already cloned

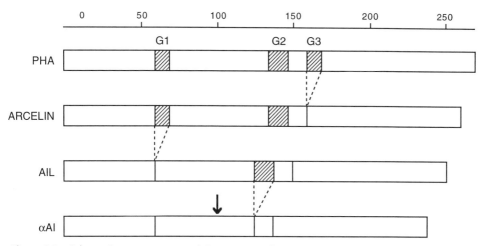

Figure 9.5 Schematic representation of the amino acid sequences of PHA, arcelin, α-amylase inhibitor like protein (AIL) and αAl. The shaded portions indicate successive deletions in the protein sequences

cDNA which was referred to as the 'bean lectin' based on its similarity to PHA (Hoffman *et al.*, 1982). That this sequence encodes an active inhibitor was demonstrated by Altabella and Chrispeels (1990) by expressing the cDNA in tobacco seeds. A standard protocol for αAl purification calls for heating the seed extract to 70°C for 15 min (Marshall and Lauda, 1975) and this procedure was followed in the work of Moreno and Chrispeels (1989). However, other methods to purify αAl have been developed, and Iguti and Lajolo (1991) showed that many varieties of bean contain two separable αAl inhibitors, one that is heat stable and one that is not. This indicates that αAl, like the homologous protein PHA, may be encoded by two different genes. The inhibitor identified by Moreno and Chrispeels (1989) is referred to as αAl-1. αAl is also found in wild accessions of the common bean as well as in related bean species including *P. acutifolius* (tepary bean), *P. coccineus* (runner bean) and *P. polyanthus* (Pueyo *et al.*, 1993). The αAl polypeptides in these species could be identified with the polyclonal antiserum against *P. vulgaris* αAl. Efforts to clone cDNAs from wild accessions and related species have yielded a number of sequences that are clearly more closely related to αAl than to either arcelin or PHA but none has as yet been shown to encode an active inhibitor (Mirkov *et al.*, 1994).

Using as starting material a wild accession of the common bean which is resistant to Mexican bean weevils, Mirkov *et al.* (1994) and Suzuki *et al.* (1994) isolated a new sequence, termed αAl-2, which has 75% amino acid sequence identity with αAl-1. Interestingly, this inhibitor is not digested by the gut enzymes of the Mexican bean weevil and unlike αAl-1, it is an effective inhibitor of Mexican bean weevil α-amylase (Suzuki *et al.*, 1993). However, when the cDNA was expressed in tobacco seeds, the resulting protein did not inhibit Mexican bean weevil α-amylase (Mirkov *et al.*, 1994). The polypeptide pattern in the transgenic tobacco seeds did not match the polypeptide pattern of αAl-2 in the bean, raising the possibility that the lack of activity was caused by incorrect processing and/or breakdown of one of the subunits. Alternatively, this gene may not encode an active inhibitor or may encode an inhibitor for which the target α-amylase has not been identified.

147

The same wild accession of *P. vulgaris* yielded another sequence identified as αAI-3 (Mirkov *et al.*, 1995a). When this cDNA was expressed in tobacco seeds, the resulting polypeptide was not proteolytically processed (Mirkov and Chrispeels, unpublished), making it very unlikely that it encodes an active amylase inhibitor. Sequence alignment of all available PHA, arcelin and αAI sequences shows that they form three groups of sequences (Mirkov *et al.*, 1995a) and that the αAI-3 sequence falls within the αAI grouping. However, the presence of a sequence in the αAI grouping apparently does not necessarily mean that the protein will be an active inhibitor. αAI-3 could represent an evolutionary intermediate between arcelin and active αAI, and should be referred to as an amylase inhibitor-like protein (Figure 9.5).

The Molecular Structure of αAI and its Binding to α-Amylase

αAI is synthesized as an initial translation product that is co-translationally and post-translationally modified to yield a holoprotein consisting of two glycoproteins, α and β, which form a complex. The processing step which gives rise to the two polypeptides occurs after protein folding has occurred, either in the trans-Golgi or in the vacuoles; it is most unlikely that the two polypeptides will again dissociate and associate as suggested by Ho *et al.* (1994). Determinations of the molecular weight of the holoprotein have ranged from 20 000 Da to 60 000 Da (see Ho *et al.*, 1994). It seems likely that there are two types of molecular structure: $\alpha\beta$ and $\alpha_2\beta_2$ and $\alpha\beta\gamma$ in which one of the $\alpha\beta$ subunits remains unprocessed. This unprocessed subunit is referred to as the γ-subunit by Yamaguchi (1993) who postulates that its presence is responsible for the heat lability of the protein. Whether this γ-subunit is the product of the same gene that produces the processed α and β subunits, or the product of a different gene is not yet clear. The sequences αAI-1 and αAI-2 have an $\alpha_2\beta_2$ structure (MW = 48 000 Da). The heat labile inhibitor has an $\alpha\beta\gamma$ structure (MW = 47 000 Da) (Yamaguchi, 1993).

It is of particular interest that the αAI protein has no inhibitory activity until it is proteolytically processed at Asn^{77} (Pueyo *et al.*, 1993). Proteolytic processing could bring about a small conformational change that creates an active site and allows αAI to bind its target enzyme. The glycans do not appear to be necessary for this activation step because mutant αAI that does not have glycosylation sites, is also processed and activated *in planta* (Mirkov and Chrispeels, unpublished). The rate of complex formation between αAI and α-amylase is slow and it takes 30–120 min to reach equilibrium depending on the conditions (Powers and Whitaker, 1977b). First, αAI binds to α-amylase to form a complex with a K_D of approx. 10^{-5}; this complex still has amylolytic activity. Subsequently, this complex is slowly converted to a new complex with a K_D of approx. 10^{-11}; this complex has no activity. Complex formation has a pH optimum around 5.5 and for this reason, αAI inhibits amylases in the acid midgut of *Coleoptera*, but not in the alkaline midgut of *Lepidoptera*. αAI is therefore particularly well suited for defense against the starch-eating bruchids.

α-Amylase inhibitors have been isolated from a large number of organisms and the amino acid sequences of a number of them have been established. The structure of one of these, the bacterial inhibitor tendamistat, has been determined by both

NMR and X-ray methods (Billeter *et al.*, 1989). Chemical modification experiments (Arai *et al.*, 1985) and sequence comparison with another bacterial amylase inhibitor, AI-3688 (Vertesy and Tripier, 1985) have established that the active site consists of the segment –Trp^{18}–Arg–Tyr–, which is in an exposed loop of the protein. We therefore examined the bean αAI sequence for segments with a sequence similar to the one found in the bacterial inhibitor. One candidate region around Trp^{188} was found, –Tyr–Gln–Trp–Ser–Tyr–, and when a model of the whole protein was made this region was close to the Asn^{77} cleavage site, and to an Arg residue, Arg^{74} (Figure 9.6). To test this site hypothesis, we created mutants of αAI, expressed them in tobacco plants with a seed specific promoter, and assayed the inhibitory activity of seed extracts toward porcine pancreatic α-amylase. Both the R74N and WSY188-190GNV mutations were inactive, as was the double mutant, supporting the conclusion that these residues, which are completely different from those involved in creating the sugar-binding site of lectins, are involved in the active site of the inhibitor.

The similarity of the active sites of αAI and tendamistat suggest that both may bind to the catalytic site of α-amylase in the same way. The Arg residue could interact with the catalytic Asp residues (Qian *et al.*, 1993) while the flanking Trp and Tyr may bind to neighboring subsites, mimicking the glucose residues of the substrate. Similar mimicry of sugar residues by aromatic amino acids was seen in peptide inhibitors of concanavalin A (Oldenberg *et al.*, 1992; Scott *et al.*, 1992) and anti-carbohydrate antibodies (Hoess *et al.*, 1993) derived from phage libraries. Mutational analysis alone cannot prove the location of the active site and crystallographic analysis of the enzyme–inhibitor complex is now in progress in collaboration with the laboratory of L. Wyns (Free University of Brussels).

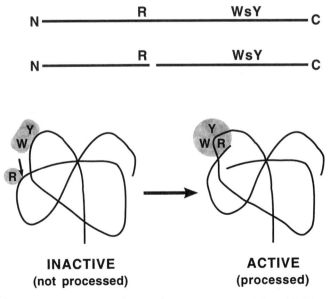

Figure 9.6   Schematic representation of αAI and its active site consisting of R, W and Y. Processing at Asn^{77} is postulated to bring R^{74} in closer proximity to W and Y to create an active site similar to the one in the bacterial α-amylase inhibitor tendamistat

Transfer of Bruchid Resistance from Beans to Peas by Genetic Engineering with the αAl Gene

Gene transfer to achieve insect resistance is a promising approach to plant protection and it is particularly well suited in this situation because the gene of interest, αAI, is present in one species, and protection is needed for a number of other species. The following developments and observations made it possible to test the hypothesis that bean αAI will inhibit the development of bruchid larvae when it is expressed in the seeds of other grain legumes:

(1) The cDNA encoding αAI was available. It had been isolated by Hoffman *et al.* (1982) and identified by Moreno and Chrispeels (1989).

(2) Seed specific promoters that are able to direct the high level expression of storage proteins and lectins in developing legume seeds had been studied for several years and were available.

(3) When the genes for seed proteins are expressed in heterologous plants, the products of the transgenes are correctly processed and targeted to vacuoles. This is particularly important in this case, because αAI is glycosylated and needs to be proteolytically processed to be active.

(4) Inhibition of larval development can be achieved with αAI levels of 0.3–0.9% in the seeds. Such levels can readily be obtained with a good seed specific promoter, such as the PHA promoter.

(5) Grain legumes are particularly recalcitrant to transformation, but Schroeder *et al.* (1993) developed a method for *Agrobacterium*-mediated transformation of peas.

(6) Several species of bruchids, including the pea weevil, the cowpea weevil and the azuki bean weevil, can be reared on peas, making it possible to test the hypothesis with peas.

The vector used for DNA transfer contained the αAI cDNA flanked by PHA 5' and 3' sequences, the Basta® resistance (BAR) gene driven by the CaMV 35S promoter and the gene encoding β-glucuronidase (GUS) also driven by the CaMV 35S promoter. This gene was added to have a screenable marker – as well as a selectable marker – but this turned out to be unnecessary. The transformed plants were produced by the Division of Plant Industry, CSIRO, Canberra, Australia according to the method of Schroeder *et al.* (1993).

Transgenic seeds of the T2 generations were tested by Murdock and Shade of Purdue University for resistance to the cowpea weevil and the azuki bean weevil (Shade *et al.*, 1994). In a blind experiment, transgenic pea seeds were bioassayed for weevil resistance by infesting them with cowpea weevils or azuki bean weevils and observing developmental times and mortalities versus untransformed seeds. The seeds were individually marked and assayed biochemically to determine the level of αAI. For the azuki bean weevil, the impact of αAI in the peas was dramatic and unequivocal: in nearly every seed where αAI was detected biochemically, all infesting larvae died.

The response of the cowpea weevil was more varied and depended on the level of αAI in the seeds. A selection of 69 of the bioassayed seeds from three transgenic lines was each assigned to one of four resistance classes – immune, resistant, susceptible to cowpea weevil, and highly susceptible – and also assayed biochemically for αAI.

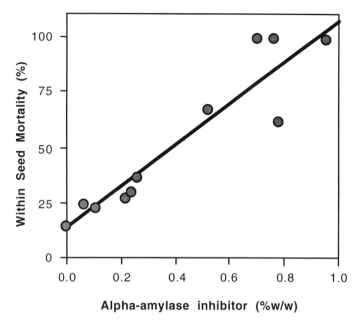

Figure 9.7 Responses of cowpea weevil larvae developing in transgenic pea seeds that express various levels of αAI. The αAI content of the seeds was determined biochemically and mortality calculated by grouping seeds in classes of highly susceptible, less susceptible, resistant and immune (see Shade *et al.*, 1994 for details)

There was excellent correspondence between the measured αAI level and the percent mortality and delay of larval development of cowpea weevils growing in these four seed classes (Figure 9.7). αAI levels of 0.8–1.0% gave complete insect mortality confirming earlier results obtained with artificial seeds (Huesing *et al.*, 1991).

The peas that express αAI in their seeds are also completely resistant to *Bruchus pisorum*, an insect that damages peas that are growing in the field (Schroeder *et al.*, 1995). The eggs hatch and the larvae eat through the seed pod and into the seed, but there they fail to develop.

It is quite likely that this simple gene transfer technology which worked so well with peas is applicable to other starchy grain legumes that suffer bruchid infestations. These include cowpeas, mungbeans, pigeon peas or chick-peas, azuki beans, broad beans and lentils. The technology is also applicable to the common bean but will require forms of αAI such as αAI-2 which inhibit the α-amylase(s) in the guts of the Mexican bean weevil and the bean weevil. A Japanese research group – Ishimoto, Sato and Kitamura – has transformed azuki beans with the same construct, and the seeds from the T2 plants are resistant to azuki bean weevils and *C. analis*. (Ishimoto *et al.*, 1996). No attempt was made to establish a quantitative relationship between the level of αAI and the degree of resistance. The major constraint to making other starchy grain legumes such as cowpeas, mungbeans and chick-peas resistant to bruchids, is the lack of suitable transformation protocols.

Can Beans be made Resistant to New World Bruchids with αAI Genes?

New world bruchids cause considerable post-harvest losses of common bean wherever it is grown, but resistance to the Mexican bean weevil and the bean weevil is found among the wild accessions of the common bean (Schoonhoven *et al.*, 1983). For resistance to the Mexican bean weevil, arcelin was identified as the likely cause of this resistance (Osborn *et al.*, 1986) and the arcelin-PHA-αAI locus was introduced into cultivated lines (Cardona *et al.*, 1990; Hartweck *et al.*, 1991; Osborn *et al.*, 1988). The locus confers resistance to the Mexican bean weevil, and purified arcelin at a high dose – 10% of total protein – inhibits the development of Mexican bean weevil larvae in artificial seeds (Osborn *et al.*, 1988). Since the locus contains not only arcelin but also αAI genes, and because the purified arcelin preparations may have been contaminated with small amounts of αAI, we checked the relationship between αAI levels and resistance in the backcrossed lines. We found that there was no relationship between resistance and αAI levels, suggesting that αAI is not involved, or that it plays only a minor role in the resistance of beans to the Mexican bean weevil (Fory *et al.*, 1996). Resistance to the bean weevil (*A. obtectus*) could not be transferred by backcrossing between resistant accessions and cultivated varieties (Kornegay and Cardona, 1991) suggesting that the entire locus plays at best a minor role in the resistance to the bean weevil. However, some wild accessions of beans contain αAI variants such as αAI-2, which strongly inhibit larval α-amylase of the Mexican bean weevil and weakly inhibit the larval α-amylase of the bean weevil. This suggests that these variant αAIs could be used for genetic engineering experiments to suppress growth of the Mexican bean weevil larvae. Beans resistant to the Mexican bean weevil have already been bred using traditional breeding methods, and this strategy therefore seems redundant given the difficulty of transforming the common bean at this time. However, these variant forms of αAI could be used to make other grain legumes such as cowpeas resistant to the Mexican bean weevil.

Will αAI-containing Legumes be Safe to Eat?

Diets based on raw legume seed meals have detrimental effects on humans and farm animals because of the presence of antinutrients or plant defense chemicals. For example, the main antinutrient protein in the common bean is PHA. When rats are fed high quality diets that contain purified PHA, the lectin induces hyperplastic and hypertrophic growth of the small intestine while the animals lose weight continuously because nutrients and energy are used to maintain this wasteful growth of the gut (Pusztai *et al.*, 1991). In soybean meal, the soybean agglutinin and trypsin inhibitors are the two principal proteinaceous antinutrients. The presence of such proteinaceous antinutrients is the main reason why legume seeds are cooked or roasted before consumption by humans. Once denatured, the proteins do not function as antinutrients any longer but provide amino acids just like seed proteins. Because αAI inactivates intraduodenal amylase in humans (Layer *et al.*, 1985), it is likely to be an antinutritional factor that must be denatured by cooking or roasting. This is confirmed by a recent study of Pusztai *et al.* (1995) who determined the effect of bean αAI on starch digestion and the growth of weanling rats. They found that the growth of rats fed 20 mg or 40 mg αAI per day in a diet containing 0.33% or 0.66% w/w αAI, was retarded and that the inhibitor reduced the utilization of dietary proteins and lipids. At the higher levels of αAI, starch digestion in the small intestine was negligible. These experiments were carried out with purified αAI that contained 8.6% protease inhibitor. Recent experiments (Eggum *et al.*, unpublished results) with the transgenic peas indicate that αAI may not pose a big problem. When rats were fed a diet consisting of 25% peas and 75% barley, there was only a small difference (about 3.5%) in the total digestibility of the diet with transgenic peas containing αAI as compared to control peas. Even this small difference may be unacceptable to the animal feed industry. We therefore need to find αAIs which inhibit the target insect, but not mammalian α-amylases.

References

Altabella, T. and Chrispeels, M. J. (1990) Tobacco plants transformed with the bean αAI gene express an inhibitor of insect α-amylase in their seeds, *Plant Physiol.* **93**, 805–810.

Arai, M., Oouchi, N., Goto, A., Ogura, S. and Murao, S. (1985) Chemical modifications of Haim, a proteinaceous α-amylase inhibitor, *Agric. Biol. Chem.* **49**, 1523–1524.

Billeter, M., Kline, A. D., Braun, W., Huber, R. and Wuthrich, K. (1989) Comparison of the high-resolution structures of the α-amylase inhibitor tendamistat determined by nuclear magnetic resonance in solution and by X-ray diffraction in single crystals, *J. Mol. Biol.* **206**, 677–687.

Cardona, C., Kornegay, J., Posso, C. E., Morales, F. and Ramirez, H. (1990) Comparative value of four arcelin variants in the development of dry bean lines resistant to the Mexican bean weevil, *Entomology* **56**, 197–206.

Chrispeels, M. J. and Raikhel, N. V. (1991) Lectins, lectin genes and their role in plant defense, *Plant Cell* **3**, 1–19.

Chrispeels, M. J. and Raikhel, N. V. (1992) Short peptide domains target proteins to plant vacuoles, *Cell* **68**, 613–616.

de Oliveira, J. T. A., Pusztai, A. and Grant, G. (1988) Changes in organs and tissues induced by feeding of purified kidney bean (*Phaseolus vulgaris*) lectins, *Nutr. Res.* **8**, 943–947.

Fory, L. F., Finardi-Filho, F., Quintero, C. M., Osborn, T. C., Cardona, C., Chrispeels, M. J. and Mayer, J. E. (1996) α-amylase inhibitors in resistance of common beans to the Mexican bean weevil and the bean weevil (Coleoptera, Bruchidae), *J. Econ. Entomol.* **89**, 204–10.

Gatehouse, A. M. R., Dewey, F. M., Dove, J., Fenton, K. A. and Pusztai, A. (1984) Effect of seed lectins from *Phaseolus vulgaris* on the development of larvae of *Callosobruchus maculatus*: mechanism of toxicity, *J. Sci. Food Agric.* **35**, 373–380.

Goossens, A., Geremia, R., Van Montagu, M. and Angenon, D. (1994) Isolation and characterisation of arcelin-5 proteins and cDNAs, *Eur. J. Biochem.* **225**, 787–795.

Hartweck, L. M., Vogelzang, R. D. and Osborn, T. C. (1991) Characterization and comparison of arcelin seed protein variants from common bean, *Plant Physiol.* **97**, 204–211.

Ho, M. F., Yin, X., Filho, F. F., Lajolo, F. and Whitaker, J. R. (1994) Naturally occurring α-amylase inhibitors: structure/function relationships, In: Yada, R. Y., Jackman, R. L. and Smith, J. L. (Eds), *Protein Structure–Function Relationships in Foods*, London: Chapman & Hall, pp. 89–119.

Hoess, R., Brinkmann, U., Handel, T. and Pastan, I. (1993) Identification of a peptide that binds to the carbohydrate-specific monoclonal antibody B3, *Gene* **100**, 43–49.

Hoffman, L. M., Ma, Y. and Barker, R. F. (1982) Molecular cloning of *Phaseolus vulgaris* lectin mRNA and use of cDNA as a probe to estimate lectin transcripts levels in various tissues, *Nucl. Acid Res.* **10**, 7819–7828.

Hoh, B., Hinz, G., Jeong, B. K. and Robinson, D. G. (1995) Protein storage vacuoles form de novo during pea cotyledon development, *J. Cell Sci.* **108**, 299–310.

Huesing, J. E., Shade, R. E., Chrispeels, M. J. and Murdock, L. L. (1991) α-Amylase inhibitor, not phytohemagglutinin explains the resistance of common bean seeds to cowpea weevil, *Plant Physiol.* **96**, 993–996.

Iguti, A. M. and Lajolo, F. M. (1991) Occurrence and purification of α-amylase isoinhibitors in bean (*Phaseolus vulgaris* L.) varieties, *J. Agric. Food Chem.* **39**, 2131–2136.

Ishimoto, M. and Chrispeels, M. J. (1996) Protective mechanisms of the Mexican bean weevil against high levels of α-amylase inhibitor in the common bean, *Plant Physiol.* **111**, 393–401.

Ishimoto, M. and Kitamura, K. (1989) Growth inhibitory effects of an α-amylase inhibitor from kidney bean, *Phaseolus vulgaris* (L.) on three species of bruchids (Coleoptera: Bruchidae), *Appl. Entomol. Zool.* **24**, 281–286.

Ishimoto, M. and Kitamura, K. (1992) Tolerance to the seed α-amylase inhibitor by the 2 insect pests of the common bean, *Zabrotes subfasciatus* and *Acanthoscelides obtectus* (Coleoptera, Bruchidae), *Appl. Entomol. Zool.* **27**, 243–251.

Ishimoto, M., Sato, J., Chrispeels, M. J. and Kitamura, K. (1996) Bruchid resistance of transgenic azuki bean expressing seed α-amylase inhibitor in the common bean, *Entomol. Exper. Appl.* **79**, 309–15.

Jaffé, W. G., Brucher, O. and Palozzo, A. (1972) Detection of four types of specific hemagglutinins in different lines of beans, *Z. Immunitaetsforsch.* **142**, 439–447.

Janzen, D. H., Juster, H. B. and Liener, I. E. (1976) Insecticidal action of the phytohemagglutinin in black beans on a bruchid beetle, *Science* **192**, 795–796.

Kornegay, J. L. and Cardona, C. (1991) Inheritance of resistance to *Acanthoscelides obtectus* in a wild common bean accession crossed to commercial bean cultivars, *Euphytica* **52**, 103–111.

Lajolo, F. M. and Finardi-Filho, F. (1985) Partial characterization of the amylase inhibitor of black beans (*Phaseolus vulgaris*), variety Rico 23, *J. Agric. Food Chem.* **33**, 132–138.

Layer, P., Carlson, G. L. and DiMagno, E. P. (1985) Partially purified white bean amylase inhibitor reduces starch digestion *in vitro* and inactivates intraduodenal amylase in humans, *Gastroenterology* **88**, 1895–1902.

Marshall, J. J. and Lauda, C. M. (1975) Purification and properties of phaseolamin, an inhibitor of α-amylase, from the kidney bean *Phaseolus vulgaris*, *J. Biol. Chem.* **250**, 8030–8037.

Minney, B. H. P., Gatehouse, A. M. R., Dobie, P., Dendy, J., Cardona, C. and Gatehouse, J. A. (1990) Biochemical bases of seed resistance to *Zabrotes subfasciatus* (bean weevil) in *Phaseolus vulgaris* (common bean): a mechanism for arcelin toxicity, *J. Insect Physiol.* **36**, 757–767.

Mirkov, T. E. and Chrispeels, M. J. (1993) Mutation of Asn[128] to Asp of *Phaseolus vulgaris* leucoagglutinin (PHA-L) eliminates carbohydrate binding and biological activity, *Glycobiology* **3**, 581–587.

Mirkov, T. E., Wahlstrom, J. E., Hagiwara, K., Finardi-Filho, F., Kjemtrup, S. and Chrispeels, M. J. (1994) Evolutionary relationships among proteins in the phytohemagglutinin-arcelin-α-amylase inhibitor family of the common bean and its relatives, *Plant Mol. Biol.* **26**, 1103–1113.

Mirkov, T. E., Evans, S. V., Wahlstrom, J., Gomez, L., Young, N. M. and Chrispeels, M. J. (1995a) Location of the active site of the bean α-amylase inhibitor and involvement of a Trp, Arg, Tyr triad, *Glycobiology* **5**, 45–50.

Mirkov, T. E., Pueyo, J. J., Mayer, J., Kjemtrup, S., Cardona, C. and Chrispeels, M. J. (1995b) Molecular and functional analysis of α-amylase inhibitor genes and proteins in the common bean *Phaseolus vulgaris*, In: Roca, W. M. (Ed.), *Proceedings of Bean Advanced Research Network Workshop*, Cali, Colombia: CIAT, pp. 296–302.

Moreno, J. and Chrispeels, M. J. (1989) A lectin gene encodes the α-amylase inhibitor of the common bean, *Proc. Natl. Acad. Sci. USA* **86**, 7885–7889.

Nodari, R. O., Tsai, S. M., Gilbertson, R. L. and Gepts, P. (1993) Towards an integrated linkage map of common bean 2. Development of an RFLP-based linkage map, *Theor. Appl. Genet.* **85**, 513–520.

Oldenberg, K. R., Loganathan, D., Goldstein, I. J., Schultz, P. J. and Gallop, M. A. (1992) Peptide ligands for a sugar-binding protein isolated from a random peptide library, *Proc. Natl. Acad. Sci. USA* **89**, 5393–5397.

Osborn, T. C., Burow, M. and Bliss, F. A. (1988) Purification and characterization of arcelin seed protein from common bean, *Plant Physiol.*, **86**, 399–405.

Osborn, T. C., Blake, T., Gepts, P. and Bliss, F. A. (1986) Bean arcelin. 2. Genetic variation, inheritance and linkage relationships of a novel seed protein of *Phaseolus vulgaris* L., *Theor. Appl. Genet.* **71**, 847–855.

Pick, K. H. and Wöber, G. (1978) Proteinaceous α-amylase inhibitor from bean (*Phaseolus vulgaris*). Purification and characterization, *Z. Physiol. Chem.* **359**, 1371–1377.

Powers, J. R. and Whitaker, J. R. (1977a) Effect of several experimental parameters on combination of red kidney bean (*Phaseolus vulgaris*) α-amylase inhibitor with porcine pancreatic α-amylase, *J. Food Biochem.* **1**, 239–260.

Powers, J. R. and Whitaker, J. R. (1977b) Purification and some physical and chemical properties of red kidney bean (*Phaseolus vulgaris*) α-amylase inhibitor, *J. Food Biochem.* **1**, 217–238.

Pueyo, J. J., Hunt, D. C. and Chrispeels, M. J. (1993) Activation of bean α-amylase inhibitor requires proteolytic processing of the pro-protein, *Plant Physiol.* **101**, 1341–1348.

Pusztai, A. (1993) Dietary lectins are metabolic signals for the gut and modulate immune and hormone functions, *Eur. J. Clin. Nutr.* **47**, 691–699.

Pusztai, A., Watt, W. B. and Stewart, J. C. (1991) A comprehensive scheme for the isolation of trypsin inhibitors and the agglutinin from soybean seeds, *J. Agric. Food Chem.* **39**, 862–866.

Pusztai, A., Grant, G., Duguid, T., Brown, D. S., Peumans, W. J., Van Damme, E. J. M. and Bardocz, S. (1995) Inhibition of starch digestion by α-amylase inhibitor reduces the efficiency of utilization of dietary proteins and lipids and retards the growth of rats, *J. Nutr.* **125**, 1554–1562.

Qian, M., Haser, R. and Payan, F. (1993) Structure and molecular model refinement of pig pancreatic α-amylase at 2.1 Å resolution, *J. Mol. Biol.* **231**, 785–799.

Rougé, P., Barre, A., Causse, H., Chatelain, C. and Porthé, G. (1993) Arcelin and α-amylase inhibitor from the seeds of common bean (*Phaseolus vulgaris* L.) are truncated lectins, *Biochem. System. Ecol.* **21**, 695–703.

Schoonhoven, A. V., Cardona, C. and Valor, J. (1983) Resistance to the bean weevil and the Mexican bean weevil (Coleoptera: Bruchidae) in noncultivated common bean accessions, *J. Econ. Entomol.* **76**, 1255–1259.

Schroeder, H. E., Schotz, A. H., Wardley-Richardson, T., Spencer, D. and Higgins, T. J. V. (1993) Transformation and regeneration of two cultivars of pea (*Pisum sativum* L.), *Plant Physiol.* **101**, 751–757.

Schroeder, H. E., Gollasch, S., Moore, A., Tabe, L. M., Craig, S., Hardie, D., Chrispeels, M. J., Spencer, D. and Higgins, T. J. V. (1995) Bean α-amylase inhibitor confers resistance to the pea weevil, *Bruchus pisorum*, in genetically engineered peas (*Pisum sativum* L.), *Plant Physiol.* **107**, 1233–1239.

Scott, M. P., Jung, R., Muntz, K. and Nielsen, N. C. (1992) A protease responsible for post-translational cleavage of a conserved Asn–Gly linkage in glycinin, the major seed storage protein of soybean, *Proc. Natl. Acad. Sci. USA* **89**, 658–662.

Shade, R. E., Schroeder, H. E., Pueyo, J. J., Tabe, L. M., Murdock, L. L., Higgins, T. J. V. and Chrispeels, M. J. (1994) Transgenic pea seeds expressing the α-amylase inhibitor of the common bean are resistant to bruchid beetles, *Bio/Technology* **12**, 793–796.

Sturm, A., Voelker, T. A., Herman, E. M. and Chrispeels, M. J. (1988) Correct glycosylation, Golgi-processing, and targeting to protein bodies of the vacuolar protein phytohemagglutinin in transgenic tobacco, *Planta* **175**, 170–183.

Suzuki, K., Ishimoto, M. and Kitamura, K. (1994) cDNA sequence and deduced primary structure of an α-amylase inhibitor from a bruchid-resistant wild common bean, *Biochim. Biophys. Acta* **1206**, 289–291.

Suzuki, K., Ishimoto, M., Kikuchi, F. and Kitamura, K. (1993) Growth inhibitory effect of an α-amylase inhibitor from the wild common bean resistant to the Mexican bean weevil

(*Zabrotes subfasciatus*), *Jap. J. Breed.* **43**, 257–265.

Talekar, N. S. (1987) Biology, damage and control of bruchid pests of mungbean, In: Shanmugasundaram, S. and McLean, B. T. (Eds), *Mungbean: Proceedings of the Second International Symposium, Bangkok, Thailand, November 16–20, 1987*, Shanhua, Taiwan: AVRDC, pp. 329–342.

Vertesy, L. and Tripier, D. (1985) Isolation and structure elucidation of an α-amylase inhibitor, AI-3688, from *Streptomyces aureofaciens*, *FEBS Lett.* **185**, 187–190.

Yamaguchi, H. (1993) Isolation and characterization of the subunits of a heat labile α-amylase inhibitor from *Phaseolus vulgaris* white kidney bean, *Biosci. Biotechnol. Biochem.* **57**, 297–302.

Acknowledgments

Work on plant defense proteins in the author's laboratory has been supported by the Department of Energy (Energy Biosciences Program). The writing of this chapter has benefited from stimulating discussions with Masao Ishimoto and Flavio Finardi-Filho. The laboratories of T. J. Higgins and H. Schroeder at CSIRO, Canberra, Australia, and of L. Murdock and R. Shade at Purdue University were equal partners in generating the transgenic peas expressing αAI. I thank Lyn Alkan for skillful production of the manuscript and the figures.

10

Proteinase Inhibitors and Resistance of Transgenic Plants to Insects

GERALD R. REECK, KARL J. KRAMER, JAMES E. BAKER, MICHAEL R. KANOST, JEFFREY A. FABRICK and CRAIG A. BEHNKE

Introduction

Proteinaceous inhibitors of proteinases have been explored for a number of years as growth retardants for insects, particularly phytophagous insects, and for other organisms as well. Many proteinase inhibitors are rather small, very stable, abundant proteins and therefore rather easy to purify. It is not surprising, then, that they have been extensively studied for many years to elucidate their structures and functional properties (Bode and Huber, 1992; Laskowski and Kato, 1980; Read and James, 1986).

Wound-inducible proteinase inhibitors have been clearly shown to be involved in plant defense, largely by Ryan and his colleagues (Ryan, 1990). This involvement of proteinase inhibitors in natural defense by plants has greatly stimulated thinking about and working toward enhancing plant defense through genetic engineering with genes that encode proteinase inhibitors. It should be noted, however, that many of the inhibitors under consideration are found in seeds and are not inducible, and their role in plant defense is more inferred than proven. For the most part, in this chapter we will not attempt to provide a comprehensive review of the literature. Instead, to illustrate important concepts we will select appropriate articles. In making these selections, we undoubtedly failed to cite many articles and we apologize to those authors whose important work we have omitted.

Overview of the Chapter

We will first provide an overview of proteinase inhibitors. We consider proteinaceous inhibitors from all biological sources, not just from plants, since in creating a transgenic plant there is no need to restrict attention to plant-derived inhibitors. We next describe the current state of knowledge of digestive proteinases in insects from physiological, biochemical, and molecular genetic viewpoints. Clearly, the most intelligent use of inhibitor genes in transgenic plants cannot be made unless the nature of the enzymes to be inhibited has been taken into account and the physiology of the digestive process considered. We then survey published studies on the inhibition of insect enzymes by proteinaceous inhibitors, both *in vitro* and in bioassays. We next examine plants that have been transformed with genes that encode inhibitors, and we close by looking at future prospects in this field of research, which we believe is likely to have a great impact in economic entomology.

Proteinase Inhibitors

Classification into Families of Related Proteins

Because proteins are genetically encoded, the most fundamental basis for classifying proteins is by phylogenetic (or evolutionary) criteria. This was first realized for inhibitors by Laskowski and Kato (1980), who recognized eight families of inhibitors based on sequence similarity.

Table 10.1 provides an updated list of families of inhibitors, or, more precisely, of protein families that contain proteinase inhibitors. We have attempted to make a comprehensive list, regardless of the biological source or function of the inhibitors. Following the suggestion of Laskowski and Kato (1980), each family in Table 10.1 is identified by the name of a prominent member of that family. We emphasize, however, that a family's name is not intended to imply either functional properties or the biological source for all members of the family. The names in Table 10.1 correspond for the most part with those in the SwissProt database, but we have included several families that do not appear in SwissProt's compilation. For other reviews on inhibitors, see Ryan (1990), Richardson (1991), and Kanost and Jiang (1996).

Evolution of Function

An evolutionary classification, properly constructed, is based solely on genetic relatedness, generally revealed by sequence similarity and disregards function. There is no need for all proteins in a family to have similar functions. In fact, some of the families in Table 10.1 contain proteins that are not proteinase inhibitors at all. For example, the 'serpin family' contains not only inhibitors, but hormone-binding proteins, ovalbumin, and an egg-white protein of no known activity (Potempa *et al.*, 1994). Thus, the name 'serpin family' means only that serpins are in the family, and not that the family contains only serpins. As another example, the 'barley trypsin inhibitor family' contains proteins that are well studied as amylase inhibitors but do not exhibit proteinase inhibitor activity (Ryan, 1990; Richardson, 1991). Obviously, one cannot infer function (activity as an inhibitor, in this case) based solely on sequence similarity. Instead, inhibitory properties of proteins must be determined directly in studies with the proteins.

Another important point about an evolutionary classification is that there is not a strict correspondence between family membership and inhibitory selectivity, or between family membership and the type of proteinase inhibited. Thus, within a particular family in Table 10.1, different proteins frequently inhibit proteinases of very different specificities. As one example among many, while most members of the potato inhibitor I family are inhibitors of chymotrypsin or elastase, a member of the family is known that inhibits trypsin (Krishnamoorthi *et al.*, 1990). Perhaps more surprising is the fact that different members of a family may inhibit proteinases of

Table 10.1 Protein families that contain proteinase inhibitors[a]

Name	Approximate domain size and number of domains
1. Bovine pancreatic trypsin inhibitor family	60 residues
2. Kazal serine protease inhibitor family	55 residues; up to 7 domains
3. Kunitz soybean trypsin inhibitor family	180 residues
4. Bowman–Birk inhibitor family	35 residues; 2 domains
5. Potato inhibitor I family	70 residues
6. Potato inhibitor II family	50 residues; 1 or 2 domains; up to 5 domains in original translation product[b]
7. Squash inhibitor family	30 residues
8. Barley trypsin inhibitor[c] family	120 residues
9. Thaumatin family	200 residues
10. *Ascaris* trypsin inhibitor family	60 residues
11. Locust inhibitor[d] family	35 residues; 2 domains in original translation product
12. Ecotin family	140 residues; 2 subunits
13. Serpin family	400 residues
14. *Streptomyces* subtilisin inhibitor family	110 residues
15. Hirudin family	65 residues
16. Cystatin family	110 residues; usually single domain but 3 inhibitory domains in kininogens and 8 in potato multicystatin
17. Calpastatin family	140 residues; 4 inhibitory domains
18. Potato carboxypeptidase inhibitor family	40 residues
19. *Ascaris* carboxypeptidase inhibitor[e] family	40 residues
20. Collagenase inhibitor[f] family	200 residues
21. *Ascaris* pepsin inhibitor[g] family	150 residues
22. α-2-Macroglobulin family	1500 residues; 2 or 4 subunits

[a] When domain number is not indicated the proteins occur as a single domain. The Ragi I-2 family listed by Richardson (1991) contains amylase inhibitors but has not been shown to contain proteinase inhibitors

[b] Atkinson *et al.* (1993)

[c] This is also called the cereal amylase/trypsin inhibitor family by SwissProt and the cereal superfamily by Richardson (1991)

[d] Boigergrain *et al.* (1992) and Kromer *et al.* (1993)

[e] Homandberg *et al.* (1989)

[f] This is also called the tissue inhibitor of metalloproteinase family by SwissProt

[g] Martzen *et al.* (1990)

different mechanistic classes. Thus, as revealed by the work of Turk and his colleagues, within the Kunitz soybean trypsin inhibitor family, there are not only proteins that inhibit serine proteinases, but also inhibitors of cysteine proteinases and aspartyl proteinases (Krizaj *et al.*, 1993). Although most members of the Bowman–

Birk family are inhibitors of serine proteinases, a family member is known (see PIR database) that inhibits a cysteine proteinase (Reddy *et al.*, 1975). Similarly, although within the serpin family most inhibitors are targeted toward serine proteinases (indeed, 'serpin' stands for SERine Proteinase INhibitor), a viral protein called CrmA of the serpin family inhibits a cysteine proteinase (Komiyama *et al.*, 1994).

Variation in Sequence Within a Family

As emphasized by Laskowski *et al.* (1987, 1990) for inhibitors that rely largely on an exposed reactive-site loop (see below), the sequence of that loop is hypervariable in evolution. Thus, the structural scaffolding that supports the loop is more highly conserved in sequence than is the reactive region of the proteins. This appears to be an evolutionary mechanism for basing a range of inhibitory selectivities on a single architectural design, since differences in sequence in the reactive-site loop, even at individual positions, can correspond to profound differences in selectivities of the inhibitors (Bigler *et al.*, 1993).

Three-dimensional Structures and Interactions of Inhibitors with Target Enzymes

There is no case where we know the three-dimensional structure of an insect proteinase interacting with an inhibitor. However, numerous structures of inhibitors in complexes with enzymes from other sources are known, and these can serve as models for insect proteinase–inhibitor complexes, which should be studied directly in the future.

In what might be called the 'genetic law of protein folding', proteins within a family exhibit similar tertiary structures or folding patterns (Chothia and Lesk, 1986). Conversely, proteins in different families generally have dissimilar tertiary structures. The compilation and analysis by Bode and Huber (1992) reaffirms this law in the case of inhibitors. Since Bode and Huber's article appeared, three-dimensional structures have been published for proteins in three additional families (Grasberger *et al.*, 1994; McGrath *et al.*, 1995; Strobl *et al.*, 1995). As a result, three-dimensional structures are now available for at least one member of 15 families that contain proteinase inhibitors. The only families listed in Table 10.1 for which representative three-dimensional structures are not available are numbers 9, 11, 17, and 19–22.

As has been recognized for some time, the basic architectural design of many (though not all) inhibitors is a globular scaffold, frequently containing several disulfide bonds, that supports an exposed loop which provides the main points of interaction with a target proteinase. Inhibitors with this design, in families 1–8, 13, and 14, resemble substrates for proteinases. In other families, the nature of the interaction is either not completely substrate-like or is not well understood.

Proteinases of Different Classes as Targets of Inhibitors

Proteinases are typically classified into four major classes: serine proteinases, cysteine proteinases, aspartyl proteinases and metallo-proteinases. At least three of these mechanistic classes are drawn from more than one protein family. As we examine the inhibition of types of proteinase, we will proceed by protein families within each mechanistic class of proteinase.

Inhibitors of Serine Proteinases

Serine proteinases fall into two genetically unrelated families: proteins related to trypsin and proteins related to the bacterial enzyme subtilisin. Inhibitors of both trypsin-related and subtilisin-related enzymes have been very thoroughly studied. For the most part, a given protein inhibits enzymes exclusively from one or the other family of serine proteinases. An interesting exception exists in the potato inhibitor I family. Barley chymotrypsin inhibitor 2 inhibits both chymotrypsin and subtilisin (Longstaff *et al.*, 1990).

Many inhibitors of serine proteinases, of either family, obey the 'standard mechanism of inhibition' as defined by Laskowski and Kato (1980). The basic feature of this scheme is that, through a combination of reversible binding steps and enzymatic cleavage of a single peptide bond in the inhibitor, an inhibitor/enzyme mixture reaches an equilibrium state in which a certain proportion of the inhibitor exists as a two-chain form. The proportion of single-chain and two-chain forms is unique for each inhibitor and typically is within an order of magnitude of each other. As a thermodynamic property this proportion is independent of the enzyme used to establish the equilibrium.

Inhibitors that obey the standard mechanism occur in numerous protein families (1–11 and 14 in Table 10.1), although detailed studies of the mechanism are still lacking for families 9–11. Although the folding patterns of such inhibitors from different families are entirely distinct, there is a remarkable similarity in local configuration in the reactive-site loop of the inhibitors, regardless of family membership. This standard or canonical, configuration of the reactive-site loop led Bode and Huber (1992) to call these 'canonical inhibitors'.

Canonical inhibitors are substrates for their target enzymes; that is, an inhibitor contains a peptide bond that is enzymatically hydrolyzed. This is the so-called scissile bond. Canonical inhibitors are in fact excellent substrates if one judges them by the standard combined kinetic constant V_{max}/K_m. The two-chain forms of canonical inhibitors are, as a thermodynamic necessity, fully inhibitory and will, when combined with a suitable enzyme, give rise to the same equilibrium state as the single-chain inhibitor. However, equilibrium may be attained rather slowly when starting with the two-chain forms of canonical inhibitors.

Among non-canonical inhibitors is ecotin, a protein from *Escherichia coli*, which inhibits serine proteinases of widely different specificities. Each ecotin homodimer inhibits two enzyme molecules, and the primary mode of interaction of the inhibitor with the enzyme is only loosely substrate-like (McGrath *et al.*, 1995).

Serpins are similar to canonical inhibitors to the extent that a scissile bond in a reactive-site loop is attacked by a target enzyme. In a serpin, however, the two-chain form is not inhibitory, evidently because of the major difference in conformation between the single-chain and two-chain serpin molecules (Bode and Huber, 1992). Exposure of an enzyme–serpin complex to sodium dodecyl sulfate results in formation of a covalent bond, presumably involving the reactive-site serine of the enzyme and the amino acid residue that contributes a carbonyl group to the scissile peptide bond. Inhibition by a serpin is treated as an irreversible inhibition, and the inter-actions of serpins with their target enzymes are described by rates of reactions rather than strengths of interaction (Potempa *et al.*, 1994).

An entirely different mechanism of inhibition of proteinases is exhibited by α2-macroglobulin. It inhibits serine proteinases by engulfing the proteinase molecules after they have cleaved sensitive peptide bonds in the inhibitor molecule (Sottrup-Jensen, 1989). The interaction of hirudin, a leech protein, with its target enzyme, thrombin, is also a special case based on a peculiar feature of thrombin (a canyon-like extension of the active site, lined with several positive charges) and a comple-mentary feature of hirudin (a highly negatively charged, unstructured region) (Bode and Huber, 1992).

Inhibitors of Cysteine Proteinases

Proteinases that rely on a cysteine as the catalytic residue in the active site fall into two families: the papain family and the calpain family (Berti and Storer, 1995). To our knowledge, a given inhibitor will not inhibit enzymes from both families of cysteine proteinases.

Proteinaceous inhibitors of papain-related cysteine proteinases occur in several families (13, 16, 17, and 22 of Table 10.1). The best studied are members of the cystatin family. The interactions of cystatins and target enzymes are reversible, but there does not appear to be cleavage of a scissile bond. The three-dimensional struc-ture of a cystatin–enzyme complex determined by X-ray crystallography provides some understanding of why no bond is cleaved (Bode and Huber, 1992). The inter-action of the inhibitor is largely at the active site of the enzyme, but nonetheless does not mimic in detail the interaction of a substrate with a cysteine proteinase. Instead, two turns on the inhibitor (along with an N-terminal segment in some cystatins) form a wedge that binds in the active site cleft (Bode and Huber, 1992).

Non-cystatin inhibitors of papain-related cysteine proteinases are known. One occurs in the soybean trypsin inhibitor (Kunitz) family, but its mechanism of action is not known in detail (Krizaj *et al.*, 1993). A member of the serpin family is known to inhibit a cysteine proteinase (Komiyama *et al.*, 1994). α2-Macroglobulin inhibits cysteine proteinases in a trapping mechanism analogous to its inhibition of serine proteinases (Sottrup-Jensen, 1989). Calpain, an intracellular mammalian cysteine proteinase (Croall and Demartino, 1991) unrelated to the papain family of cysteine proteinases, is inhibited by a multidomain protein called calpastatin. The calcium-dependent interaction of calpain and inhibitory domains from calpastatin is not understood in detail, but calpastatin binding is not competitive with substrate binding and therefore is not substrate-like (Ma *et al.*, 1994).

Inhibitors of Aspartyl Proteinases

Relatively few proteins are known that inhibit pepsin-related proteinases, which rely on aspartyl residues in their active sites. However, low molecular weight inhibitors of the HIV protease, an aspartyl proteinase, have been studied extensively (Wlodawer and Erickson, 1993).

Aspartyl proteinase inhibitors are known in the Kunitz soybean trypsin inhibitor family (Ritonja *et al.*, 1990), and an *Ascaris* aspartyl proteinase inhibitor has been described (Martzen *et al.*, 1990). In neither case is the mechanism of inhibition well characterized.

Inhibitors of Metallo-proteinases

Metallo-proteinases include aminopeptidases, carboxypeptidases, and metallo-endoproteinases (collagenases), which fall into several protein families themselves. Among these, only the interaction of carboxypeptidase A with a 39-residue potato inhibitor (from family number 18) has been studied in detail. This inhibitor forms a substrate-like interaction with the enzyme (Rees and Lipscomb, 1982), in which the C-terminal amino acid residue of the inhibitor is hydrolyzed, but the products of hydrolysis are not released from the enzyme.

Insect Digestive Proteinases and Protein Digestion in Phytophagous Insects

Proteinaceous inhibitors expressed in transgenic plants and ingested through normal feeding by an insect will be exposed to a series of complex biochemical and physiological processes in the midgut region of the intestinal tract, a primary inter-face between phytophagous insects and their trophic environment. The degree to which the ingested inhibitor can interact with its target enzyme within this complex system will affect its usefulness as a pest control agent. In addition to information obtained from direct *in vitro* and *in vivo* studies of enzyme inhibitor interactions, a detailed knowledge of the overall digestive process itself, in a given pest insect species, will maximize the likelihood of a successful use of these inhibitors in a pest management program. The small size of many insects complicates the analysis of food polymer breakdown and nutrient utilization. Nevertheless, there has been excellent recent progress in our understanding of the biochemistry and physiology of many aspects of insect digestion. Excellent reviews include Applebaum (1985), Chapman (1985a,b), Dadd (1985), Slansky and Scriber (1985), Terra (1988, 1990), Terra and Ferreira (1994), and Turunen (1985).

Passage of Food through the Intestinal Tract

Ingestion of plant material by phytophagous insects initiates a series of physiologi-cal and biochemical events that result in the conversion of protein, carbohydrate, and lipid present in the food into utilizable nutrients. In most phytophagous insect species with chewing mouthparts, this conversion process typically occurs within the foregut and midgut regions of the intestinal tract and results from the action of enzymes originating from or present in the midgut.

After the initiation of feeding, salivary secretions are mixed with masticated plant tissue in the buccal cavity of most chewing insects (Chapman, 1985a). These secre-tions are thought to moisten the food bolus and facilitate its passage through the esophagus. There is little evidence for the presence of significant levels of proteinase activity in salivary secretions from orthopterous (Ferreira *et al.*, 1990) or lepidopter-ous (Lenz *et al.*, 1991; Santos *et al.*, 1983) species. Salivary glands are poorly devel-oped in coleopterans (Wigglesworth, 1972), and details about possible salivary proteinases in these species are lacking.

Ingested food material is subsequently passed through the esophageal region of the foregut into the midgut region. Because of numerous morphological modifications in the foregut and midgut regions of different insect taxa (Chapman, 1985a; Terra, 1990), details of physiological and biochemical aspects of protein digestion can vary widely between species. In species where the esophageal region of the foregut is modified into a crop or storage organ, proteinases passed forward from the midgut can initiate protein digestion (Baker *et al.*, 1984a; Ferreira *et al.*, 1990). Nevertheless, the midgut is the primary site of protein digestion in most insect species, and even in species where initial protein digestion can occur in the crop, intermediate and final stages of protein digestion occur in the midgut (Ferreira *et al.*, 1990).

Peristalsis drives undigested food material present in the midgut into the hindgut. Some orthopterous species have uniquely modified hindguts that contribute significantly to protein digestion (Thomas and Nation, 1984). However, in most species only low levels of proteinases are present in the hindgut region (Ferreira *et al.*, 1990; Santos *et al.*, 1983; Terra *et al.*, 1979). The presence of only low levels of proteinases in fecal material has been interpreted as evidence that a mechanism for the recirculation or recovery of digestive enzymes exists in the midgut (Terra *et al.*, 1979).

Compartmentalization of the Midgut

Ingested protein is sequentially hydrolyzed to oligomers, dimers, and monomers through a spatially arranged enzyme cascade in the insect midgut. The peritrophic membrane, a chitin–protein matrix in midguts of many insects (Chapman, 1985a; Richards and Richards, 1977), effectively partitions the midgut into endoperitrophic and ectoperitrophic regions (Terra and Ferreira, 1981, 1994). Typically, endoproteinases are secreted and pass through the peritrophic membrane into the endoperitrophic space and are responsible for the initial stages of protein hydrolysis. There is also evidence that some endoproteinase activity is immobilized in the peritrophic membrane of some lepidopterans (Ferreira *et al.*, 1994). As digestion proceeds, oligomers produced in the endoperitrophic space pass through the peritrophic membrane into the ectoperitrophic space where oligopeptide and dipeptide hydrolases continue the hydrolytic process. Final digestion occurs by peptidases present in the ectoperitrophic fluid, either bound to the midgut cell microvilli or present in midgut cells.

Enzyme localization also occurs along the length of the midgut. For example, different classes of endoproteinases can occur in anterior and posterior regions of the endoperitrophic space (Thie and Houseman, 1990). Exopeptidases are often differentially bound to epithelial cells located in the posterior midgut region (Baker *et al.*, 1984b).

Recirculation of Digestive Enzymes

Terra's group has used the models for countercurrent flows developed by Berridge (1970) and Dow (1981) to hypothesize an endo–ecto circulation of digestive enzymes in the insect midgut. In the proposed model (Terra, 1990), proteinases in the endoperitrophic space within the posterior region of the midgut pass through the peritrophic membrane into the ectoperitrophic space and move anteriorly in a countercurrent fashion, subsequently reentering the endoperitrophic space in the anterior region of the midgut. This recirculation, a mechanism that would prevent loss of enzymes as undigested food is passed into the hindgut and expelled, is thought to be driven by water secreted into the posterior region and absorbed in the anterior region of the midgut. There is no information on the ability of proteinase inhibitors themselves, or the complexes formed between proteinases and proteinase inhibitors, to pass through pores in the peritrophic membrane. The increased molecular size of the complex, or its configuration, may slow or prevent movement.

Occurrence of Multiple Proteinases in Phytophagous Insects

Terra and Ferreira (1994) have provided a comprehensive review of digestive proteinases and their properties in different insect taxa. In characterizing the complement of digestive enzymes, the most typical approach is to use inhibitors that are diagnostic for specific mechanistic classes of proteinases. Diisopropylfluorophosphate and phenylmethylsulfonylfluoride are taken as diagnostic inhibitors for serine proteinases, E-64 (L-trans-epoxysuccinylleucylamido-(4-guanidino)-butane) for cysteine proteinases, pepstatin for aspartyl proteinases and chelating agents for metallo-proteinases. Within a mechanistic class, specificities of the contributing proteinases are typically judged by relative activities against low molecular weight substrates that are preferentially cleaved by mammalian proteinases of known specificities.

Based on numerous studies, we can conclude that multiple proteinases with different cleavage site specificities are present in midguts of most phytophagous species (Wolfson and Murdock, 1990a). In the alkaline midguts of lepidopterans, serine proteinases are common and most species examined have enzymes of more than one specificity, for instance, trypsin-like, chymotrypsin-like or elastase-like proteinases (Ahmad *et al.*, 1980; Christeller *et al.*, 1992; Houseman and Chin, 1995; Houseman *et al.*, 1989; Johnston *et al.*, 1995; Rymerson and Bodnaryk, 1995). In coleopterans, midgut pH can range from slightly acidic to very alkaline and the predominant proteinase classes can vary greatly among species. Some species, in particular bruchids, have active cysteine proteinases (Murdock *et al.*, 1987) that exist in multiple forms (Gillikin *et al.*, 1992; Michaud *et al.*, 1995) or that are present in combination with aspartyl proteinases (Blanco-Labra *et al.*, 1996; Houseman and Thie, 1993; Silva and Xavier-Filho, 1991). Many other coleopterans (cerambycids, scarabs, dermestids, elaterids, carabids, tenebrionids, and curculionids) depend on a complement of serine proteinases for protein digestion (Baker, 1982; Bian *et al.*, 1996; Chen *et al.*, 1992; Christeller *et al.*, 1989; Levinsky *et al.*, 1977).

Although a diverse complement of proteinases is typically present in most insect midguts, it is generally not known if the individual enzymes are components of a common secretion, if they are secreted independently, or how the control of secretion of a given enzyme is regulated. Lehane *et al.* (1995) and Terra and Ferreira (1994) have reviewed current hypotheses concerning the secretory process and the control mechanisms thought to regulate secretion. More information about the control of proteinase gene expression and secretion in phytophagous species is needed, especially since there is some evidence that the proteinase complement in insects can be altered by ingestion of dietary proteinase inhibitors (Bolter and Jongsma, 1995; Broadway, 1995; Jongsma *et al.*, 1995; see below).

Gaps in Our Knowledge of Inhibition of Insect Digestion

When a phytophagous insect ingests plant material containing an inhibitory protein targeted to a digestive proteinase, the effective binding of the inhibitor to the enzyme can be influenced by biochemical factors in the intestinal tract including:

(1)   solubility of inhibitor in the endoperitrophic fluid;

(2)   pH, ionic strength, and redox potential of the endoperitrophic fluid as well as temperature;

(3)   susceptibility of the inhibitor to proteolytic cleavage by non-target proteinases;

(4)   selectivity of the inhibitor and competition with other substrates for the target proteinase;

(5)   presence of inhibitor binding proteins in the diet or endoperitrophic fluid; and

(6)   non-specific interactions of the inhibitor with other proteins such as phenoloxidases, which can catalyze the cross-linking of proteins and interfere with inhibitor–proteinase interactions (Constabel *et al.*, 1995).

Physiological factors within an insect species that also could influence the effectiveness of the inhibitors include:

(1)   changes in feeding rates of the insect, i.e. change in the rate of ingestion of the inhibitor;

(2)   rate of passage of food bolus through the intestinal tract;

(3)   secretory rate of target proteinase;

(4)   permeability of the peritrophic membrane to the inhibitor as well as the inhibitor–enzyme complex; and

(5)   compensatory responses of the insect to adjust both feeding rates and the complement of proteinases within the gut.

Answers to most of these questions, with respect to major pest insect–crop associations, are lacking. They would certainly be helpful in assessing the potential success or failure of pest resistance factors such as proteinase inhibitors in transgenic plants.

Molecular Genetics of Insect Digestive Proteinases

Despite the power of recombinant DNA methods and their suitability for investigating insect proteins, the number of cloned DNAs that encode digestive enzymes of insects is still quite small. In the sequence database of the National Center for Biotechnology Information, there are cloned digestive proteinase sequences from only two species of phytophagous insects. Both of these are lepidopterans, *Manduca sexta* (tobacco hornworm), from which three trypsin-like and one chymotrypsin-like cDNA sequences have been cloned (Peterson *et al.*, 1994, 1995), and *Choristoneura fumiferana* (spruce budworm), from which a trypsin-like enzyme's cDNA has been isolated (Wang *et al.*, 1993). A serine proteinase's sequence has also been determined by Edman degradation of the purified protein from midgut of *Bombyx mori* (Sasaki *et al.*, 1993). These serine proteinase sequences contain a large number of arginine residues and a relatively low number of lysine residues. The arginine residues remain charged at high pH, which may contribute to the extreme stability of the proteinases in the alkaline lepidopteran midgut.

The only other insect order from which cloned proteinase gene sequences are available is Diptera. There has been some interest in studying proteinases from blood-feeding flies, which must digest large amounts of protein taken in sporadic, but large meals. Serine proteinase sequences have been obtained from two mosquito species, *Aedes aegypti* (Barrillas-Mury *et al.*, 1991; Noriega *et al.*, 1996) and *Anopheles gambiae* (Muller *et al.*, 1993), a blackfly species, *Simulium vittatum* (Ramos *et al.*, 1993) and the hornfly, *Haematobia irritans* (Elvin *et al.*, 1993). All of these species appear to contain families of serine proteinase genes. Perhaps having many serine proteinase genes is an adaptation for rapidly induced expression of a large amount of proteinase upon blood-feeding. The trypsin-like activity of the mosquito *Aedes aegypti* (and presumably the other blood-feeding flies as well) is susceptible to inhibition by proteinase inhibitors (serpins and α2-macroglobulin) which occur at high concentration in mammalian plasma (Barillas-Mury and Kanost, unpublished data). Thus, these blood-feeding insects can face a problem similar to plant-feeding species, overcoming proteinase inhibitors present in their food. The mosquitoes may produce an excess of a proteinase, such that after titration of the plasma inhibitors, the remaining active proteinase is able to digest the meal.

Serine proteinase encoding gene clones have also been isolated from the flesh flies, *Lucilia cuprina* and *Hypoderma lineatum* (Casu *et al.*, 1994; Moire *et al.*, 1994), and the fruit fly, *Drosophila melanogaster* (Davis *et al.*, 1985). These three species also contain families of serine proteinase genes.

All of these insect serine proteinase sequences contain typical hydrophobic signaling (secretion) sequences, followed by sequences of 7–40 amino acid residues prior to the sequence of the amino-terminus of the mature enzyme. The sequences contain an Arg residue at the putative activation cleavage site. Although the existence of inactive forms of digestive enzymes is not well documented in insects (Applebaum, 1985), the nucleotide sequences suggest that, like mammalian serine proteinases, at least some insect digestive enzymes are produced as zymogens, and are activated by proteolytic (perhaps tryptic) removal of an amino-terminal peptide.

A putative cysteine proteinase encoding genomic DNA sequence from *D. melanogaster*, apparently expressed in the midgut, has been determined (Matsumoto *et al.*, 1995). The sequence does not appear to contain a translation initiation site and no corresponding protein has been isolated. Therefore, the biological significance of this gene is as yet unclear.

Although molecular genetic data are not yet available to document their existence directly, it is clear from biochemical studies described above that, within the insect class, there are genes encoding serine, cysteine, aspartyl, and metallo-proteinases. Genes for all of these classes of proteinases might well exist in individual insect species. The relative levels of expression of the genes would determine the complement of enzymes that one studies under any physiological condition.

In vitro Inhibition of Insect Proteinases by Proteinaceous Inhibitors

Because only a few insect proteinases have been purified to homogeneity (Terra and Ferreira, 1994), virtually all studies of the effects of inhibitors on insect proteinases have been carried out on extracts or partially purified enzymes. This fact greatly limits the conclusions that can be drawn from these studies. In this section, we will first examine studies that focused principally on canonical inhibitors of serine proteinases, giving special emphasis to the revealing work of Christeller and his colleagues. We then describe studies that use cystatins, mainly from plants.

Canonical Inhibitors of Serine Proteinases

The earliest work on inhibition of insect proteinases by proteinaceous inhibitors was by Fraenkel, Birk, Applebaum, and colleagues, who examined the effects of soybean inhibitor preparations on proteolytic activity in extracts from the yellow mealworm, *Tenebrio molitor*, and confused flour beetle, *Tribolium castaneum* (Birk and Applebaum, 1960; Birk *et al.*, 1962; Lipke *et al.*, 1954). They found that a substantial portion of the caseinolytic activity in each species was inhibited by those soybean inhibitor preparations, which represent what are now called the Kunitz soybean trypsin inhibitor and Bowman–Birk inhibitor families. Birk and her colleagues also documented the inhibition of a partially purified trypsin from *T. molitor* and of purified trypsin-like and chymotrypsin-like enzymes from the African migratory locust, *Locusta migratoria*, by soybean inhibitors (Levinsky *et al.*, 1977; Sakal *et al.*, 1988, 1989). This important work verified that canonical inhibitors that had been well characterized by their interactions with mammalian proteinases appear to interact comparably with digestive proteinases from insects. Canonical inhibitors, including the soybean inhibitors, ovomucoid, and bovine pancreatic trypsin inhibitor (aprotinin), have been used by numerous other investigators to inhibit proteolytic activity in extracts of various insect species (e.g., see Baker, 1982; Lee and Anstee, 1995; Purcell *et al.*, 1992; Thie and Houseman, 1990; Wolfson and Murdock, 1990a). Perhaps the best way to summarize this body of work is to say that, on a case-by-case basis, canonical inhibitors of mammalian serine proteinases can be effective as inhibitors of insect proteolytic activity. Of course, the activity that is inhibited will be limited to the serine proteinase portion of a particular insect species' repertoire of digestive enzymes.

Christeller and his collaborators have published a series of papers that provide several conclusions important to future work on creating transgenic plants for increased resistance to insect pests. Their work has been notable in that they examined the effects of as many as 26 different inhibitors (Christeller and Shaw, 1989; Christeller *et al.*, 1989, 1992, 1994a; McGhie *et al.*, 1995), including eglin C, $\alpha 2$-macroglobulin and a serpin, none of which had been exploited previously by other workers. A striking result of Christeller and Shaw (1989) is that strong inhibitors of bovine trypsin vary widely in their strengths of inhibition (over a 100 000-fold range) of the trypsin-like enzyme from larvae of the grass grub (*Costelytra zealandica*). Thus, a protein that inhibits bovine trypsin effectively might or might not be a strong inhibitor of a particular insect trypsin-like enzyme. A similar conclusion can be drawn from the study of Houseman *et al.* (1989), who found that trypsin-like enzymes in the European corn borer, *Ostrinia nubilalis*, were not inhibited by ovomucoid, a strong inhibitor of bovine trypsin.

Christeller's use of eglin C, an elastase inhibitor in the potato I family, established the importance of elastase-like proteinases in the digestive system of numerous lepidopteran species (Christeller *et al.*, 1992). They have recently explored the interaction of eglin C with partially purified elastase-like enzymes from *Cydia pomonella* and *Helicoverpa punctigera* (Christeller *et al.*, 1994a).

The investigations of Christeller and his collaborators emphasize the importance of examining many inhibitors to find those that strongly inhibit a selected insect proteinase, rather than simply assuming, for instance, that an inhibitor of bovine trypsin will inhibit an insect's trypsin-like enzyme. Another important insight from their work is that inhibitors of elastase should be examined as potentially useful agents for adversely affecting insect growth and development.

Cystatins

In contrast to their levels of serine proteinase inhibitors, plants contain quite low levels of cystatins. Thus, in examining proteolytic activity in insect extracts, many studies have relied on the commercially available peptide derivative E-64 for class-specific inhibition of cysteine proteinases. The availability in the last several years of purified cystatins from plants (e.g., Abe *et al.*, 1987; Liang *et al.*, 1991) and high-level expression of recombinant forms of cystatin (Abe *et al.*, 1992; Chen *et al.*, 1992) has allowed examination of the effects of plant cystatins on proteolytic activity in insects.

Since many coleopteran species have cysteine proteinases as the major component of their digestive proteinases (Chen *et al.*, 1992; Liang *et al.*, 1991; Purcell *et al.*, 1992; Wolfson and Murdock, 1990a), it is not surprising that cystatins are quite effective in inhibiting a substantial portion of the proteolytic activity in several coleopteran species that have been studied. We demonstrated the effectiveness of cystatin purified from rice seeds (Liang *et al.*, 1991) and a recombinant form of oryzacystatin I (Chen *et al.*, 1992) in inhibiting most of the caseinolytic activity in the rice weevil, red flour beetle, and cadelle beetle. On the other hand, less than half the caseinolytic activity in gut extracts from the dark mealworm and yellow mealworm, as well as from the Indianmeal moth, was inhibited by recombinant oryzacystatin. Yelle and his colleagues have demonstrated the efficacy of two rice cystatins in inhibiting cysteine proteinase activity from midgut extracts of the Colorado potato beetle, *Leptinotarsa decemlineata* (Michaud *et al.*, 1993, 1995). Gillikin *et al.* (1992) and Orr *et al.* (1994) have demonstrated that cystatins can inhibit a high percentage of the proteolytic activity from gut extracts of two species of *Diabrotica*, the southern and western corn rootworms. The work of Orr *et al.* (1994) is particularly notable in that those authors have used as an inhibitor the potato 'multicystatin', which consists of eight cystatin domains contained in a continuous polypeptide chain.

Bioassays using Proteinase Inhibitors

Knowledge of the effects of specific proteinase inhibitors on the physiological processes and behavioral responses of insects is needed to advance our understanding of insect–plant relationships. Such information will be required in the development of resistant plant cultivars with proteinase inhibitors acting as the defense mechanism. There are many reports in the literature about bioassays of proteinase inhibitors and their effects on insect growth and development. The potential for a proteinase inhibitor to be active against an insect depends primarily on the quantity and classes of proteinases present in the insect's midgut, and the affinity of the inhibitor for a proteinase that is critical for the insect's growth and development. Many of the bioassay studies have concentrated on the serine and cysteine protein-ases with limited reports on the presence of aspartyl proteinases (see Oppert *et al.*, 1993 and references therein).

Bioassays might underestimate the potency that an inhibitor could have in tissues of living plants. Many artificial diets are of very high nutritional quality and optimized for growth such that an individual inhibitor might have little effect on insect development when added as a supplement to such a diet. In a living tissue, in which stresses, including plant defense mechanisms, are operating on an insect, the additional stress imposed by small amounts of an individual inhibitor could have substantial effects. Artificial diets with limited protein quality are thus preferred for bioassays of proteinase inhibitors.

For best results in determining which kinds of proteinase inhibitor would be effective against a target insect when expressed in transgenic plants, the gut protein-ases of that target insect should be characterized and inhibitors selective for those enzymes should be identified. In practice, however, random or shotgun screening of inhibitors has been a more common approach to identifying proteinase inhibitors that adversely affect insect populations. The effects of specific proteinase inhibitors on selected insects have been determined by their *in vitro* activity on gut homoge-nates or by measuring their effects on growth, development and mortality using artificial diets supplemented with proteinase inhibitors (Murdock *et al.*, 1987; Oppert *et al.*, 1993, and references therein). The first of many of these reports was published in 1954 when it was observed that crude preparations of soybean trypsin inhibitors in diets inhibited growth of *Tribolium confusum* larvae (Lipke *et al.*, 1954). Limited *in vivo* bioassays, conducted by applying inhibitors to an insect's host plant, have also been conducted to substantiate reported *in vitro* biochemical assays and to provide a more valid assessment of the effectiveness of specific inhibitors (Wolfson and Murdock, 1987). Several *in vivo* bioassays have been developed to screen proteinase inhibitors for their ability to inhibit growth and development in insects and to determine which classes of inhibitors are active. Effects on larval growth, foliar feeding, pupation, adult emergence, and mortality have been reported. For example, insects susceptible to the effects of cysteine proteinase inhibitors in their diets include *Hypera postica* (Elden, 1995), *Leptinotarsa decemlineata* (Michaud *et al.*, 1995), *Acanthoscelides obtectus* (Hines *et al.*, 1990), and *Diabrotica unde-cimpunctata* (Edmonds *et al.*, 1996). Insects susceptible to serine proteinase inhibi-tors include *Melanoplus sanguinipes* (Hinks and Hupka, 1995), *Spodoptera litura* (McManus and Burgess, 1995), *Cydia pomonella* (Markwick *et al.*, 1995), *Apis mellif-era* (Malone *et al.*, 1995), *Costelytra zealandica* (Dymock *et al.*, 1992), *Helicoverpa armigera* (Johnston *et al.*, 1993) and *Teleogryllus commodus* (Burgess *et al.*, 1991). Mixtures of inhibitors have also been tested, and combinations of cysteine and serine proteinase inhibitors in diets can be toxic to insects at levels where individual inhibitors are non-toxic (Oppert *et al.*, 1993).

The mechanism by which proteinase inhibitors slow growth and cause mortality probably involves creating a deficiency in one or more amino acids essential for development. Dietary supplements of either a single amino acid or a mixture of amino acids rescued several species of insects from the detrimental effects of protein-ase inhibitors in their diets (Broadway and Duffey, 1986; Hines *et al.*, 1990; Oppert *et al.*, 1993). Another possible mode of action of proteinase inhibitors is to cause hyperproduction of digestive proteinases (to titrate the proteinase inhibitors). This, together with insufficient dietary availability of amino acids needed for proteinase synthesis, can cause growth inhibition (Broadway and Duffey, 1986).

Correlation of Proteinase Inhibitors with Host-plant Resistance and Interactions with other Proteinaceous Resistance Factors

Attempts to correlate levels of proteinase inhibitors found in different cultivars and the resistance/susceptibility to insect predation have been met with mixed success. Measurements of trypsin inhibitor content of seeds of six cowpea cultivars indicated that there is no correlation between trypsin inhibitor content and resistance to *Callosobruchus maculatus* (Gatehouse and Boulter, 1983; Xavier-Filho *et al.*, 1989). Correlation analysis between proteinase inhibitor levels or amylase inhibitor levels in 20 cowpea lines and extent of insect attack showed that neither type of inhibitor separately explained the resistance to the weevil (Piergiovanni *et al.*, 1991). Interestingly, however, bruchid-resistant lines had high levels of both inhibitors, whereas lines with low levels of trypsin and amylase inhibitors were bruchid susceptible. In other words, there was a correlation of resistance to *combined* trypsin inhibitor and amylase inhibitor levels. This result indicates that combinations of different types of digestive enzyme inhibitors may be effective for resistance against this insect pest. Because it is likely that the bruchid beetle has cysteine proteinases as one or more of its major digestive enzymes, the cowpea lines should be reexamined for cysteine proteinase inhibitors to determine whether a correlation exists between cysteine proteinase inhibitor levels and resistance.

Proteinase inhibitors might also affect the activity of other proteinaceous resistance factors in transgenic plants. For example, proteinase inhibitors could prevent the activation of the protoxin from *Bacillus thuringiensis* or inhibit toxin degradation. There have been conflicting reports about the effects of proteinase inhibitors on the efficacy of the insecticidal endotoxin. A serine proteinase inhibitor gene fused to a truncated *B. thuringiensis* toxin gene in transgenic tobacco was reported to enhance the activity of the endotoxin against lepidopteran species (MacIntosh *et al.*, 1990). However, serine proteinase inhibitors added to semi-artificial diets failed to synergize *B. thuringiensis* endotoxins toward *Plutella xylostella* (Tabashnik *et al.*, 1992). More work is needed to determine whether and how proteinase inhibitors influence the activity of *B. thuringiensis* endotoxins and other proteinaceous resistance factors toward insects.

Induced Proteinase Inhibitors and Host-plant Resistance

The effects of proteinase inhibitors induced by wounding host plants on insect development have also been investigated (Wolfson, 1991; Wolfson and Murdock, 1990b). Although other factors besides inhibitor levels in plants might also have been affected by wounding, insect growth rates were decreased by wounding, indicating that the induced proteinase inhibitors were suppressing growth, even though comparable concentrations of proteinase inhibitors did not affect larval growth rates when tested in artificial diets. For example, there is a significant inverse correlation between the concentration of wound-inducible proteinase inhibitors in tomato and cabbage foliage and the growth of lepidopterans (Broadway and Colvin, 1992;

Broadway *et al.*, 1986; Johnson *et al.*, 1989). However, these results are only corre-lational and there is no direct evidence of causation.

Insect Adaptation or Resistance to Proteinase Inhibitors

Several insect species have been reported to adapt to proteinase inhibitors by alter-ing their complement of secreted proteinases. For example, *Spodotera exigua* appar-ently adapted to high levels of serine proteinase inhibitors in tobacco leaves by induction of gut proteinases that were not affected by the inhibitors (Jongsma *et al.*, 1995), and larvae of *Leptinotarsa decemlineata* apparently compensated for inhibited gut proteolytic activity during chronic intake of cysteine proteinase inhibitors by secreting insensitive proteinase(s) (Bolter and Jongsma, 1995). The relative propor-tion of digestive proteinases can be shifted in response to ingestion of proteinase inhibitors such that insensitive enzymes are predominant in the complement of midgut proteinases (Broadway, 1995). While these studies provide evidence for a rapid compensatory response to the effects of ingested proteinase inhibitors in several phytophagous species, additional studies are needed to determine the pos-sible effects of sensitive proteinases that might be overproduced with only a portion becoming bound to proteinase inhibitors in the diet.

Another factor, host range, appears to influence the susceptibility of herbivorous insects to non-host plant proteinase inhibitors in that polyphagous insects that feed on numerous species of host plants are better adapted than more specialized feeders to tolerate non-host proteinase inhibitors (Broadway and Villani, 1995). Overall, these studies indicate the necessity for additional knowledge of the physiological and biochemical responses of insects to their dietary components. These responses will have to be understood and accounted for in strategies developed to delay adap-tation by pest insects to specific proteinase inhibitors.

Plants Transformed with Proteinase Inhibitor Genes

A good deal of work, much of it quite elegant, is currently underway with transgenic plants to probe the nature of the wound inducibility of members of the potato inhibitor II family in tomato, potato, and tobacco (e.g., see McGurl *et al.*, 1994; Keil *et al.*, 1990). There have also been several studies aimed at creating transgenic plants with increased resistance to insect pests. Early papers from two groups served to establish the legitimacy of this line of investigation. The pioneering study of Hilder *et al.* (1987) generated a good deal of excitement. Those workers transformed tobacco with a gene that encoded a cowpea trypsin inhibitor (in the Bowman–Birk family) and reported that the transgenic plants had heightened resistance to feeding by the tobacco budworm, *Heliothis virescens*. The same transformants were shown in a subsequent study to produce increased mortality to the corn earworm, *Helico-verpa zea*. In a seminal paper, Ryan and his co-workers transformed tobacco with genes encoding either the potato or tomato proteinase inhibitor II and observed a reduced growth rate of the tobacco hornworm feeding on the transgenic plants (Johnson *et al.*, 1989). In addition, Ryan's group has inhibited the wound-responsive defense system by transforming tomato plants with an antisense prosystemin gene (systemin is a wound-inducible peptide that facilitates the expression of proteinase inhibitors). The transgenic plants allowed the growth of *M. sexta* larvae (Orozco-

Cardenas *et al.*, 1993). This work provides particularly strong evidence for the role of wound-inducible proteinase inhibitors in resistance to insects.

Subsequent to the formative papers of Hilder *et al.* (1987) and Johnson *et al.* (1989), there have been several reports of the creation of transgenic plants with genes for foreign proteinase inhibitors. Carbonero *et al.* (1992) made a preliminary report of plants transformed by a barley trypsin inhibitor (barley trypsin inhibitor family), which increased the mortalities of the black cutworm, *Agrotis ipsilon* and the armyworm, *Spodoptera littoralis*. McManus *et al.* (1994) transformed tobacco with potato inhibitor II, which inhibits chymotrypsin. Larvae of *Chrysodeixis eriosma*, the green looper, grew more slowly on leaf tissue from the transgenic plants than from non-transgenic plants, whereas no differences were observed in the growth rates of *Spodoptera litura* or *Thysanoplusia orichalcea* larvae fed on leaves from transgenic or non-transgenic plants. The work of McManus and co-workers thus makes the important point that increasing resistance to a particular insect pest may require a customized transgenic system utilizing a foreign gene that encodes an inhibitor effective against the digestive enzymes of the insect in question. Recently, we transformed tobacco with a gene encoding the 13 kDa bifunctional corn inhibitor, which inhibits both trypsin-like serine proteinases and certain insect amylases (Masoud *et al.*, 1996), and observed both activities of the expressed inhibitor. The use of bifunctional inhibitors might, for an appropriate insect, provide a more potent effect than use of a monofunctional inhibitor alone. It is noteworthy that several proteinase inhibitors also inhibit insect amylases (Richardson, 1991).

More recently, plants other than tobacco have been transformed with genes that encode inhibitors of serine proteinases. Newell *et al.* (1995) created transgenic sweet potato containing the gene for cowpea trypsin inhibitor, but no studies were reported on the susceptibility of plant tissues to attack by insects. Thomas *et al.* (1994) introduced a gene for an insect serpin into alfalfa and observed a reductive effect on thrip predation. Klopfenstein *et al.* (1993) have taken this area of investigation to new heights (!) by transforming trees (*Populus* hybrids) with the gene for potato proteinase inhibitor II. Studies of the effects on insect attack on the transgenic trees are planned. Several transgenic systems are being explored in rice (Rockefeller Foundation, 1994), but these studies have not yet appeared in the peer-reviewed literature. Among these is the work of Zhao *et al.* (1996), who expressed two insecticidal genes in the same tobacco plant. One encoded cowpea trypsin inhibitor and the other encoded a toxin gene from *B. thuringiensis*. Bioassays showed that plants that contained both genes had enhanced toxicity to *Helicoverpa armigera* larvae compared to plants that contained either of the individual genes. These data support the hypothesis originally proposed by Macintosh *et al.* (1990) that proteinase inhibitors can synergize the insecticidal activity of *B. thuringiensis* toxins.

Recent work of Thomas *et al.* (1995a,b) illustrates the important point that we need not restrict our attention to genes that encode canonical inhibitors, or to inhibitors that in their natural context inhibit digestive enzymes, or to wild-type inhibitors. In their work, Thomas and co-workers transformed tobacco and cotton plants with a gene that encodes a serpin from the tobacco hornworm. This serpin is normally expressed in the insect's hemolymph. The wild-type inhibitor is an inhibitor of elastase. Mutant forms of the inhibitor that inhibited trypsin or chymotrypsin were created by site-directed mutagenesis and also used for creating transgenic plants. Plants expressing the serpins were found to suppress the reproduction of the sweetpotato whitefly type B (*Bemisia tabaci*).

Because of the importance of cysteine proteinases in digestion in many insect species, it is not surprising that a spate of papers has appeared in which plants are transformed by genes that inhibit cysteine proteinases (Benchekroun *et al.*, 1995; Hosoyama *et al.*, 1994; Leple *et al.*, 1995; Masoud *et al.*, 1993; Urwin *et al.*, 1995). These studies were in potato, rice, poplars, tobacco, and tomato, respectively. Only in two of the cases was tissue from transformed plants tested for its ability to support the growth of a pest. Urwin *et al.* (1995) tested the growth of two nematodes, *Caenorhabditis elegans*, which is free living, and *Gobodera pallida*, which is a plant parasite, on roots of transformed tomato plants. This group worked with plants transformed with genes that encoded either wild-type oryzacystatin-I or oryzacystatin variants created by site-directed mutagenesis. The most profound detrimental effect was seen on the growth and fecundity of *G. pallida* in plants transformed with a mutant form of oryzacystatin-I. Leple *et al.* (1995) demonstrated the toxicity toward larvae of the poplar leaf beetle *Chrysomela tremulae* of leaves of poplars transformed with a gene for oryzacystatin I.

Summary and Prospects

After an initial burst of enthusiasm and optimism, we have grown more realistic and sophisticated as a scientific community in the past decade in thinking about the prospects for using proteinase inhibitor genes to engineer increased plant resistance to insect attack. We have realized, for instance, that an individual proteinase inhibitor is not going to be universally useful as an insect growth retardant. Instead, transgenic systems must be customized: an inhibitor should be selected based on *in vitro* measurements and, if possible, bioassays employing the insect that is targeted for control. The ability of individual insects, for at least some species, to adapt physiologically to the presence of inhibitors in their diets indicates that inhibitors may simply impose a stress on insects. To have a strong effect on insect growth and development, this stress, which may be relatively minor in individual cases, should probably be applied in combination with stresses from products of other transgenes. One useful strategy of combined transgenes may be to use genes that encode inhibitors of more than one mechanistic class of proteinase. One might use inhibitors of some of the initial complement of proteinases along with inhibitors of the proteinases that arise adaptively due to the presence of inhibitors of the initial proteinases. Finally, the development of resistance within an insect population to the stresses imposed by transgenes for proteinase inhibitors will clearly have to be studied and understood. As genes for effective inhibitors are transferred into crop plants, probably in combination with other genes, we need to grow increasingly sophisticated in the way these genes are expressed. Tissue-specific expression or wound-inducible expression will obviously be highly desirable in some applications.

In searching for particularly effective inhibitors, we need to expand our horizons beyond trypsin inhibitors, which were the focus of most of the early work. Eglin C and cystatins are already receiving attention, but we can certainly think beyond them to proteins like ecotin and α2-macroglobulin, which have broader selectivities. Genes that encode inhibitory fragments of monoclonal antibodies directed toward insect proteinases should be explored. Mutant forms of inhibitors should be created in searches for inhibitors with greater affinity or selectivity. A particularly promising approach is the phage-display of libraries of mutant inhibitors coupled with selection of a variant with desired properties (Roberts *et al.*, 1992).

As such approaches are pursued, we will certainly rely heavily on recombinant DNA methods for studies of both inhibitors and insect proteinases. Cloning of genes or cDNAs for inhibitors is useful for purposes of expression (as a source of material for bioassays) and is essential for creating mutant inhibitors. By our count, approximately 40 different proteinase inhibitors from 10 protein families have been successfully expressed in microbial systems (bacteria or yeast). Thus, it appears that successful microbial expression of an inhibitor can be expected to be straightforward.

Purified proteinases are needed to study the effect of individual inhibitors as potential insect growth retardants, and purified insect proteinases are essential in order to create and select mutant inhibitors with desired binding characteristics. Unfortunately, insect digestive enzymes have been difficult to purify by conventional biochemical procedures, and this has undoubtedly slowed progress in identifying appropriate inhibitors for use in insect control. Recombinant DNA methods should, however, provide a powerful alternative to conventional purification methods as more genes for insect proteinases are cloned and expressed. The expression of active proteinases has proven difficult in bacteria, and eukaryotic systems, such as baculovirus expression systems, may well be needed for the successful expression of active insect proteinases.

We believe that there is great promise in the use of genes that encode proteinase inhibitors for insect pest control in transgenic plants. Inhibitors should, however, be selected carefully, with consideration of the physiology and biochemistry of protein digestion in the insect to be controlled. In addition, combinations of inhibitors with other stress inducing or growth suppressing proteins should be exploited.

References

Abe, K., Kondo, H. and Arai, S. (1987) Purification and characterization of a rice cysteine proteinase inhibitor, *Agric. Biol. Chem.* **51**, 2763–2768.

Abe, M., Abe, K., Kuroda, M. and Arai, S. (1992) Corn kernel cysteine proteinase inhibitor as a novel cystatin superfamily member of plant origin: molecular cloning and expression studies, *Eur. J. Biochem.* **209**, 933–937.

Ahmad, Z., Saleemuddin, M. and Siddi, M. (1980) Purification and characterization of three alkaline proteases from the gut of the larva of army worm, *Spodoptera litura, Insect Biochem.* **10**, 667–673.

Applebaum, S. (1985) Biochemistry of digestion, In: Kerkut, G. A. (Ed.), *Comprehensive Insect Physiology, Biochemistry and Pharmacology*, Vol. 4, New York: Pergamon Press, pp. 279–311.

Atkinson, A. H., Heat, R. L., Simpson, R. J., Clarke, A. E. and Anderson, M. A. (1993) Proteinase inhibitors in *Nicotiana alata* stigmas are derived from a precursor protein which is processed into five homologous inhibitors, *Plant Cell* **5**, 203–213.

Baker, J. (1982) Digestive proteinases of *Sitophilus* weevils (Coleoptera: Curculionidae) and their response to inhibitors from wheat and corn flour, *Can. J. Zool.* **60**, 3206–3214.

Baker, J., Woo, S. and Byrd, R. (1984a) Ultrastructural features of the gut of *Sitophilus granarius* (L.) (Coleoptera: Curculionidae) with notes on distribution of proteinases and amylases in crop and midgut, *Can. J. Zool.* **62**, 1251–1259.

Baker, J., Woo, S. and Mullen, M. (1984b) Distribution of proteinases and carbohydrases in the midgut of larvae of the sweetpotato weevil *Cylas formicarius elegantulus* and response of proteinases to inhibitors from sweet potato, *Entomol. Exper. Appl.* **36**, 97–105.

Barillas-Mury, C., Graf, R., Hagedorn, H. and Wells, M. (1991) cDNA and deduced amino acid sequence of a blood meal-induced trypsin from the mosquito, *Aedes aegypti, Insect Biochem.* **21**, 825–831.

Benchekroun, A., Michaud, D., Nguyen-Quoc, B., Overney, S., Desjardins, Y. and Yelle, S. (1995) Synthesis of active oryzacystatin I in transgenic potato plants, *Plant Cell Rep.* **14**, 585–588.

Berridge, M. (1970) A structural analysis of intestinal absorption, In: Neville, A. E. (Ed.), *Insect Ultrastructure, Symp. R. Ent. Soc. Lond.* **5**, 135–151.

Berti, P. and Storer, A. (1995) Alignment/phylogeny of the papain superfamily of cysteine proteases, *J. Mol. Biol.* **246**, 273–283.

Bian, S., Shaw, B. D., Han, Y. and Christeller, J. T. (1996) Midgut proteinase activities in larvae of *Anoplophora glabripennis* (Coleoptera: Cerambycidae) and their interaction with proteinase inhibitors, *Arch. Insect Biochem. Physiol.* **31**, 23–37.

Bigler, T., Lu, W., Park, S., Tashiro, M., Wieczorek, M., Wynn, R., Birk, Y. and Applebaum, S. (1960) Effect of soybean trypsin inhibitors on the development and midgut proteolytic activity of *Tribolium castaneum* larvae, *Enzymologia*, **22**, 318–326.

Bigler, T., Lu, W., Park, S., Tashiro, M., Wieczorek, M., Wynn, R. and Laskowski, M. (1993) Binding of amino acid side chains to preformed cavities: interaction of serine proteinases with turkey ovomucoid third domains with coded and noncoded P1 residues, *Protein Sci.* **2**, 786–799.

Birk, Y. and Applebaum, S. (1960) Effect of soybean trypsin inhibitors on the development and midgut proteolytic activity of *Tribolium castaneum* larvae, *Enzymologia*, **22**, 318–326.

Birk, Y., Harpaz, I., Ishaaya, I. and Bondi., A. (1962) Studies on the proteolytic activity of the beetles *Tenebrio* and *Tribolium, J. Insect Physiol.* **8**, 417–429.

Blanco-Labra, A., Martinez-Gallardo, N. A., Sandoval-Cardoso, L. and Delano-Frier, J. (1996) Purification and characterization of a digestive cathepsin D proteinase isolated from *Tribolium castaneum* larvae (Herbst), *Insect Biochem. Mol. Biol.* **26**, 95–100.

Bode, W. and Huber, R. (1992) Natural protein proteinase inhibitors and their interaction with proteinases, *Eur. J. Biochem.* **204**, 433–451.

Boigergrain, R., Mattras, H., Brehelin, M., Paroutaud, P. and Coletti-Previere, M. (1992) Insect immunity: two proteinase inhibitors from hemolymph of *Locusta migratoria, Biochem. Biophys. Res. Commun.* **189**, 790–793.

Bolter, C. and Jongsma, M. (1995) Colorado potato beetles (*Leptinotarsa decemlineata*) adapt to proteinase inhibitors induced in potato leaves by methyl jasmonate, *J. Insect Physiol.* **41**, 1071–1078.

Broadway, R. (1995) Are insects resistant to plant proteinase inhibitors? *J. Insect Physiol.* **41**, 107–116.

Broadway, R. and Colvin, A. (1992) Influence of cabbage proteinase inhibitors *in situ* on the growth of larval *Trichoplusia ni* and *Pieris rapae, J. Chem. Ecol.* **18**, 1009–1024.

Broadway, R. and Duffey, S. (1986) Plant proteinase inhibitors: mechanism of action and effect on the growth and digestive physiology of larval *Heliothis zea* and *Spodoptera exigua, J. Insect Physiol.* **32**, 827–833.

Broadway, R. and Villani, M. (1995) Does host range influence susceptibility of herbivorous insects to non-host plant proteinase inhibitors? *Entomol. Exper. Appl.* **76**, 303–312.

Broadway, R., Duffey, S., Pearce, G. and Ryan, C. (1986) Plant proteinase inhibitors: a defense against herbivorous insects? *Entomol. Exper. Appl.* **41**, 33–38.

Burgess, E., Stevens, P., Keen, G., Laing, W. and Christeller, J. (1991) Effects of protease inhibitors and dietary protein level on the black field cricket *Teleogryllus commodus, Entomol. Exper. Appl.* **61**, 123–130.

Carbonero, P., Royo, J., Diaz, I., Garcia-Maroto, F., Gonzalez-Hidalgo, E., Gutierrez, C. and Castanera, P. (1992) Cereal inhibitors of insect hydolases (α-amylases and trypsin): genetic control, transgenic expression and insect tests, In: *Third International Workshop on Pathogenesis-Related Proteins in Plants*, Arolla, Switzerland, 16–20 August 1992.

Casu, R., Jarmey, J., Elvin, C. and Eisemann, C. (1994) Isolation of a trypsin-like serine protease gene family from the sheep blowfly *Lucilia cuprina, Insect Mol. Biol.* **3**, 159–170.

Chapman, R. (1985a) Structure of the digestive system, In: Kerkut, G. A. (Ed.), *Comprehensive Insect Physiology, Biochemistry and Pharmacology*, Vol. 4, New York: Pergamon Press, pp. 165–211.

Chapman, R. (1985b) Coordination of digestion, In: Kerkut, G. A. (Ed.), *Comprehensive Insect Physiology, Biochemistry and Pharmacology*, Vol. 4, New York: Pergamon Press, pp. 213–240.

Chen, M., Johnson, B., Wen, L., Muthukrishnan, S., Kramer, K., Morgan, T. and Reeck, G. (1992) Rice cystatin: bacterial expression, purification, cysteine proteinase inhibitory activity and insect growth suppressing activity of a truncated form of the protein, *Protein Expr. Purif.* **3**, 41–49.

Chothia, C. and Lesk, A. (1986) The relation between the divergence of sequence and structure in proteins, *EMBO J.* **5**, 823–826.

Christeller, J. and Shaw, B. (1989) The interaction of a range of serine proteinase inhibitors with bovine trypsin and *Costelytra zealandica* trypsin, *Insect Biochem.* **19**, 233–241.

Christeller, J., Gatehouse, A. and Laing, W. (1994a) The interaction of the elastase inhibitor, eglin c, with insect digestive endopeptidases: effect of pH on the dissociation constants, *Insect Biochem. Mol. Biol.* **24**, 103–109.

Christeller, J., Markwick, N. and Burgess, P. (1994b) Midgut proteinase activities of three keratinolytic larvae, *Hofmannophila pseudospretella, Tineola bisselliella*, and *Anthrenocerus australis*, and the effect of proteinase inhibitors on proteolysis, *Arch. Insect Biochem. Physiol.* **25**, 159–173.

Christeller, J., Laing, W., Markwick, N. and Burgess, E. (1992) Midgut protease activities in 12 phytophagous lepidopteran larvae: dietary and protease inhibitor interactions, *Insect Biochem. Mol. Biol.* **22**, 735–746.

Christeller, J., Shaw, B., Gardiner, S. and Dymock, J. (1989) Partial purification and characterization of the major proteases of grass grub larvae (*Costelytra zealandica*, Coleoptera: Scarbaeidae), *Insect Biochem.* **19**, 221–231.

Constabel, C., Bergey, D. and Ryan, C. (1995) Systemin activates synthesis of wound-inducible tomato leaf polyphenol oxidase via the octadecanoid defense signaling pathway, *Proc. Natl. Acad. Sci. USA* **92**, 407–411.

Croall, D. and Demartino, G. (1991) Calcium-activated neutral protease (calpain) system: structure, function, and regulation, *Physiol. Rev.* **71**, 813–847.

Dadd, R. (1985) Nutrition: organisms, In: Kerkut, G. A. (Ed.), *Comprehensive Insect Physiology, Biochemistry and Pharmacology*, Vol. 4, New York: Pergamon Press, pp. 313–390.

Davis, C., Riddler, D., Higgins, M., Holden, J. and White, B. (1985) A gene family in *Drosophila melanogaster* coding for trypsin-like enzymes, *Nucl. Acid Res.* **13**, 6605–6619.

Dow, J. (1981) Countercurrent flows, water movements and nutrient uptake in locust midgut, *J. Insect Physiol.* **27**, 579–585.

Dymock, J., Laing, W., Shaw, B., Gatehouse, A. and Christeller, J. (1992) Behavioral and physiological responses of grass grub larvae (*Costelytra zealandica*) feeding on protease inhibitors, *NZ J. Zool.* **19**, 123–131.

Edmonds, H. S., Gatehouse, L. M., Hilder, V. A. and Gatehouse, J. A. (1996) The inhibitory effects of the cysteine proteinase inhibitor, oryzacystatin, on digestive proteinases and on larval survival and development of the southern corn rootworm (*Diabrotica undecimpunctata howardi*), *Entomol. Exper. Appl.* **78**, 83–94.

Elden, T. (1995) Selected proteinase inhibitor effects on alfalfa weevil growth and development, *J. Econ. Entomol.* **88**, 1586–1590.

Elvin, C., Whan, V. and Riddles, P. (1993) A family of serine protease genes expressed in adult buffalo fly (*Haematobia irritans exigua*), *Mol. Gen. Genet.* **240**, 132–139.

Ferreira, C., Oliveria, M. and Terra, W. (1990) Compartmentalization of the digestive process in *Abracris flavolineata* (Orthoptera: Acrididae) adults, *Insect Biochem.* **20**, 267–274.

Ferreira, C., Capella, A., Sitnik, R. and Terra, W. (1994) Digestive enzymes in midgut cells, endo- and ectoperitrophic contents, and peritrophic membranes of *Spodoptera frugiperda* (Lepidoptera) larvae, *Arch. Insect Biochem. Physiol.* **26**, 299–313.

Gatehouse, A. and Boulter, D. (1983) Assessment of the antimetabolic effects of trypsin inhibitors from cowpea (*Vigna unguiculata*) and other legumes on development of the bruchid beetle *Callosobruchus maculatus, J. Sci. Food Agric.* **34**, 345–350.

Gillikin, J., Bevilacqua, S. and Graham, J. (1992) Partial characterization of digestive tract proteinases from western corn rootworm larvae, *Diabrotica virgifera, Arch. Insect Biochem. Physiol.* **19**, 285–298.

Grasberger, B., Clore, G. and Gronenborn, A. (1994) High-resolution structure of *Ascaris* trypsin inhibitor in solution: direct evidence for a pH-induced conformational transition in the reactive site, *Structure* **2**, 669–678.

Hilder, V., Gatehouse, A., Sheerman, S., Barker, R. and Boulter, D. (1987) A novel mechanism of insect resistance engineered into tobacco, *Nature* **330**, 160–163.

Hines, M., Nielsen, S., Shade, R. and Pomeroy, M. (1990) The effect of two proteinase inhibitors, E-64 and the Bowman–Birk inhibitor, on the developmental time and mortality of *Acanthoscelides obtectus, Entomol. Exper. Appl.* **57**, 201–207.

Hinks, C. and Hupka, D. (1995) The effects of feeding leaf sap from oats and wheat, with and without soybean trypsin inhibitor, on feeding behavior and digestive physiology of adult males of *Melanoplus sanguinipes, J. Insect Physiol.* **41**, 1007–1015.

Homandberg, G., Litwiller, R. and Peanaksy, R. (1989) Carboxypeptidase inhibitors from *Ascaris suum*: the primary structure, *Arch. Biochem. Biophys.* **270**, 153–161.

Hosoyama, H., Irie, K., Abe, K. and Arai, S. (1994) Oryzacystatin exogenously introduced into protoplasts and regeneration of transgenic rice, *Biosci. Biotechnol. Biochem.* **58**, 1500–1505.

Houseman, J. and Chin, P. (1995) Distribution of digestive proteinases in the alimentary tract of the European corn borer *Ostrinia nubilalis* (Lepidoptera: Pyralidae), *Arch. Insect Biochem. Physiol.* **28**, 103–111.

Houseman, J. and Thie, N. (1993) Difference in digestive proteolysis in the stored maize beetles: *Sitophilus zeamais* (Coleoptera: Curculionidae) and *Prostephanus truncatus* (Coleoptera: Bostrichidae), *J. Econ. Entomol.* **86**, 1049–1054.

Houseman, J., Philogene, B. and Downe, A. (1989) Partial characterization of proteinase activity in the larval midgut of the European corn borer, *Ostrinia nubilalis* Hubner (Lepidoptera: Pyralidae), *Can. J. Zool.* **67**, 864–868.

Johnson, R., Narvaez, J., An, G. and Ryan, C. (1989) Expression of proteinase inhibitors I and II in transgenic tobacco plants: effects on natural defense against *Manduca sexta* larvae, *Proc. Natl. Acad. Sci. USA* **86**, 9871–9875.

Johnston, K., Gatehouse, J. and Anstee, J. (1993) Effects of soybean protease inhibitors on the growth and development of larval *Helicoverpa armigera, J. Insect Physiol.* **39**, 657–664.

Johnston, K., Lee, M., Brough, C., Hilder, V., Gatehouse, A. and Gatehouse, J. (1995) Protease activities in the larval midgut of *Heliothis virescens*: evidence for trypsin and chymotrypsin-like enzymes, *Insect Biochem. Mol. Biol.* **25**, 375–383.

Jongsma, M., Bakker, P., Peters, D., Bosch, D. and Stiekema, W. (1995) Adaptation of *Spodoptera exigua* larvae to plant proteinase inhibitors by induction of gut proteinase activity insensitive to inhibition, *Proc. Natl. Acad. Sci. USA* **92**, 8041–8045.

Kanost, M. and Jiang, H. (1996) Proteinase inhibitors in invertebrate immunity, In: Soderhall, K., Iwanaga, S. and Vasta, G. (Eds), *New Directions in Invertebrate Immunology*, Fair Haven, NJ: SOS Publications.

Keil, M., Sanchez-Serrano, J., Schell, J. and Willmitzer, L. (1990) Localization of elements important for the wound-inducible expression of a chimeric potato proteinase inhibitor II-CAT gene in transgenic tobacco plants, *Plant Cell* **2**, 62–70.

Klopfenstein, N., McNabb, H., Hart, E., Hall, R., Hanna, R., Heuchelin, S., Allen, K., Shi, N. and Thornburg, R. (1993) Transformation of *Populus* hybrids to study and improve pest resistance, *Silvae Genet.* **42**, 86–90.

Komiyama, T., Ray, C., Pickup, D., Howard, A., Thornberry, N., Peterson, E. and Salvesen, G. (1994) Inhibition of interleukin-1B converting enzyme by the cowpox virus Serpin CrmA, *J. Biol. Chem.* **269**, 19331–19337.

Krishnamoorthi, R., Gong, Y. and Richardson, M. (1990) A new protein inhibitor of trypsin and activated Hageman factor from pumpkin (*Cucurbita maxima*) seeds, *FEBS Lett.* **273**, 163–167.

Krizaj, I., Drobnic-Kosorok, M., Brzin, J., Jerala, R. and Turk, V. (1993) The primary structure of inhibitor of cysteine proteinases from potato, *FEBS Lett.* **333**, 15–20.

Kromer, E., Nakakura, N. and Lagueux, M. (1993) Cloning of a *Locusta* cDNA encoding a precursor peptide for two structurally related proteinase inhibitors, *Insect Biochem. Mol. Biol.* **24**, 329–331.

Laskowski, M. (1993) Binding of amino acid side chains to preformed cavities: interaction of serine proteinases with turkey ovomucoid third domains with coded and noncoded P1 residues, *Protein Sci.* **2**, 786–799.

Laskowski, M. and Kato, I. (1980) Protein inhibitors of proteinases, *Annu. Rev. Biochem.* **49**, 593–626.

Laskowski, M., Kato, I., Ardelt, W., Cook, J., Denton, A., Empie, M., Kohr, W., Park, S., Parks, K., Schatzley, B. *et al.* (1987) Ovomucoid third domains from 100 avian species: isolation, sequences and hypervariability of enzyme-inhibitor contact residues, *Biochemistry* **26**, 202–221.

Laskowski, M., Apostol, I., Ardelt, W., Cook, J., Giletto, A., Kelly, C., Lu, W., Park, S., Qasim, M., Whatley, H. *et al.* (1990) Amino acid sequences of ovomucoid third domain from 25 additional species of birds, *J. Protein Chem.* **9**, 715–725.

Lee, M. and Anstee, J. (1995) Endoproteases from the midgut of larval *Spodoptera littoralis* include a chymotrypsin-like enzyme with an extended binding site, *Insect Biochem. Mol. Biol.* **25**, 49–61.

Lehane, M., Blakemore, D., Williams, S. and Moffatt, M. (1995) Regulation of digestive enzyme levels in insects, *Comp. Biochem. Physiol.* **110B**, 285–289.

Lenz, C., Kang, J., Rice, W., McIntosh, A., Chippendale, G. and Schubert, K. (1991) Digestive proteinases of larvae of the corn earworm *Heliothis zea*: characterization, distribution and dietary relationships, *Arch. Insect Biochem. Physiol.* **16**, 201–212.

Leple, J.C., Bonade-Bottino, M., Augustin, S., Pilate, G., Dumanois Le Tan, V., Delplanque, A., Cornu, D. and Jouanin, L. (1995) Toxicity to *Chrysomela tremulae* (Coleoptera: Chrysomelidae) of transgenic poplars expressing a cystein proteinase inhibitor, *Mol. Breeding* **1**, 319–328.

Levinsky, H., Birk, Y. and Applebaum, S. (1977) Isolation and characterization of a new trypsin-like enzyme from *Tenebrio molitor* L. larvae, *Int. J. Peptide Protein Res.* **10**, 252–264.

Liang, C., Brookhart, G., Feng., G., Reeck, G. and Kramer, K. (1991) Inhibition of digestive proteinases of stored grain coleoptera by oryzacystatin, a cysteine proteinase inhibitor from rice seed, *FEBS Lett.* **278**, 139–142.

Lipke, H., Fraenkel, G. and Liener, I. (1954) Effect of soybean inhibitors on growth of *Tribolium confusum*, *J. Agric. Food Chem.* **2**, 410–414.

Longstaff, C., Campbell, A. and Fersht A. (1990) Recombinant chymotrypsin inhibitor 2: expression, kinetic analysis of inhibition with α-chymotrypsin and wild-type and mutant subtilisin BPN', and protein engineering to investigate inhibitory specificity and mechanism, *Biochemistry* **29**, 7339–7347.

Ma, H., Yang, H., Takano, E., Hatanaka, M. and Maki, M. (1994) Amino-terminal conserved region in proteinase inhibitor domain of calpastatin potentiates its calpain inhibitory activity by interacting with calmodulin-like domain of the proteinase, *J. Biol. Chem.* **269**, 24430–24436.

Macintosh, S., Kishore, G., Perlak, F., Marrone, P., Stone, T., Sims, S. and Fuchs, R. (1990)

Potentiation of *Bacillus thuringiensis* insecticidal activity by serine protease inhibitors, *J. Agric. Food Chem.* **38**, 1145–1152.

Malone, L., Giacon, H., Burgess, E., Maxwell, J., Christeller, J. and Laing, W. A. (1995) Toxicity of trypsin endopeptidase inhibitors to honey bees, *J. Econ. Entomol.* **88**, 46–50.

Markwick, N., Reid, S., Laing, W. and Christeller, J. (1995) Effects of dietary protein and protease inhibitors on codling moth (Lepidoptera: Tortricidae), *J. Econ. Entomol.* **88**, 33–39.

Martzen, M., McMullen, B., Smith, N., Fujikawa, K. and Peanasky, R. (1990) Primary structure of the major pepsin inhibitor from the intestinal parasitic nematode *Ascaris suum*, *Biochemistry* **29**, 7366–7372.

Masoud, S., Johnson, L., White, F. and Reeck, G. (1993) Expression of a cysteine proteinase inhibitor (oryzacystatin-I) in transgenic tobacco plants, *Plant Mol. Biol.* **21**, 655–663.

Masoud, S., Ding, X., Johnson, L., White, F. and Reeck, G. (1996) Expression of a corn bifunctional inhibitor of serine proteases and insect α-amylases in transgenic tobacco plants, *Plant Sci.* **115**, 59–69.

Matsumoto, I., Watanabe, H., Abe, K., Arai, S. and Emori, Y. (1995) A putative digestive cysteine proteinase from *Drosophila melanogaster* is predominantly expressed in the embryonic and larval midgut, *Eur. J. Biochem.* **227**, 582–587.

McGhie, T., Christeller, J., Ford, R. and Allsopp, P. (1995) Characterization of midgut proteinase activities of white grubs: *Lepidiota noxia*, *Lepidiota negatoria* and *Antitrogus consanguineus* (Scarabaeidae, Melolonthini), *Arch. Insect Biochem. Physiol.* **28**, 351–363.

McGrath, M., Gillmor, S. and Fletterick, R. (1995) Ecotin: lessons on survival in a protease-filled world, *Protein Sci.* **4**, 141–148.

McGurl, B., Orozco-Cardenas, M., Pearce, G. and Ryan, C. (1994) Overexpression of the prosystemin gene in transgenic tomato plants generates a systemic signal that constitutively induces proteinase inhibitor synthesis, *Proc. Natl. Acad. Sci. USA* **91**, 9799–9802.

McManus, M. and Burgess, E. (1995) Effects of the soybean (Kunitz) trypsin inhibitor on growth and digestive proteases of larvae of *Spodoptera litura*, *J. Insect Physiol.* **41**, 731–738.

McManus, M., White, D. and McGregor, P. (1994) Accumulation of a chymotrypsin inhibitor in transgenic tobacco can affect the growth of insect pests, *Transgen. Res.* **3**, 50–58.

Michaud, D., Nguyen-Quoc, B. and Yelle, S. (1993) Selective inhibition of Colorado potato beetle cathepsin H by oryzacystatins I and II, *FEBS Lett.* **331**, 173–176.

Michaud, D., Bernier-Vadnais, N., Overney, S. and Yelle, S. (1995) Constitutive expression of digestive cysteine proteinase forms during development of the Colorado potato beetle, *Leptinotarsa decemlineata* Say (Coleoptera: Chrysomelidae), *Insect Biochem. Mol. Biol.* **25**, 1041–1048.

Moire, N., Bigot, Y., Periquet, G. and Boulard, C. (1994) Sequencing and gene expression of hypodermis A, B, C in larval stages of *Hypoderma lineatum*, *Mol. Biochem. Parasitol.* **66**, 233–240.

Muller, H., Crampton, J., della Torre, A., Sinden, R. and Crisanti, A. (1993) Members of a trypsin gene family in *Anopheles gambiae* are induced in the gut by blood meal, *EMBO J.* **12**, 2891–2900.

Murdock, L., Brookhart, G., Dunn, P., Foard, D., Kelley, S., Kitch, L., Shade, R., Shukle, R. and Wolfson, J. (1987) Cysteine digestive proteinases in coleoptera, *Comp. Biochem. Physiol.* **87B**, 783–787.

Newell, C., Lowe, J., Merryweather, A., Rooke, L. and Hamilton, W. (1995) Transformation of sweet potato (*Ipomoea batatas* (L.) Lam.) with *Agrobacterium tumefaciens* and regeneration of plants expressing cowpea trypsin inhibitor and snowdrop lectin, *Plant Sci.* **107**, 215–227.

Noriega, F., Wang, X., Pennington, J., Barillas-Mury, C. and Wells, M. (1996) Early trypsin, a female-specific midgut protease in *Aedes aegypti*: isolation, amino-terminal sequence determination, and cloning and sequencing of the gene, *Insect Biochem. Mol. Biol.* **26**,

119–126.

Oppert, B., Morgan, T., Cubertson, C. and Kramer, K. (1993) Dietary mixtures of cysteine and serine proteinase inhibitors exhibit synergistic toxicity toward the red flour beetle, *Tribolium castaneum, Comp. Biochem. Physiol.* **105C**, 379–385.

Orozco-Cardenas, M., McGurl, B. and Ryan, C. A. (1993) Expression of an antisense prosystemin gene in tomato plants reduces resistance toward *Manduca sexta* larvae, *Proc. Natl. Acad. Sci. USA* **90**, 8273–8276.

Orr, G., Strickland, J. and Walsh, T. (1994) Inhibition of *Diabrotica* larval growth by multicystatin from potato tubers, *J. Insect Physiol.* **40**, 893–900.

Peterson, A., Barillas-Mury, C. and Wells, M. (1994) Sequence of three cDNAs encoding an alkaline midgut trypsin from *Manduca sexta, Insect Biochem. Mol. Biol.* **24**, 463–471.

Peterson, A., Fernando, G. and Wells, M. (1995) Purification, characterization and cDNA sequence of an alkaline chymotrypsin from the midgut of *Manduca sexta, Insect Biochem. Mol. Biol.* **25**, 765–774.

Piergiovanni, A., Sonnante, G., Gatta, C. and Perrino, P. (1991) Digestive enzyme inhibitors and storage pest resistance in cowpea (*Vigna unguiculata*) seeds, *Euphytica* **54**, 191–194.

Potempa, J., Korzus, E. and Travis, J. (1994) The Serpin superfamily of proteinase inhibitors: structure, function and regulation, *J. Biol. Chem.* **269**, 15957–15960.

Purcell, J., Greenplate, J. and Sammons, D. (1992) Examination of midgut luminal proteinase activities in six economically important insects, *Insect Biochem. Mol. Biol.* **22**, 41–47.

Ramos, A., Mahowald, A. and Jacobs-Lorena, M. (1993) Gut-specific genes from the black fly *Simulium vittatum* encoding trypsin-like and carboxypeptidase-like proteins, *Insect Mol. Biol.* **1**, 149–163.

Read, R. and James, M. (1986) Introduction to the protein inhibitors: X-ray crystallography, In: Barrett, A. and Salvesen, G., (Eds) *Proteinase Inhibitors*, Amsterdam: Elsevier, pp. 301–336.

Reddy, M. N., Keim, P. S., Heinrikson, R. L. and Kezdy, F. J. (1975) Primary structural analysis of sulfhydryl protease inhibitors from pineapple stem, *J. Biol. Chem.* **250**, 1741–1750.

Rees, D. and Lipscomb, W. (1982) Refined crystal structure of the potato inhibitor complex of carboxypeptidase A at 2.5 A resolution, *J. Mol. Biol.* **160**, 475–498.

Richards, A. and Richards, P. (1977) The peritrophic membranes of insects, *Annu. Rev. Entomol.* **22**, 219–240.

Richardson, M. (1991) Seed storage proteins: the enzyme inhibitors, *Meth. Plant Biochem.* **5**, 259–305.

Ritonja, A., Krizaj, I., Mesko, P., Kopitar, M., Lucovnik, P., Strukelj, B., Pungercar, J., Buttle, D., Barrett, A. and Turk, V. (1990) The amino acid sequence of a novel inhibitor of cathepsin D from potato, *FEBS Lett.* **267**, 13–15.

Roberts, B., Markland, W., Ley, A., Kent, R., White, D., Guterman, S. and Ladner, R. (1992) Directed evolution of a protein: selection of potent neutrophil elastase inhibitors displayed on M13 fusion phage, *Proc. Natl. Acad. Sci. USA* **89**, 2429–2433.

Rockefeller Foundation (1994) *Seventh Meeting of the International Program on Rice Biotechnology*, Bali, Indonesia, 16–20 May 1994.

Ryan, C. (1990) Protease inhibitors in plants: genes for improving defenses against insects and pathogens, *Annu. Rev. Phytopathol.* **28**, 425–449.

Rymerson, R. and Bodnaryk, R. (1995) Gut proteinase activity in insect pests of canola, *Can. Entomol.* **127**, 41–48.

Sakal, E., Applebaum, S. and Birk, Y. (1988) Purification and characterization of *Locusta migratoria* chymotrypsin, *Int. J. Peptide Prot. Res.* **32**, 590–598.

Sakal, E., Applebaum, S. and Birk, Y. (1989) Purification and characterization of trypsins from the digestive tract of *Locusta migratoria, Int. J. Peptide Prot. Res.* **34**, 498–505.

Santos C., Ferreira, C. and Terra, W. (1983) Consumption of food and spatial organization of digestion in the cassava hornworm, *Erinnyis ello, J. Insect Physiol.* **29**, 707–714.

Sasaki, T., Hishida, T., Ichikawa, K. and Asari, S. (1993) Amino acid sequence of alkaliphilic serine protease from silkworm, *Bombyx mori*, larval digestive juice, *FEBS Lett.* **320**, 35–37.

Silva, C. and Xavier-Filho, J. (1991) Comparison between the levels of aspartic and cysteine proteinases of the larval midguts of *Callosobruchus maculatus* (L.) and *Zabrotes subfasciatus* (Boh.) (Coleoptera: Bruchidae), *Comp. Biochem. Physiol.* **99B**, 529–533.

Slansky Jr, F. and Scriber, J. (1985) Food consumption and utilization. In: Kerkut, G. A. (Ed.), *Comprehensive Insect Physiology, Biochemistry and Pharmacology*, Vol. 4, New York: Pergamon Press, pp. 87–163.

Sottrup-Jensen, L. (1989) α-Macroglobulins: structure, shape and mechanism of proteinase complex formation, *J. Biol. Chem.* **264**, 11539–11542.

Strobl, S., Muhlhahn, R., Bernstein, R., Wiltscheck, R., Maskos, K., Wunderlich, M., Huber, R., Glockshuber, R. and Holak, T. (1995) Determination of the three-dimensional structure of the bifunctional α-amylase/trypsin inhibitor from ragi seeds by NMR spectroscopy, *Biochemistry* **34**, 8281–8293.

Tabashnik, B., Finson, N. and Johnson, M. (1992) Two protease inhibitors fail to synergize *Bacillus thuringiensis* in diamondback moth, *J. Econ. Entomol.* **85**, 2082–2087.

Terra, W. (1988) Physiology and biochemistry of insect digestion: an evolutionary perspective, *Braz. J. Med. Biol. Res.* **21**, 675–734.

Terra, W. (1990) Evolution of digestive systems of insects, *Annu. Rev. Entomol.* **35**, 181–200.

Terra, W. and Ferreira, C. (1981) The physiological role of the peritrophic membrane and trehalase: digestive enzymes in the midgut and excreta of starved larvae of *Rhynchosciara*, *J. Insect Physiol.* **27**, 325–331.

Terra, W. and Ferreira, C. (1994) Insect digestive enzymes: properties, compartmentalization and function, *Comp. Biochem. Physiol.* **109B**, 1–62.

Terra, W., Ferreira, C. and De Bianchi, A. (1979) Distribution of digestive enzymes among the endo- and ectoperitrophic spaces and midgut cells of *Rhynchosciara* and its physiological significance, *J. Insect Physiol.* **25**, 487–494.

Thie, N. and Houseman, J. (1990) Cysteine and serine proteolytic activities in larval midgut of yellow mealworm, *Tenebrio molitor* L. (Coleoptera: Tenebrionidae), *Insect Biochem.* **20**, 741–744.

Thomas, K. and Nation, J. (1984) Protease, amylase and lipase activities in the midgut and hindgut of the cricket, *Gryllus rubens* and mole cricket, *Scapteriscus acletus*, *Comp. Biochem. Physiol.* **79A**, 297–304.

Thomas, J., Wasmann, C., Echt, C., Dunn, R., Bohnert, H. and McCoy, T. (1994) Introduction and expression of an insect proteinase inhibitor in alfalfa (*Medicago sativa* L.), *Plant Cell Rep.* **14**, 31–36.

Thomas, J., Adams, D., Keppenne, V., Wasmann, C., Brown, J., Kanost, M. and Bohnert, H. (1995a) *Manduca sexta* encoded protease inhibitors expressed in *Nicotiana tabacum* provide protection against insects, *Plant Physiol. Biochem.* **33**, 611–614.

Thomas, J., Adams, D., Keppenne, V., Wasmann, C., Brown, J., Kanost, M. and Bohnert, H. (1995b) Protease inhibitors of *Manduca sexta* expressed in transgenic cotton, *Plant Cell Rep.* **14**, 758–762.

Turunen, S. (1985) Absorption, In: Kerkut, G. A. (Ed.), *Comprehensive Insect Physiology, Biochemistry and Pharmacology*, Vol. 4, New York: Pergamon Press, pp. 241–277.

Urwin, P., Atkinson, H., Waller, D. and McPherson, M. (1995) Engineered oryzacystatin-I expressed in transgenic hairy roots confers resistance to *Globodera pallida*, *Plant J.* **8**, 121–131.

Wang, S., Magoulas, C. and Hickey, D. (1993) Isolation and characterization of a full-length trypsin-encoding cDNA clone from the Lepidopteran insect, *Choristoneura fumiferana*, *Gene* **136**, 375–376.

Wigglesworth, V. (1972) *The Principles of Insect Physiology*, 6th ed., London: Methuen.

Wlodawer, A. and Erickson, J. (1993) Structure-based inhibitors of HIV-1 protease, *Annu.*

Rev. Biochem. **62**, 543–585.

Wolfson, J. (1991) The effects of induced plant proteinase inhibitors on herbivorous insects, In: Tallamy, D. W. and Raupp, M. J. (Eds), *Phytochemical Induction by Herbivores*, New York: John Wiley & Sons, pp. 223–243.

Wolfson, J. and Murdock, L. (1987) Suppression of larval Colorado potato beetle growth and development by digestive proteinase inhibitors, *Entomol. Exper. Appl.* **44**, 235–240.

Wolfson, J. and Murdock, L. (1990a) Diversity in digestive proteinase activity among insects, *J. Chem. Ecol.* **16**, 1089–1102.

Wolfson, J. and Murdock, L. (1990b) Growth of *Manduca sexta* on wounded tomato plants, role of induced proteinase inhibitors, *Entomol. Exper. Appl.* **54**, 257–264.

Xavier-Filho, J., Campos, F., Ary, M., Silva, C., Carvalho, M., Macedo, M., Lemos, F. and Grant, G. (1989) Poor correlation between the levels of proteinase inhibitors found in seeds of different cultivars of cowpea (*Vigna unguiculata*) and the resistance/susceptibility to predation by *Callosobruchus maculatus*, *J. Agric. Food Chem.* **37**, 1139–1143.

Zhao, R., Shi, X., Wang, J. and Fan, Y. (1996) Transgenic tobacco plants expressed both B.t. and CpTI genes and their homozygotes, *Rice Biotechnol. Quart.* **25**, 35–36.

Acknowledgments

Work in our laboratories that is related to the material in this chapter has been supported by the Agricultural Research Service of the USDA, the Kansas Agricultural Experiment Station, and grants from the following sources: US-Israel BARD program (to GR, KK, Y. Birk and S. Applebaum); the Rockefeller Foundation (to GR and S. Muthukrishnan); NIH-GM41247 (to MK); Pioneer Hi-Bred International (to GR and KK) and the USDA National Research Initiatives Competitive Grant program (GR and J. Reese). We thank Drs Brenda Oppert, S. Muthukrishnan, Lowell Johnson, and Frank White for many discussions and insights related to the material in this chapter, and Drs John Christeller, Igor Kregar, and Clarence Ryan for valuable suggestions based on their reading of a draft of this chapter. Contribution no. 96-306-B of the Kansas Agricultural Experiment Station.

11

Chitinases for Insect Control

KARL J. KRAMER, SUBBARATNAM MUTHUKRISHNAN, LOWELL JOHNSON and FRANK WHITE

Introduction

Chitin, the unbranched homopolymer of 2-acetamido-2-deoxy-D-glucopyranoside (N-acetylglucosamine, GlcNAc) in a β-1,4 linkage (Flach *et al.*, 1992), occurs as a structural component in the cuticles and shells of arthropods, in the cell walls of fungi and some algae, in nematodes, in mollusks, and in many other types of organisms (Cabib, 1987; Gooday, 1990; Kramer and Koga, 1986). This polysaccharide ranks among the most abundant biopolymers. Because of the importance of chitin and chitinolytic enzymes in insect growth and development, the latter are receiving attention in regard to their development as biopesticides or chemical defense proteins in transgenic plants and microbial biological control agents.

Chitinases are defined as enzymes (EC 3.2.1.14) with a specific hydrolytic activity directed against the homopolymer chitin. However, some chitinases hydrolyze other related polymers, such as cell wall polysaccharides containing β,1-4 linked N-acetylglucosamines and N-acetylmuramates, at significant rates. Enzymatic cleavage occurs randomly at internal locations over the entire length of the chitin microfibril. The final products formed via chitinase catalysis are soluble low molecular mass multimers of GlcNAc such as chitotetraose, chitotriose and chitobiose, with the latter being predominant. These oligosaccharides are substrates for another type of chitinolytic enzyme, β-N-acetylglucosaminidase (EC 3.2.1.30), which cleaves off GlcNAc units sequentially from the non-reducing end (Kramer and Koga, 1986). Both types of enzyme have been detected in a variety of organisms, including those that contain chitin, such as insects, crustaceans, yeasts, and fungi, and also organisms that do not contain chitin, such as bacteria, higher plants, and vertebrates.

In arthropods, chitinases are involved in molting and digestion. Insects periodically shed their old cuticles and resynthesize new ones. This process is mediated by the elaboration of chitinases in the molting fluid that accumulates in the space between the old cuticle and the epidermis. The products of hydrolysis are recycled for the synthesis of the new cuticle. Often the larva will ingest the old cuticle. Apparently, chitinases found in the gut have a digestive function in addition to their role in breaking down chitin present in the gut lining. In fungi, chitinases apparently help to degrade and mobilize organic matter and perhaps to antagonize the growth of other fungi. In yeast, chitinases are important for cell separation. Mutants lacking chitinolytic activity tend to aggregate at their septum regions (Kuranda and Robbins, 1991). Furthermore, addition of allosamidin, a chitinase inhibitor, causes a similar effect on yeast cell aggregation (Sakuda *et al.*, 1990).

Considerable interest in the physical, chemical, kinetic, and biocidal properties of chitinases has been stimulated by their possible involvement as defensive agents against chitinous pathogenic or pestiferous organisms such as fungi and insects (Carr and Klessig, 1989; Linthorst, 1991; Sahai and Manocha, 1993). Resistance to organisms can be imparted by the degradation of vital structures such as the peritrophic membrane or cuticle of insects, the cell wall of fungal pathogens, or by liberation of substances that subsequently elicit other defensive responses (Boller, 1987).

The peritrophic membrane and exoskeleton of insects act as physicochemical barriers to environmental hazards and predators. However, entomopathogenic fungi such as *Metarhizium anisopliae, Beauveria bassiana, Nomuraea rileyi* and *Aspergillus flavus* apparently overcome these kinds of barriers by producing multiple extracellular enzymes including chitinolytic and proteolytic enzymes that help to penetrate the cuticle and facilitate infection (El-Sayed *et al.*, 1989; St Leger *et al.*, 1986, 1991a,b). Some insect venoms also contain chitinolytic enzymes that might serve to facilitate the entry of venomous components into prey (Krishnan *et al.*, 1994). Similarly, the nematode, *Brugia malayi*, utilizes a chitinase to break down a protective chitinous extracellular sheath and/or the peritrophic membrane to gain entry into the mosquito host (Huber *et al.*, 1991; Sahabuddin *et al.*, 1993). Baculoviruses also contain genes for chitinases, but their precise role(s) in infection of the host is unclear (Ayres *et al.*, 1994). A role in putrefaction of the host has been proposed. Nevertheless, hydrolytic enzymes used by insects, fungi, and other organisms for molting or barrier penetration are potentially useful in pest management, because their physiological action is to destroy vital structures such as the exoskeleton or peritrophic membrane of insects.

The role of plant chitinases in disease resistance is well documented (see Graham and Sticklen, 1994, for a recent comprehensive review). Numerous plant and microbial chitinase genes or cDNAs have been cloned. Some of these have been introduced into plants under the control of constitutive promoters, resulting in an enhancement of resistance of the host plant to fungal pathogens (Broglie *et al.*, 1991, 1993; Lin *et al.*, 1995; Vierheilig *et al.*, 1993). However, the role of various chitinases in mediating plant resistance to insects is less well understood. Furthermore, genes coding for chitinolytic enzymes from insect sources have not been reported until recently (Kramer *et al.*, 1993; Krishnan *et al.*, 1994; Nagamatsu *et al.*, 1995; Zen *et al.*, 1996). This review summarizes recent progress in the area of the use of microbial chitinases for insect control, the cloning of insect chitinolytic enzyme genes, and their expression in insect tissues and in cell lines infected with baculoviruses containing an insect chitinase gene. We also report the results of some recent studies involving the introduction of an insect chitinase gene into plants. Our current understanding of the possible function of chitinases as plant defensive proteins against insects is also presented.

Microbial Chitinases for Pest Control

Chitinases have been used in mixing experiments to increase the potency of ento-mopathogenic microorganisms. Bacterial chitinolytic enzymes have been used to enhance the activity of microbial insecticides including *Bacillus thuringiensis* (*Bt*) and a baculovirus. Larvae of spruce budworm, *Choristoneura fumiferana*, died more rapidly when exposed to a chitinase–*Bt* mixture than when exposed to the enzyme or bacterium alone (Lysenko, 1976; Morris, 1976; Smirnoff, 1973, 1974). In another study, mortality of gypsy moth (*Lymantria dispar*) larvae was enhanced when chitin-ase was added with *Bt* compared to treatment with *Bt* alone (Dubois, 1977), and this effect was correlated with enzyme levels (Gunner *et al.*, 1985). The larvicidal activity of a nuclear polyhedrosis virus toward gypsy moth larvae was increased about five-fold when it was coadministered with bacterial chitinase (Shapiro *et al.*, 1987). Hypothetically, chitinases cause perforations in the gut peritrophic membrane, which facilitate entry of the pathogens into the hemocoel of susceptible insects (Brandt *et al.*, 1978).

Chitinases also appear to be involved in the penetration of host cuticle by ento-mopathogenic fungi (Coudron *et al.*, 1989; St Leger *et al.*, 1991a,b). Chitinases and β-N-acetylglucosaminidases are secreted when the entomopathogens *M. anisopliae*, *B. bassiana*, and *Verticilium lecanii* are grown on insect cuticles (St Leger *et al.*, 1986). At the time of cuticle penetration, virulent isolates of *N. rileyi* exhibit substan-tially higher levels of chitinase activity than avirulent strains (El-Sayed *et al.*, 1989). Chitinase gene expression in entomopathogenic fungi is believed to be controlled by a repressor–inducer system in which chitin or the oligomeric products of degrada-tion serve as inducers (St Leger *et al.*, 1986). However, bacterial chitinases were ineffective in assays in which insects were fed a diet containing the enzymes. No mortality of the nymphal stages of the rice brown plant hopper, *Nilaparvata lugens*, occurred when 0.09% w/v *Streptomyces griseus* chitinase was added to an artificial diet (Powell *et al.*, 1993). Similarly, *Serratia* and *Streptomyces* chitinases at 1–2% levels in the diet of the merchant grain beetle, *Oryzaephilus mercator*, caused no mortality (see below).

Plant Chitinases for Pest Control

We know of no reports of the successful use of a plant chitinase in controlling insect pests. Cereal grains have substantial levels of chitinases – 10–100 µg/g are typical (Leah *et al.*, 1987; Molano *et al.*, 1979; Wadsworth and Zikakis, 1984). Yet stored grains are susceptible to insect attack, suggesting that stored product insects have evolved to overcome the effects of plant chitinases. In further support of this conclu-sion, we have found that transgenic rice plants expressing relatively high levels of a rice chitinase (0.05% total protein) have no detrimental effects on the growth of the fall armyworm, *Spodoptera frugiperda* (unpublished data).

Insect Chitinases for Pest Control

Chitinases from several insect species, including the tobacco hornworm, *Manduca sexta*, and the silkworm, *Bombyx mori*, have been characterized (Koga *et al.*, 1992; Kramer and Koga, 1986). The insect chitinases have a size range of 40 to 85 kDa, which is typically larger than that of plant chitinases (25 to 40 kDa) and bacterial chitinases (21 to 58 kDa). The insect enzymes are associated primarily with the molting process, but they might also have a digestive role in the gut (Terra and Ferreira, 1994). A cDNA encoding the major molting fluid chitinase of *M. sexta* was characterized by DNA sequencing (Kramer *et al.*, 1993). As expected for secreted proteins, the protein encoded in the cDNA contains a hydrophobic 19 amino acid-long leader peptide followed by a sequence corresponding to the mature N-terminus of the molting fluid chitinase. The length of the mature protein predicted from this clone is 554 amino acids, corresponding to a molecular mass of 62 kDa. Glycosylation results in a protein with an apparent molecular mass of 85 kDa. The sequence of *M. sexta* chitinase has very little similarity with sequences of class I and class II plant chitinases but shows limited similarity to sequences of class III and class V plant and microbial chitinases. In particular, two highly conserved regions, including one proposed to be a part of the active site of the enzyme, are present in *M. sexta* chitinase. On the other hand, *M. sexta* chitinase shows a significantly greater level of sequence identity with a recently reported chitinase encoded by a cDNA from the venom gland of an endoparasitic wasp (*Chelonus* sp.) and to a lesser extent to several oviductins, which are glycoproteins with an affinity for carbohydrate compounds (Krishnan *et al.*, 1994; Malette *et al.*, 1995).

Expression and Regulation of *Manduca sexta* Chitinase

The *M. sexta* chitinase gene is not active during larval feeding behavior, but it is expressed in a narrow time frame just prior to larval–larval and larval–pupal molting. The activity of this gene is regulated positively by ecdysteroid and controlled negatively by juvenile hormone (Kramer *et al.*, 1993). The tight developmental and hormonal regulation of chitinase expression suggests that this enzyme might be detrimental to insect growth, if expressed at an inappropriate time or if not expressed at an appropriate time. Thus, plants constitutively expressing an insect chitinase might be resistant to insects that feed on them, because exposure to this enzyme might damage the gut lining of these herbivores.

Transgenic Plants Expressing an Insect Chitinase

Chimeric gene constructs were prepared by inserting the 1.8 kb EcoRI fragment from a cDNA clone containing the entire coding region for *M. sexta* chitinase (Kramer *et al.*, 1993) into binary vectors between single or double CaMV 35S promoter and either the *nos* or *pinII* polyadenylation signal sequences (Ding, 1995). This gene was introduced into tobacco (*Nicotiana tabacum* var. *xanthi*) and tomato (*Lycopersicon esculentum*) plants using *Agrobacterium tumefaciens*-mediated transfer techniques of Masoud *et al.* (1993). T_0 plants were shown by immunoblot analysis to express the insect chitinase. The size of the protein was only 46 kDa instead of 85 kDa, which is the mass of the native enzyme in molting fluid. However, the truncated protein was found to be enzymatically active. T_0 plants expressing chitinase were self-fertilized, and T_1 progeny were assayed for expression of insect chitinase by immunoblot analysis and glycolchitin hydrolytic activity. Progeny from several plants segregated with insect chitinase expression in a ratio of approximately 3 positive : 1 negative, which was expected for an active gene at a single locus. Leaves were excised from chitinase-positive and chitinase-negative plants and fed to first instar larvae of the tobacco budworm (*Heliothis virescens*). After three weeks, the total mass of larvae surviving on chitinase-negative leaves was 966 mg, whereas that on the chitinase-positive leaves was only 177 mg, a reduction of more than 80%. Reductions in larval biomass and feeding damage were also observed using intact T_1 plants derived by transformation with a 35S promoter. Larvae reared on plants lacking the *Manduca* chitinase gene consumed substantially more leaf tissue than did larvae feeding on plants expressing the gene. Thus, constitutive expression of the truncated insect chitinase in tobacco protected the plants from extensive feeding damage by the budworm.

In order to determine whether the *Manduca* chitinase from transgenic tobacco and several chitinases from other sources are directly toxic to insects, a beetle feeding study was conducted using enzymes purified from transgenic tobacco as well as two bacterial and one plant species. A relatively small species, the merchant grain beetle, *O. mercator*, was utilized for the bioassay, because not enough recombinant insect chitinase from tobacco could be purified to bioassay larger insects such as the budworm *in vitro*. Recombinant *Manduca* chitinase from transgenic tobacco and chitinases from *Serratia*, *Streptomyces* (bacteria), and *Hordeum* (barley) species were incorporated into a ground wheat germ diet at a 1–2% level and fed to neonate beetle larvae. Whereas growth and mortality of larvae consuming the bacterial and plant chitinase-supplemented diets were the same as those of larvae consuming the untreated diet, all of the larvae consuming the recombinant insect chitinase-supplemented diet were dead a few days after egg hatch (Ding, 1995). These data indicate that insect chitinase is a potential host plant resistance factor in transgenic plants and might be more potent than chitinases from other sources.

Baculovirus Expressing *M. sexta* Chitinase has Enhanced Larvicidal Activity

M. sexta chitinase has also been shown to increase the killing rate of a recombinant baculovirus (Gopalakrishnan *et al.*, 1995). A recombinant non-occluded baculovirus, *Autographa californica* nuclear polyhedrosis virus (AcMNPV) containing the 1.8 kb DNA fragment from an *M. sexta* chitinase cDNA (Kramer *et al.*, 1993) under the control of the polyhedrin gene promoter, expressed a glycosylated chitinase with a size of 85 kDa, which was indistinguishable in size from the enzyme in molting fluid. This chitinase was secreted into the medium when insect cell lines were infected with the virus. When the recombinant virus was injected in *M. sexta* larvae, chitinase was found in the hemolymph, where it does not normally occur. The recombinant baculovirus expressing the chitinase killed larvae of fall armyworms (*S. frugiperda*) in approximately three-quarters of the time required for the wild-type virus to kill the larvae ($LT_{50} = 65$ h versus 88 h; Gopalakrishnan *et al.*, 1995). The results indicate that insect chitinase, by increasing the killing rate of insect pathogens, might be useful as an enhancing factor for biological control agents. The occluded form of the virus expressing the insect chitinase might be more useful as a biocontrol agent, because the virus is effective as a larvicide when ingested.

N-Acetylglucosaminidases for Pest Control

Another type of chitinolytic enzyme from insects, β-N-acetylglucosaminidase, has been characterized (Kramer and Koga, 1986; Nagamatsu *et al.*, 1995; Zen *et al.*, 1996). At least two forms of this enzyme, with apparent molecular masses of 55 and 62 kDa, exist in molting fluid of *M. sexta*. cDNAs for the 62 kDa enzyme have been cloned from the tobacco hornworm and silkworm, *B. mori* (Nagamatsu *et al.*, 1995; Zen *et al.*, 1996). However, none of these enzymes have yet been assayed for insecticidal activity. Like chitinases, β-N-acetylglucosaminidases are secreted by fungal pathogens upon infection of insects (Bidochka *et al.*, 1993; St Leger *et al.*, 1991b), and these enzymes might have biocidal properties. Furthermore, a combination of chitinases and β-N-acetylglucosaminidases has been shown to degrade chitin at a rate six times faster than that of either enzyme alone (Fukamizo and Kramer, 1985). Therefore, the possibility of genetically engineering both chitinase and β-N-acetyl-glucosaminidase genes into crop plants as a strategy to control insect pests is quite appealing.

Concluding Remarks

Manipulation of the levels and/or the integrity of chitin in insect cuticles and peri-trophic membranes offers an excellent strategy for selective pest control that has not been well explored. A certain amount of chitin must be maintained during each developmental stage, and an excess or deficiency can produce deleterious effects. The likelihood of biotechnological development of a selective chitin-targeted insecticide that is relatively non-toxic to other organisms should be high, because chitin is not present in higher animals or plants. Chitinolytic enzymes might have unique properties such that the enzymes themselves possess selective insecticidal activities. The enzymes also might facilitate the uptake or enhance the activity of other insecticides that are orally or topically active. We are still working to develop insect chitinase as a biocide for insects and plant pathogens and so far have made some progress with one plant and a baculovirus expressing this enzyme. Research also needs to be conducted on improving expression levels of the enzyme and using the chitinase gene to transform other plants and microbial biocontrol agents that target other pest species.

Chitin metabolism has been presumed to be a relatively unique target for insect control agents, because vertebrates do not use chitin as a significant structural element. However, several reports now document the existence of vertebrate chitinases and families of related proteins (Choi *et al.*, 1996; Hakala *et al.*, 1993; Malette *et al.*, 1995; Renkema *et al.*, 1995). Therefore, prior to their use as pest control agents, the relationship between the structure and function of insect chitinases and their vertebrate homologues should be determined.

In nature, most host plant resistance is based on multiple genes and a diverse set of resistance factors (Pimental, 1991). A diverse polygenic resistance system probably prevents plant-feeding species from overcoming resistance by the host plant. Adding chitinolytic enzyme genes – whose encoded proteins have the potential to affect insect pests and microbial pathogens adversely – to the repertoire of other defense genes in plants and entomopathogens might enhance the effectiveness of biological control strategies. Our hope is that a more thorough understanding of chitin catabolism and its regulation will help to facilitate the development of novel ways to control insect pests.

References

Ayres, M. D., Howard, S. C., Kuzio, J., Lopez-Ferber, M. and Possee, R. D. (1994) The complete DNA sequence of *Autographa californica* nuclear polyhedrosis virus, *Virology* **202**, 586–605.

Bidochka, M. J., Tong, K. I. and Khachatourians, G. G. (1993) Partial purification and characterization of two extracellular N-acetyl-D-glucosaminidases produced by the entomopathogenic fungus, *Beauveria bassiana, Can. J. Microbiol.* **39**, 40–45.

Boller, T. (1987) Hydrolytic enzymes in plant disease resistance, In: Kosuge, T. and Nester, E. W. (Eds), *Plant–Microbe Interactions: Molecular and Genetic Perspectives*, Vol. 2, New York: MacMillan, pp. 384–414.

Brandt, C. R., Adang, M. J. and Spence, K. D. (1978) The peritrophic membrane: ultrastructural analysis and function as a mechanical barrier to microbial infection in *Orgyia pseudotsugata, J. Invert. Pathol.* **32**, 12–24.

Broglie, R., Broglie, K., Roby, D. and Chet, I. (1993) Production of transgenic plants with enhanced resistance to microbial pathogens, In: Kung, S. D. and Wu, R. (Eds), *Transgenic Plants*, Vol. 1, San Diego, CA: Academic Press, pp. 265–276.

Broglie, K., Chet, I., Holliday, M., Cressman, R., Biddle, P., Knowlton, S., Mauvais, J. and Broglie, R. M. (1991) Transgenic plants with enhanced resistance to the fungal pathogen, *Rhizoctonia solani, Science* **254**, 1194–1197.

Cabib, E. (1987) The synthesis and degradation of chitin, *Adv. Enzymol.* **59**, 59–101.

Carr, J. P. and Klessig, D. F. (1989) In: Setlow, J. K. (Ed.), *Genetic Engineering – Principles and Methods*, Vol. 11, New York: Plenum Press, p. 65.

Choi, H. K., Choi, K. H., Kramer, K. J. and Muthukrishnan, S. (1996) Isolation and characterization of a genomic clone for the gene of an insect molting enzyme, chitinase, *Insect Biochem. Molec. Biol.* (in press).

Coudron, T. A., Kroha, M. J. and Ignoffo, C. M. (1989) Levels of chitinolytic activity during development of three entomopathogenic fungi, *Comp. Biochem. Physiol.* **79B**, 339–348.

Ding, X. (1995) *Manduca* chitinase-mediated resistance to tobacco budworm (*Heliothis virescens*) and tobacco hornworm (*Manduca sexta*) larvae in transgenic tobacco plants, PhD dissertation, Kansas State University, Manhattan.

Dubois, N. R. (1977) Pathogenicity of selected resident microorganisms of *Lymantria dispar* after induction for chitinase. PhD dissertation, University of Massachusetts, Amherst.

El-Sayed, G. N., Coudron, T. A. and Ignoffo, C. M. (1989) Chitinolytic activity and virulence associated with native and mutant isolates of the entomopathogenic fungus, *Nomuraea ileyi, J. Invert. Pathol.* **54**, 394–403.

Flach, J., Pilet, P. E. and Jolles, P. (1992) What's new in chitinase research? *Experientia* **48**, 701–716.

Fukamizo, T. and Kramer, K. J. (1985) Mechanism of chitin hydrolysis by the binary chitinase system in insect moulting fluid, *Insect Biochem.* **15**, 141–145.

Gooday, G. (1990) The ecology of chitin degradation, *Microbiol. Ecol.* **10**, 387–431.

Gopalakrishnan, B., Muthukrishnan, S. and Kramer, K. J. (1995) Baculovirus-mediated expression of a *Manduca sexta* chitinase gene: properties of the recombinant protein, *Insect Biochem. Mol. Biol.* **25**, 255–265.

Graham, L. S. and Sticklen, M. B. (1994) Plant chitinases, *Can. J. Bot.* **72**, 1057–1083.

Gunner, H. B., Met, M. Z. and Berger, S. (1985) Chitinase-producing *Bt* strains, In: Grimble, D. G. and Lewis, F. B (Eds), *Microbial Control of Spruce Budworms and Gypsy Moths*, US For. Serv. GTR-NE-100, pp. 102–108.

Hakala, B. E., White, C. and Recklies, A. D. (1993) Human cartilage gp-39, a major secretory product of articular chondrocytes and synovial cells, is a mammalian member of a chitinase protein family, *J. Biol. Chem.* **268**, 25803–25810.

Huber, M., Cabib, E. and Miller, L. H. (1991) Malaria parasite chitinase and penetration of the mosquito peritrophic membrane, *Proc. Natl. Acad. Sci. USA* **88**, 2807–2810.

Koga, D., Funakoshi, T., Mizuki, K., Ide, A., Kramer, K. J., Zen, K. C., Choi, H. K. and Muthukrishnan, S. (1992) Immunoblot analysis of chitinolytic enzymes in integument and molting fluid of the silkworm, *Bombyx mori*, and the tobacco hornworm, *Manduca sexta*, *Insect Biochem. Mol. Biol.* **22**, 305–311.

Kramer, K. J. and Koga, D. (1986) Insect chitin: physical state, synthesis, degradation and metabolic regulation, *Insect Biochem.* **16**, 851–877.

Kramer, K. J., Corpuz, L. M., Choi, H. and Muthukrishnan, S. (1993) Sequence of a cDNA and expression of the gene encoding epidermal and gut chitinases of *Manduca sexta*, *Insect Biochem. Mol. Biol.* **23**, 691–701.

Krishnan, A., Nair, P. N. and Jones, D. (1994) Isolation, cloning and characterization of a new chitinase stored in active form in chitin-lined venom reservoir, *J. Biol. Chem.* **269**, 20971–20976.

Kuranda, M. J. and Robbins P. W. (1991) Chitinase is required for cell separation during growth of *Saccharomyces cerevisiae*, *J. Biol. Chem.* **266**, 19758–19767.

Leah, R., Mikkelsen, J. D., Mundy, J. and Svendsen, I. (1987) Identification of a 28 000 dalton endochitinase in barley endosperm, *Carlsberg Res. Comm.* **52**, 31–37.

Lin, W., Anuratha, C. S., Datta, K., Potrykus, I., Muthukrishnan, S. and Datta, S. K. (1995) Genetic engineering of rice for resistance to sheath blight, *Bio/Technology* **13**, 686–691.

Linthorst, H. J. M. (1991) Pathogenesis-related proteins of plants. *Crit. Rev. Plant Sci.* **10**, 123–150.

Lysenko, O. (1976) Chitinase of *Serratia marcescens* and its toxicity to insects, *J. Invert. Path.* **27**, 385–386.

Malette, B., Paquette, Y., Merlen, Y. and Bleau, G. (1995) Oviductins possess chitinase- and mucin-like domains: a lead in the search for the biological function of these oviduct-specific ZP-associating glycoproteins, *Molec. Reprod. Develop.* **41**, 384–397.

Masoud, S. A., Johnson, L. B., White, F. F. and Reeck, G. R. (1993) Expression of a cysteine proteinase inhibitor (oryzacystatin-I) in transgenic tobacco plants, *Plant Molec. Biol.* **21**, 655–663.

Molano, J., Polacheck, I., Duran, A. and Cabib, E. (1979) An endochitinase from wheat germ, *J. Biol. Chem.* **254**, 4901–4907.

Morris, O. N. (1976) A 2-year study of the efficacy of *Bacillus thuringiensis*–chitinase combinations in spruce budworm (*Choristoneura fumiferana*) control, *Can. Entomol.* **108**, 3225–3233.

Nagamatsu, Y., Yanagisawa, I., Kimoto, M., Okamoto, M. and Koga, D. (1995) Purification of a chitooligosaccharidolytic β-N-acetylglucosaminidase from *Bombyx mori* larvae during metamorphosis and the nucleotide sequence of its cDNA, *Biosci. Biotech. Biochem.* **59**, 219–225.

Pimental, D. (1991) Diversification of biological control strategies in agriculture, *Crop Prot.* **10**, 243–253.

Powell, K. S., Gatehouse, A. M. R., Hilder, V. A. and Gatehouse, J. A. (1993) Antimetabolic effects of plant lectins and plant and fungal enzymes on the nymphal stages of two important rice pests, *Nilaparvata lugens* and *Nephotettix cinciteps*, *Entomol. Exp. Appl.* **66**, 119–126.

Renkema, G. H., Boot, R. G., Muijsers, A. O., Donker-Koopman, W. E. and Aerts, J. M. F. G. (1995) Purification and characterization of human chitotriosidase, a novel member of the chitinase family of proteins, *J. Biol. Chem.* **270**, 2198–2202.

Sahabuddin, M., Toyoshima, T., Aikawa, M. and Kaslow, D. (1993) Transmission blocking activity of a chitinase inhibitor and activation of malarial parasite chitinase by mosquito protease, *Proc. Natl. Acad. Sci. USA* **90**, 4266–4270.

Sahai, A. S. and Manocha, M. S. (1993) Chitinases of fungi and plants: their involvement in morphogenesis and host-parasite interaction, *FEMS Microbiol. Rev.* **11**, 317–338.

Sakuda, S., Nishinato, Y., Ohi, M., Watanabe, M., Takayama, S., Isogai, A. and Yamada, Y. (1990) Effects of demethylallosamidin, a potent yeast chitinase inhibitor, on the cell division of yeast, *Agric. Biol. Chem.* **54**, 1333–1335.

Shapiro, M., Preisler, H. K. and Robertson, J. L. (1987) Enhancement of baculovirus activity on gypsy moth (Lepidoptera: Lymantriidae) by chitinase, *J. Econ. Entomol.* **80**, 1113–1116.

Smirnoff, W. A. (1973) Results of test with *Bacillus thuringiensis* and chitinase on larvae of spruce bud worm, *J. Invert. Pathol.* **21**, 116–118.

Smirnoff, W. A. (1974) Three years of aerial field experiments with *Bacillus thuringiensis* plus chitinase formulation against the spruce bud worm, *J. Invert. Pathol.* **24**, 344–348.

St Leger, R. J., Cooper, R. M. and Charnley, A. K. (1986) Cuticle-degrading enzymes of entomopathogenic fungi: regulation of production of chitinolytic enzymes, *J. Gen. Microbiol.* **132**, 1509–1517.

St Leger, R. J., Cooper, R. M. and Charnley, A. K. (1991a) Characterization of chitinase and chitobiase produced by the entomopathogenic fungus *Metarhizium anisopliae*, *J. Invert. Pathol.* **58**, 415–426.

St Leger, R. J., Staples, R. C. and Roberts, D. W. (1991b) Entomopathogenic isolates of *Metarhizium anisopliae*, *Beauveria bassiana* and *Aspergillus flavus* produce multiple extracellular chitinase isozymes, *J. Invert. Pathol.* **61**, 81–84.

Terra, W. T. and Ferreira, C. (1994) Insect digestive enzymes: properties, compartmentalization and function, *Comp. Biochem. Physiol.* **109B**, 1–62.

Vierheilig, H., Alt, M., Neuhas, J-M., Boller, T. and Wiemken, A. (1993) Colonization of transgenic *N. sylvestris* plants, expressing different forms of *N. tabacum* chitinase, by the root pathogen, *Rhizoctonia solani*, and by the mycorrhizal symbiont, *Glomus mosseau*, *Molec. Plant Microbe Interac.* **6**, 261–264.

Wadsworth, S. A. and Zikakis, J. P. (1984) Chitinase from soybean seeds: purification and some properties of the enzyme system, *J. Agric. Food Chem.* **32**, 1284–1288.

Zen, K. C., Choi, H. K., Nandigama, K., Muthukrishnan, S. and Kramer, K. J. (1996) Cloning, expression, and hormonal regulation of an insect β-N-acetylglucosaminidase gene, *Insect Biochem. Molec. Biol.*, **26**, 435–444.

Acknowledgments

Research on chitinolytic enzymes in our laboratories was supported in part by USDA-NRI grants. Agricultural Research Service, USDA is an equal opportunity/ affirmative action employer and all agency services are available without discrimination. Contribution no. 96-212-B of the Kansas Agricultural Experiment Station.

12

The Role of Peroxidase in Host Insect Defenses

PATRICK F. DOWD and L. MARK LAGRIMINI

Peroxidase Properties

Many excellent reviews of peroxidases exist (e.g., Campa, 1991; Gaspar *et al.*, 1982, 1992; Kolattukudy *et al.*, 1992; Marañón and van Huystee, 1994; Penel *et al.*, 1992). We will concentrate on the properties of peroxidases with regard to host insect defenses, and the factors that may influence their effectiveness when introduced into transgenic plants.

Peroxidases are a subclass of oxidoreductase (EC 1.11.1) that use a peroxide such as H_2O_2 as an oxygen acceptor (IUB, 1992). In addition to the guaiacol peroxidases in plants (EC 1.11.1.7), there are several other peroxidases and related enzymes in this subclass, including NADH peroxidase (EC 1.22.1.1), cytochrome-c peroxidase (EC 1.11.1.4), catalase (EC 1.11.1.6), glutathione peroxidase (EC 1.11.1.9), L-ascorbate peroxidase (EC 1.11.1.11) and manganese peroxidase (EC 1.11.1.13) (IUB, 1992). For the sake of the present discussion, we will concentrate on the guaiacol peroxidases. However, in many cases what will be said about these plant peroxidases may be applicable to the others as well.

Plant peroxidases are monomeric proteins, generally have a common heme (protoporphyrin IX) group, bound Ca^{2+}, and some degree of glycosylation (Marañón and van Huystee, 1994). There is also some evidence that Mn^{2+} is required in some forms (Marañón and van Huystee, 1994). For these respective functionalities, there are conserved regions for the peptide sequence (Welinder, 1991). The heme group is bound by a histidyl imidazole and an aspartic acid residue (Marañón and van Huystee, 1994). The Ca^{2+} is bound by eight ligands in the calcium binding loop of lignin peroxidase (Poulos *et al.*, 1993); however, it has not been determined how Ca^{2+} is bound in plant peroxidases (Marañón and van Huystee, 1994; Penel, 1986). The glycosyl groups can range from eight – in horse-radish (Fujiyama *et al.*, 1990), potato (Roberts *et al.*, 1988) and tomato (Roberts and Kolattukudy, 1989) – to zero for wheat (Rebmann *et al.*, 1991). The carbohydrate moieties are covalently attached to the peptide chain through asparagine residues (Welinder and Gajhede, 1993). It has been shown that Mn^{2+} stimulates the production of H_2O_2 from NAD(P)H (Halliwell, 1978). There appears to be a Mn^{2+} in cationic peanut peroxidase but as yet evidence suggests it does not play a catalytic role (as opposed to its role in manganese peroxidase) (Marañón and van Huystee, 1994). The prosthetic groups for peroxidase have different functions. The protoporphyrin group is involved in catalysis. It has been proposed that Ca^{2+} stabilizes the heme moiety (Ogawa *et al.*, 1979; Xu and van Huystee, 1993). The presence of the glycosyl groups may help stabilize the peroxidase by decreasing the turnover rate (Marañón and van Huystee, 1994). Plant peroxidases are highly complex enzymes whose activities are closely regulated by the plant.

As of 1993, peroxidases have been cloned from tobacco (Lagrimini *et al.*, 1987), potato (Roberts *et al.*, 1988), horseradish (Fujiyama *et al.*, 1988, 1990), tomato (Roberts and Kolattukudy, 1989), peanut (Buffard *et al.*, 1990), cucumber (Morgens *et al.*, 1990), *Arabidopsis* (Intapruk *et al.*, 1991), wheat (Hertig *et al.*, 1991; Rebmann *et al.*, 1991), barley (Rasmussen *et al.*, 1991; Theilade and Rasmussen, 1992), rice (Reimmann *et al.*, 1992), and maize (Hwang, 1993). Plant peroxidases are members of a superfamily of proteins, with as many as 42 distinct isoenzyme forms in one plant species (Welinder, 1991). Peroxidase isoenzymes can differ by as much as 30% within the same species, and by as much as 70% among different plant species (Welinder, 1991). Peroxidases are often grouped into anionic, cationic, and neutral forms based on their migration on native or isoelectric focusing gels. However, artifacts may also occur in preparations, due to binding of phenolics or the hydrolysis of glycosides by β-glucosidases (Marañón and van Huystee, 1994). In situations where properties have been investigated by isoelectric focusing, a continuum across at least a pH range of 3.5 to 9.0 exists (Dowd, 1994b; Dowd and Norton, 1995; Lagrimini and Rothstein, 1987). This is further complicated by the number of alleles of a particular peroxidase gene that may exist – up to six for some in maize (Brewbaker and Hasegawa, 1975; Brewbaker *et al.*, 1985). Overall, 12 isoenzymes are considered to be present in tobacco (Lagrimini and Rothstein, 1987; Mäder *et al.*, 1977), 24 in maize (Brewbaker *et al.*, 1985), and 42 isoenzymes/forms from preparations of horseradish (Hoyle, 1977) (but see above caveats).

Although peroxidases are generally considered to have wide substrate specificity, (Campa, 1991; Gaspar *et al.*, 1982, 1992; Penel *et al.*, 1992), there does appear to be substrate 'preference' for different isoenzymes (see also following discussion on peroxidase functions). For example, when horseradish peroxidase C and the tobacco anionic peroxidase were compared, their activity toward different substrates varied by as much as tenfold (Gazaryan and Lagrimini, 1996a,b). Mäder *et al.*, 1977 compared different tobacco peroxidase isoenzymes for reactivity toward various substrates and found activity also varied by as much as tenfold. Of the several maize isoenzymes investigated with natural and artificial substrates, a relative preference rate of 1–4 was noted, depending on the substrate and isoenzyme (Brewbaker and Hasegawa, 1975). A more recent examination of many natural substrates indicated very strong preference for some natural substrates (Dowd, 1994b; Dowd and Norton, 1995). This leads into a consideration of what the natural function of peroxidases includes.

Peroxidase Functions

Hydrogen Peroxide Catalysis

By definition, plant peroxidases act on a peroxide as an oxygen acceptor, e.g. H_2O_2 (IUB, 1992). Although H_2O_2 is generated by general cellular processes (see discussion by Felton and Duffey, 1991), its level can be mediated by the presence of peroxidase, catalase, glutathione peroxidase and superoxide dismutase (Allen, 1995). Where peroxidases occur, H_2O_2 may be generated from the quenching of super-oxide radicals generated during quinone formation by superoxide dismutase (Appel, 1993; Elstner, 1982), glycolate oxidation (Elstner, 1982), polyamine/diamine oxidase (Angelini and Federico, 1989), and peroxidases themselves using NAD(P)H as an electron source (Campa, 1991; Elstner and Heupel, 1976; Mäder and Füssl, 1982). Thus, in any particular tissue or cellular compartment, there appear to be multiple mechanisms for providing the necessary H_2O_2 for rapid peroxidase function.

Hydrogen peroxide is considered to be a toxin at ca. 1 mM for *Drosophila melanogaster* (Orr and Sohal, 1992) and *Dalbulus maidis* at 100 ppm (Dowd and Vega, 1996). Plants are known to produce H_2O_2 and other active oxygen species in response to stress, e.g. chilling, wounding, disease, etc. (Levine *et al.*, 1994; Mehdy, 1994). Endogenous H_2O_2 generation that exceeds toxic levels (more than 2 mM) can result in cellular damage. Although more specific enzymes are recognized for remov-ing H_2O_2 (Allen, 1995), the presence of peroxidases at H_2O_2 generation sites can serve as a 'detoxification' for the peroxide. However, depending on the hydrogen donor, free radicals more toxic than H_2O_2 may be formed (Appel, 1993) (see below).

Regulation of Endogenous Hormone Levels

Indole-3-acetic acid (IAA) – one of the most studied of the plant growth regulators – has a very complicated catabolism which has been the subject of intensive investiga-tion. It has been known for considerable time that peroxidase catalyzes the oxida-tive decarboxylation of indole-3-acetic acid to methylene-oxindole and indole-3-aldehyde (Dunford, 1991; Epstein *et al.*, 1980). This reaction is unusual for peroxi-dases in that it does not consume hydrogen peroxide; however, it has been reported that hydrogen peroxide stimulates the oxidation of IAA (Gazaryan and Lagrimini, 1996a,b; Pfanz, 1993). There are indications that some isoenzymes from maize (Brewbaker and Hasegawa, 1975) have greater specificity toward IAA than other isoenzymes from the same plant. Dwarf forms of maize and pea have increased levels of peroxidase (McCune and Galston, 1959). When treated with gibberellic acid, dwarf forms grew more normally and peroxidase differences declined (McCune and Galston, 1959). The gibberellic acid appears to suppress peroxidase secretion (Fry, 1980). Factors other than peroxidase are also involved in IAA regulation – including rate of synthesis, conjugation, and non-peroxidative degradation (Epstein *et al.*, 1980), so the significance of IAA metabolism by peroxidase probably depends on the plant species and the developmental stage of the organism.

Polymerization

Peroxidases are critical in the biosynthesis of plant cell walls. Peroxidases promote the peroxidative polymerization of the monolignols coniferyl, *p*-coumaryl and sinapyl alcohol into lignin (Greisbach, 1981). Different plant species will have varying ratios of the monolignol species assembled in a semi-random fashion (Hwang *et al.*, 1991). Some of the best evidence to date that peroxidases participate in the formation of lignin can be found using transgenic plants over-expressing the tobacco anionic peroxidase (Lagrimini, 1991; Lagrimini *et al.*, 1992).

Although there appears to be little information on how relative levels of iso-enzymes can affect lignin composition, work with maize peroxidases, using corresponding acid precursors for the alcohols, suggests some isoenzyme specificity (Dowd, 1994b; Dowd and Norton, 1995). However, in tobacco, there is no indication of substrate specificity of lignin precursors by lignifying peroxidases based on model substrates (Pang *et al.*, 1989). Although the brown midrib (bm) mutants of corn appear to have differing lignin compositions, so far this is only attributed to the variation in levels of one of the alcohol participants due to an inactivated enzyme in the pathway. Although some of these lignin mutants appear to be useful as forage due to greater digestibility (Barriere and Argillier, 1993), increased susceptibility to insects (Barriere and Argillier, 1993) and disease (Barriere and Argillier, 1993; Nicholson *et al.*, 1976) have also been noted with these mutants. Other forms of lignin may be produced by peroxidases, including that from ferulic acid (Stafford, 1960).

Lignification serves to strengthen and reinforce cell walls (Fincher and Stone, 1986). The overall result is a toughening of the plant tissue (Fincher and Stone, 1986; Gaspar *et al.*, 1992). Lignin composition differs among monocots (Davin and Lewis, 1992; Fincher and Stone, 1986) and dicots (Fincher and Stone, 1986; Gross, 1978).

Other Forms of Polymerization

Similar to lignification is suberization, generally considered to be a wound-healing process. Suberin is composed of a peroxidase-generated polymer of hydroxy-cinnamic acids (Bernards and Lewis, 1992) esterified with long-chain fatty acids (C_{16}–C_{26}) (Kolattukudy and Dean, 1974; Riley and Kolattukudy, 1975). An acidic peroxidase isoenzyme from tomato has been associated with formation of suberin in wounded or infected plants (Espelie *et al.*, 1986). The production of polyphenolic acids has been known to occur via an oxidative mechanism in the presence of polyphenol oxidase (Mayer, 1987). However, peroxidase has also been shown to promote the condensation of phenolic acids in the presence of H_2O_2 (Lagrimini, 1991). Also, condensed tannins are the result of the peroxidase-catalyzed polymerization of flavan-3-ols and flavan-3,4-diols (Swain, 1979).

Crosslinking

A number of cell wall polymers are joined together by peroxidases. There are instances where large molecules of a non-phenolic nature are crosslinked via a free-radical mechanism initiated by a peroxidase. For example, it has been shown that polygalacturonase residues are crosslinked in a peroxidase-dependent fashion through ferulic acid residues in the graminaceous monocots (Fry, 1979, 1986; Markwalder and Newkom, 1976). Extensins or hydroxyproline-rich glycoproteins are known to be crosslinked by a peroxidase through tyrosine residues (Fry, 1982; Lamport, 1989; O'Connel *et al.*, 1990).

198

Conglomerization

The formation of covalent bonds described in the various polymerization and cross-linking reactions just discussed often occur in association with one another. Thus, composites, or conglomerates of multiple structural materials may occur. For example, the crosslinked extensins may subsequently be further bound to lignin during its formation (Vance *et al.*, 1980). The ferulic acid polysaccharides may be further crosslinked to cell walls (Fry, 1986) and their lignin (Iiyama *et al.*, 1994). Phenolics and proteins that contain groups subject to attack by peroxidases (such as ferulic acid or tyrosine, respectively; see Dowd, 1994b) would further be expected to be bound to polymerizing materials. The result can be a formidable barrier to insect consumption and digestion.

Defensive Function

The defensive function of peroxidases is somewhat controversial. Different systems yield different results. Laboratory studies may not reflect field observations. Iso-enzyme numbers/levels may not reflect physical parameters they are thought to produce. Certainly pH can influence the activity of the enzymes (Gazaryan and Lagrimini, 1996a,b; Mader *et al.*, 1977), and the identity of the products (Appel, 1993; Fry, 1986; Fulcrand *et al.*, 1994; Patzlaff and Barz, 1978). Substrate to enzyme ratio may also affect product identity (Fry, 1982). The difficulty in interpreting per-oxidase involvement is that the system is highly dynamic and interactive. Defensive effects of peroxidases are for the most part indirect and dependent on the effects of enzyme products. As complexity increases, the predictability decreases. We will begin by discussing peroxidase activities that may be interpreted as having a defensive role, and then cite specific examples of peroxidase involvement (both for and against), and relevant caveats.

Crosslinking and Polymerization

It should be obvious from the prior discussion on polymerization and crosslinking reactions that the products of these reactions can play a significant defensive role. Lignin serves as a defense material (Swain, 1979), providing both a physical barrier (Wainhouse *et al.*, 1990) and chemical deterrent to insect attack (Stamoupoulos, 1988). As a general rule, tougher materials are often more resistant to insect feeding (Bergvinson *et al.*, 1994, 1995a; Coley, 1983; Raupp, 1985). Highly crosslinked materials appear to provide enhanced resistance to disease (e.g., O'Connell *et al.*, 1990) and insects (Arnason *et al.*, 1994; Dowd, 1994b; Dowd and Norton, 1995). However, there are apparently levels of activity that must be obtained to achieve an effective defense response. Lignin content and crosslinking were considered to be the main reason older corn leaves of resistant varieties were more resistant to *O. nubilalis*, while the allelochemical DIMBOA was thought to determine resistance in younger leaves (Bergvinson *et al.*, 1995b). Condensed tannins are known to promote resistance to several insect species (Swain, 1979). However, chemically polymerized catechin was less toxic to the aphid *Macrosiphum rosae* compared to catechin (Peng and Miles, 1988).

Tying Up of Nutrients

Due to the highly reactive nature of the compounds activated by peroxidases, such as quinones (see below), and the nature of functional groups on some nutrients, binding of activated compounds to nutrients can interfere with their utilization. Binding of quinones derived from phenolics to proteins has been reported to reduce digestibility to insects (Felton *et al.*, 1989, 1992) and is presumed to occur with pathogens as well (Dowd *et al.*, 1996). Binding of activated phenolics generated by peroxidases to sugars/carbohydrates/starch has also been reported (Stafford and Ibrahim, 1992). Due to the substrate specificity of digestive enzymes, these modifications often make the nutrient more resistant to attack by the enzymes (Felton *et al.*, 1992). In some cases toxic forms of amino acids may also be generated (Felton *et al.*, 1992).

Inhibition of Enzymes

Essentially enzyme inhibition is interrelated to the nutrient tie-up just discussed. Digestive enzymes will appear to be inhibited by quinones generated by peroxidases when the products bind to substrates and limit accessibility to the active sites of enzymes. There is also evidence that reactive products bind directly to enzymes and inhibit them, which would be expected from the protein-binding properties previously discussed. Examples of quinone inhibition include proteases (Hoffman-Ostenhof, 1947), amylases (Sissler and Cox, 1960) and P-450 monooxygenases (Kostyuk, 1987).

Generation of Highly Reactive, Toxic Species

Highly reactive chemical species are often generated as a result of the activity of peroxidases. Reactive species may be generated during the different steps of the catalytic process. The reaction of H_2O_2 with peroxidase generates the Compound I form of peroxidase which reacts readily with electron donors such as *o*-aromatic alcohols, *p*-aromatic alcohols, corresponding amides, esters, and ethers (Gaspar *et al.*, 1982). This can result in the generation of highly reactive quinones. Subsequently, these electrophilic quinones can react with groups such as -SH, -NH, and -OH which are common to biological macromolecules (Appel, 1993; Felton *et al.*, 1989, 1992). During quinone formation, highly reactive superoxide anion radicals may be formed (Appel, 1993; Kanofsky, 1991; Summers and Felton, 1994), and additional radicals subsequently generated (Appel, 1993). The net result is the anti-nutritive effect from the modification of plant proteins, carbohydrates, etc., and the generation of numerous cytotoxic species which are damaging to pathogens and insects.

Examples of Peroxidase Involvement in Disease Resistance

Peroxidases are generally considered to be involved in disease resistance (Bell, 1981). Some of the evidence for peroxidase involvement in disease resistance is the result of laboratory studies involving the generation of oxidative products by peroxidases or polyphenol oxidases (which may or may not generate the same products). The H_2O_2 which may be generated in peroxidase associated activities is only a weak sporicide at room temperature (Shin *et al.*, 1994). However, the H_2O_2 can apparently mediate the rate of crosslinking of extensins as part of a rapid defense response to fungi (Bradley *et al.*, 1992). Transgenic plants expressing a gene for glucose oxidase generation of H_2O_2 had higher levels of H_2O_2 and greater resistance to bacterial and fungal pathogens (Shah *et al.*, 1995). Several plant-derived phenolic compounds oxidized by peroxidase were significantly more toxic to *Xanthomonas phaseoli* compared to the original phenolics (Urs and Dunleavy, 1975). Catechin oxidized by peroxidase strongly inhibited the growth of *Xanthomonas malvacearum* (Venere, 1980). However, the activity may depend on the timing of the generation of reactive species. Caffeic acid oxidized by polyphenol oxidase inhibited the growth of *Phytophtora infestans* more effectively than caffeic acid only within the first few minutes of the reaction (Metlinskii *et al.*, 1972). Several maize allelochemicals and related compounds were active against the maize pathogen *Fusarium graminearum* in a generally dose dependent manner as far as peroxidase, H_2O_2, and the allelochemical (Dowd *et al.*, 1996). Other evidence is dependent on the use of histochemical reactions to determine activity of peroxidase at the site of pathogen invasion. Evidence for specific interactions has been described for maize (Cardeña-Gomez and Nicholson, 1987; Macri *et al.*, 1974) and rice (Young *et al.*, 1995).

Distribution of particular isoenzymes of peroxidase has also been used as an indication that particular isoenzymes are involved in disease resistance, although as indicated above, the presence of the appropriate substrates is required for efficacy. These substrates have seldom been characterized or quantified in conjunction with peroxidases. Lines isogeneic except for different peroxidase isoenzyme profiles have not been investigated. Thus, evidence along these lines must be viewed with these caveats in mind. Maize lines resistant to *F. moniliforme* were found to have additional cationic peroxidases compared to susceptible lines (Brad *et al.*, 1974). Some maize lines resistant to *Aspergillus flavus* also have increased numbers of isoenzymes compared to susceptible lines (Dowd, 1992, 1994b). Overall, a better understanding of peroxidase involvement in disease resistance may occur if both toxic and barrier effects are considered (Moerschbacher, 1992). Associated elevated levels of H_2O_2 may regulate action of salicylic acid in systemic acquired resistance (Chen *et al.*, 1993).

Examples of Peroxidase Involvement in Insect Resistance

The browning reaction that results when tissues are damaged has often been used as an indicator that polyphenol oxidases and/or peroxidases are involved in disease/insect resistance or susceptibility. Although peroxidase alone has little effect on insects (Dowd and Vega, 1996; Felton and Duffey, 1991; Rahbé and Febvay, 1993), there is some evidence that peroxidases can contribute to insect resistance. Oxidation products of several corn allelochemicals produced by peroxidases were more toxic than precursors when produced in sucrose solutions fed to the maize leaf-hopper *Dalbulus maidis* (Dowd and Vega, 1996). Many proteins oxidized by poly-phenol oxidases were of poorer nutritional value to *Spodoptera exigua* (Felton *et al.*, 1989, 1992). Induction of peroxidases in tomato may depend on the type of insect and damage (Stout *et al.*, 1994). Also, antioxidants present in diets containing maize silks with high levels of the flavonoid maysin (a luteolin analog) were less toxic to *H. zea* compared to diets without antioxidants (Wiseman and Isenhour, 1993).

Enhanced peroxidase isozyme activity was associated with the hypersensitive response of *Solanum dulcamara* to mites (Bronner *et al.*, 1991). An inbred of maize with kernel resistance to *Aspergillus flavus* (Scott and Zummo, 1988) and the dusky sap beetle *Carpophilus lugubris* (Dowd, 1994a) compared to a susceptible inbred shows a rapid browning response when damaged (Dowd, 1994a,b, 1995). Sub-sequent analysis indicated the resistant variety had several additional peroxidase isoenzymes compared to the susceptible inbred (Dowd, 1992, 1994b). Cold-shocking the Ch mutant of maize prevented the browning of the pericarp (Dowd, 1994b). The cold-shocked pericarps did not brown nearly to the same degree as those that were not cold-shocked, and were more susceptible to feeding by the maize weevil *Sitophilus zeamais* (Dowd, 1994b). The cold-shocked pericarps had fewer peroxidase iso-enzymes compared to the normal pericarps (Dowd, 1994b). Maize callus derived from the resistant inbred that was brown was more resistant to feeding by *H. zea* (Dowd and Norton, 1995), and the brown material was found to have additional peroxidase isoenzymes capable of oxidizing ferulic acid and sinapic acid, but little difference in banding pattern was seen with *p*-coumaric acid as a substrate (Dowd and Norton, 1995). Taken together, these studies suggest that increases in peroxi-dase activity or isoenzyme type can be associated with resistance to insects.

However, there is also evidence that enhanced peroxidase activity can increase susceptibility to insects. The salivary peroxidase in aphids appears to detoxify cate-chin through polymerization (Miles and Peng, 1989). *In vitro* polymerization of catechin by peroxidase appears to detoxify the material and suggests enzymatic polymerization would have the same effect (Peng and Miles, 1988). Other gut enzymes may also have a similar effect (Appel, 1993). Catalase activity in the midgut of *H. zea* may mediate the effects of peroxidases in plant tissue once they enter the insect (Felton and Duffey, 1991). Overall polyphenol oxidase activity was about twice the level in the southwestern corn borer (*Diatraea grandiosella*) susceptible varieties of maize as in resistant varieties (Hedin *et al.*, 1984). Silks of maize varieties that browned rapidly were generally more susceptible to *H. zea* compared to those that did not (Byrne *et al.*, 1989). Although the presence of different isoenzymes was not investigated in this particular study, past studies have indicated lack of sub-strate, and not lack of peroxidase activity is responsible for the non-browning of silks when damaged (Levings and Stuber, 1971). When a population with browning silks that contained high levels of maysin was selected for browning and non-browning silks, those with non-browning silks had silks with significantly lower levels of maysin that better supported corn earworm growth compared to the browning silk line; associated changes in levels of other phenolics may also be involved (Byrne *et al.*, 1996). Insects also have several enzyme systems capable of dealing with reactive oxygen forms that may be generated by peroxidases or other enzymes (Felton and Summers, 1995). Overall, predicting the effect of enhanced enzymatic oxidation on insect feeding is likely to be difficult (Appel, 1993; Miles and Oertli, 1993).

Potential for Producing Insect Resistance with Transgenics Expressing Novel Forms or Levels of Peroxidase

The complexity of interactions involved with peroxidases, and the number of necessary conditions required for the production of functional peroxidases, makes predicting the effects of overexpression or *de novo* expression of a particular peroxidase, or peroxidases, rather difficult. If the overall literature is considered, it appears that increased resistance to insects may be expected provided that the enhanced activity exceeds some particular threshold for a particular insect species and size. There is more evidence for this with chewing insects, especially caterpillars, than other species of insect. However, based on these complexities, it would not be surprising to find that enhanced peroxidase activity would not increase resistance to a particular species or size of insect.

As discussed previously, peroxidases are indicated in the polymerization of monolignols into lignin. Therefore, it would be expected that elevated peroxidase activity in transgenic plants would result in increased lignin content. This lignified barrier could deter insects from feeding. A similar general benefit would be assumed where peroxidases generate reactive species that are acutely toxic, inhibit digestive enzymes, or tie up nutrients. However, the net effect on plant allelochemicals that can act as peroxidase substrates and that are already present as defensive factors may be difficult to determine. Even for a particular compound, such as a flavonoid, the material may initially become more toxic through quinone formation, subsequent rearrangement may make it less so, or further activity by peroxidase may result in highly toxic benzoquinones (Barz and Hoesel, 1978; Berlin and Barz, 1975; Patzlaff and Barz, 1978).

Closely related plants such as tomato and tobacco, have common defenses in the form of sticky trichomes (Stipanovic, 1983), alkaloids (Robinson, 1979), the flavonoid rutin (Sheen, 1974) and the phenolic chlorogenic acid (Sheen, 1974). However, the defensive compounds scopoletin (Reigh *et al.*, 1973), chlorogenic acid (Felton *et al.*, 1992) and rutin (Dowd and Vega, 1996) are all substrates for peroxidase. Thus, peroxidases could potentially activate or detoxify the defensive chemicals. Since the chemistry of the two plants is similar, the effects might be expected to be similar. It would be more difficult to predict the effect of introducing a Solanaceae-derived peroxidase gene into a non-related plant such as maize. The nature of lignin formation in maize has already been discussed. Although tissue distribution varies, more common maize defensive chemicals such as phenolics (e.g. ferulic, Serratos *et al.*, 1987), hydroxamic acids (Meyer, 1988), luteolin-derived flavonoids (e.g. maysin, Waiss *et al.*, 1979), and rutin (Styles and Ceska, 1977) occur, but no trichome defense has been reported. Although these compounds or their analogs are also peroxidase substrates (Dowd, 1994b; Dowd and Norton, 1995; Dowd and Vega, 1996; Dowd *et al.*, 1996; Stafford, 1960; Venere, 1980), predicting the overall effect when introducing a foreign peroxidase gene is not as comparable as if it had come from the same plant family. Again using the tobacco/tomato example, the secondary chemistry of maize is much more distant, thus making predictability more difficult.

The optimum situation would be to have identified the particular peroxidase involved in enhancing insect resistance, and having the gene of this enzyme available for use. Effects may be more predictable for the plant species from which the gene was obtained, but less so the greater the taxonomic distance of the gene source from the target plant species.

Properties of Transgenic Tobacco, Tomato, and Sweetgum Plants Overexpressing Peroxidase

The tobacco anionic peroxidase (pI 3.5) is loosely associated with the cell walls of lignifying vessels and the shoot epidermis of immature plants. The anionic peroxidase is also expressed in the endodermis and storage parenchyma of older roots and shoots (Klotz and Lagrimini, unpublished results). This isoenzyme is the predominant form found in healthy aboveground tissues (more than 80% of total peroxidase activity); however, the anionic peroxidase is expressed only at trace levels in older roots and not at all in younger roots (Lagrimini and Rothstein, 1987). Overall, the tobacco anionic peroxidase is constitutively expressed in healthy growing plants (shoots) and is not induced by disease, wounding or other forms of stress (Lagrimini and Rothstein, 1987). This is in contrast to other peroxidase isoforms which are expressed at low levels until induced by stress (Lagrimini and Rothstein, 1987). On a biochemical level, the purified anionic peroxidase has been shown to form lignin *in vitro* (Mäder and Füssl, 1982), and catalyze the oxidative decarboxylation of indole-3-acetic acid (Gazaryan and Lagrimini, 1995).

A number of years ago a cDNA of the anionic peroxidase was cloned sequenced, and its expression was characterized (Lagrimini *et al.*, 1987). Shortly after this the coding sequence for the tobacco anionic peroxidase was fused to the Cauliflower Mosaic Virus 35S promoter and transformed into *N. tabacum* and *N. sylvestris* (Lagrimini *et al.*, 1990). These plants constitutively overexpress the tobacco anionic peroxidase – by as much as tenfold in leaf tissue. The most striking phenotype of peroxidase overexpression is chronic wilting which begins at approximately the time of flowering (Lagrimini *et al.*, 1990). In addition, these plants are retarded in growth, have smaller, compacted cells (Lagrimini, 1992), and brown rapidly in response to wounding (Lagrimini, 1991).

Tomato plants were also transformed with the same tobacco peroxidase construct with similar results (Lagrimini *et al.*, 1992). These plants were also found to wilt severely after flowering, and showed excessive browning and reduced fruit size (Lagrimini *et al.*, 1993). *Liquidamber styraciflua* (sweet gum) was also transformed with the CaMV 35S/peroxidase chimeric gene (Sullivan and Lagrimini, 1993). These trees are over three years old and have not flowered or wilted.

Insect Resistance of Transgenic Plants Expressing Peroxidase

Tobacco

Initial studies dealt with the tobacco species *Nicotiana sylvestris*. Plants were grown in the greenhouse, and standard fertilizer treatments were supplemented with iron in the hope that this would reduce or eliminate wilting (which it did). The insects used in this study were larvae of the corn earworm, *Helicoverpa zea*, which is an important pest of this plant (Metcalf and Metcalf, 1993). Leaf and stem tissue were examined. Third instar larvae (ca. 5 mg) were caged on fully expanded, ca. 30 cm leaves (#2–4 from unexpanded, rosette stage). Larvae were weighed after feeding for about two days, and leaf areas consumed were determined. In three separate series of experiments, there were no significant differences in the leaf material consumed or weights of the larvae (except for one experiment where insects were held on leaves an extra eight hours, and weights of larvae caged on transgenic leaves were ca. 30% higher than for those caged on wild type leaves) (Dowd and Lagrimini, 1996a; Dowd *et al.*, 1993). Terminal and basal stem sections from bolting plants were removed from plants, and newly hatched larvae were caged with them. There were no significant differences in mortality or weights of those caged with terminal stem sections. However, a significantly greater number of larvae caged with basal stem sections from the transgenic plants died (90%) compared to wild type plants (40%). Larvae tended to bore into the pith of the stems as opposed to feeding on other parts of the stem (Figure 12.1).

Another insect, the dusky sap beetle (*Carpophilus lugubris*) was also tested with basal stem sections. This insect feeds on a wide variety of fruit and vegetable material (Connell, 1956), but is not known to feed to any extent on tobacco. With adults of *C. lugubris*, effects were seen similar to those with *H. zea* on basal stems. Mortality was significantly higher, and penetration of basal stem sections was significantly lower (Figure 12.2) for transgenic plants (48% mortality, 12% penetration) compared to wild type plants (25% mortality, 88% penetration).

At the end of this study after *N. sylvestris* plants were being held for seed production, an infestation of green peach aphids (*Myzus persicae*) was noted. Although general observation indicated numbers on leaves were similar and over 100 per leaf, quantitation of numbers feeding on the major leaf veins indicated the wild type had significantly higher numbers (3.1) feeding on these veins compared to the transgenics (0.7).

A second series of studies was run with *N. sylvestris* once underproducers became available. In order to examine effects under more controlled conditions, studies were run with seedlings in growth chambers. Leaf disk assays were conducted with older, mature leaves (#3–4, ca. 6 cm) and growing leaves (#8–10, ca. 9 cm) of week-old plants using newly hatched *H. zea* larvae. There was no significant difference in feeding or mortality of larvae placed on any of the three types of mature leaves. There were also no significant differences in mortality of larvae placed on any of the three types of young leaves. However, the amount of leaf tissue from overproducer transgenic plants consumed by the larvae was significantly less (about twofold) com-

Figure 12.1 Corn earworm (*Helicoverpa zea*) on tobacco. Left, wild type tobacco; right, transgenic tobacco (overproducer)

Figure 12.2   Dusky sap beetle (*Carpophilus lugubris*) on tobacco. Left, wild type tobacco; right, transgenic tobacco (overproducer)

pared to that consumed from leaves of wild type and underproducers after both 1 and 2 days (Figure 12.3). Although the amount of leaf tissue consumed by larvae was somewhat less for normal plants compared to underproducers, it was not significantly so.

Overall, overexpression of peroxidase in *N. sylvestris* generally increases resistance to insects, provided that the insect is small enough and depending on the maturity of the tissue. There appears to be some sort of threshold effect working. Terminal stems produce lower levels of peroxidase than basal stems (Thorpe *et al.*,

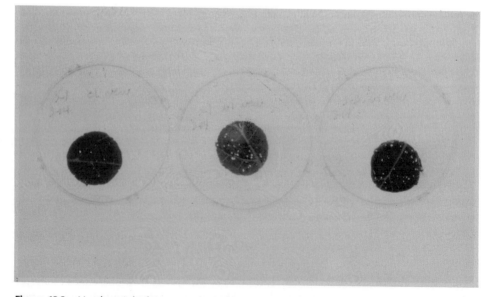

Figure 12.3   *N. sylvestris* leaf tissue consumed by *H. zea* larvae. Left, overproducer; center, wild type; right, underproducer

1978). We also noted this for the tobacco anionic peroxidase expressed in terminal and basal stems in wild type vs. transgenic *N. sylvestris* (Dowd and Lagrimini, 1996a; Dowd *et al.*, 1993). The terminal stems of the transgenics were not resistant to feeding by *H. zea* compared to the wild type, but the corresponding basal stems were resistant. There was no difference in resistance for the older leaves, but there was for younger leaves when newly hatched larvae were used. When older larvae were caged on leaves, there was no resistance. Young leaves of underexpressors were not significantly more susceptible to newly hatched caterpillars compared to wild type leaves.

Tomato

A study similar to the initial tests with *N. sylvestris* was performed with tomato *Lycopersicon esculentum*. Results with *H. zea* larvae caged on leaves (#3–4 fully expanded) of plants in the greenhouse were similar to those seen for *N. sylvestris* in that there was no significant difference in mortality or leaf consumption between the wild type and transgenic plants in two different experiments (Dowd and Lagrimini, 1996a; Dowd *et al.*, 1993). In two studies with basal stems (terminal stems were not tested), higher mortality occurred with stems from transgenic plants (60% and 70%) compared to wild type (35% and 35%), as was seen with tobacco. The same trend was noted when adults of *C. lugubris* were caged with basal stem sections, in that mortality was significantly higher for beetles caged with transgenic stem sections in the study carried out for ten days (80%) compared to those caged with wild type stem sections (30%). Penetration by adults was significantly lower in seven- and ten-day studies for those caged with transgenic stems (30% and 30%, respectively) compared to wild type plants (68% and 60%, respectively). Again, feeding/penetration of both insects was into the pith (Figure 12.4).

Green tomato fruit removed from plants was also tested with adult *C. lugubris* and both third instar and newly hatched larvae of *H. zea*. Adults of *C. lugubris* would not feed on either type of tomato. There was no mortality of third instar *H. zea* caged with either type of fruit, and no difference in weights in two different studies. However, the number of holes chewed through the skin of the fruit was about two times lower for the third instar larvae caged with the transgenic vs. wild type fruit. Mortality of newly hatched *H. zea* larvae caged with transgenic tomato fruit in two different studies was significantly greater (90% and 93%) compared to mortality of larvae caged on wild type fruit in the corresponding studies (40% and 54%).

A field test with the wild type and transgenic tomato was run in 1994 in Columbus, Ohio. Plants were not treated with any pesticides. Although aphid (probably *M. persicae*) numbers were lower on the transgenic plants compared to the wild type ones at the first sample, the numbers on both plant types declined to more equivalent levels over time. There were no significant differences in numbers of adult lady beetles over time. However, overall numbers of lady beetle larvae were significantly lower (by a factor of about two) on the transgenic plants compared to the wild type (Dowd and Lagrimini, 1996a). When total amount of aphids were estimated based on aphids seen, and aphids presumably consumed by lady beetle adults and larvae seen (based on figures from prior studies), calculated aphid numbers were significantly lower on transgenic compared to wild type plants. Insect predators and prey often interact in a sine curve fashion, with the predator lagging behind the prey.

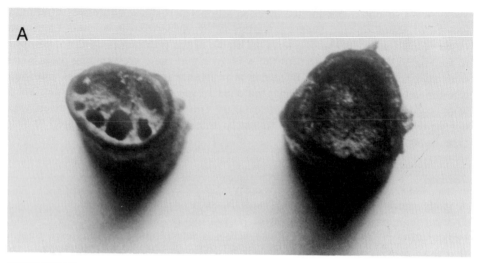

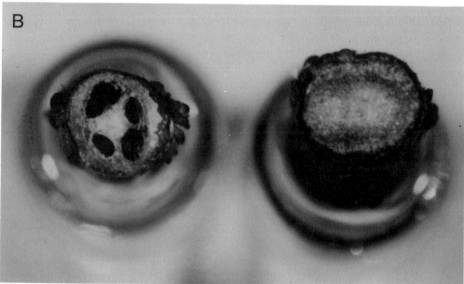

Figure 12.4 **Top**: corn earworm (*Helicoverpa zea*) on tomato stems. Left, wild type tomato; right, transgenic tomato (overproducer). **Bottom**: dusky sap beetle (*Carpophilus lugubris*) on tomato. Left, wild type tomato; right, transgenic tomato (overproducer)

Although a full curve was not obtained from the data, the partial curve suggested this effect was occurring. Thus, the lady beetle larvae were controlling many of the aphids. It is common for the number of eggs laid by lady beetles to reflect the prey population, thus, larger numbers of larvae would be present where larger prey populations occur. Examining our data in this light, it appears that the adult lady beetles had laid significantly fewer eggs on the transgenic plants compared to wild type plants prior to the first sample due to fewer aphids being present, resulting in significantly fewer lady beetle larvae at the time of the first sample. Thus, the number of aphids present prior to sampling was significantly different enough to the lady beetles to result in significantly fewer eggs being laid, and by the time the first sample occurred, significantly fewer lady beetle larvae were seen.

As was seen with tobacco, the stage and part of plant and age and species of insect may determine whether resistance is seen in plants overexpressing the tobacco anionic peroxidase. This was best illustrated in the tomato with the first instar *H. zea* larvae caged with the tomato fruit. The transgenic fruit was more resistant to first instar larvae compared to the wild type. Although not statistically significant, the third instar larvae made fewer holes in the transgenic fruit. This suggests the skin is serving as a barrier, that can be more readily overcome by larger than smaller larvae. Although not clear cut, the laboratory and field data suggest that leaves may be more resistant to the smaller insects (newly hatched caterpillars and aphids) compared to the larger caterpillars tested with leaves and fruit.

Sweet Gum

Timing of studies with sweet gum (*Liquidamber styraciflua*) has been based on the seasonal availability of two tree-feeding insects, the eastern tent caterpillar *Malacosoma americana* and the fall webworm (*Hyphantria cunea*) in Illinois. Although cherry (*Prunus* spp.) and apple (*Malus* spp.) are preferred hosts of these insects, their host range can also include *L. styraciflua* (Baker, 1972; Metcalf and Metcalf, 1993). Initial studies in 1994 were with *H. cunea*. Larvae collected from cherry (ca. 1.5 cm, 30 mg) were caged with one leaf (2–4 mature leaves from terminal) in plastic bags held in growth chambers. Mean damage by larvae was only 0.1% for transgenic plant leaves compared to 31% for wild type leaves (Dowd and Lagrimini, 1996b). Mortality for larvae on transgenic plant leaves at the end of the study was also significantly higher (44%) compared to those caged with wild type leaves (22%). The first (from the terminal) fully expanded leaves were also tested in studies with fall armyworm *Spodoptera frugiperda* and the cigarette beetle *Lasioderma serricorne*. *S. frugiperda* feeds on a wide variety of plants although not sweet gum (Metcalf and Metcalf, 1993). *L. serricorne* is a pest of a wide variety of stored materials in addition to being the major pest of stored tobacco (Metcalf and Metcalf, 1993). Both newly hatched and second instar larvae of *S. frugiperda* were tested. No significant mortality of either instar of *S. frugiperda* larva occurred with either type of leaf disk during the course of the study. Although the amount of damage caused by the two instars of *S. frugiperda* was generally less for the transgenic vs. wild type plants, it was not significantly so. With *L. serricorne*, adults caused significantly less damage to transgenic leaf disks (10%) as compared to wild type leaf disks (34%). A second study was performed in the spring of 1995 due to the availability of *M. americana*. Larvae (3 cm, 200 mg) were collected from apple. In this case, approximately half grown immature leaves (appropriate at season) were placed in plastic bags with two larvae. There were no significant differences in the amount of leaf tissue consumed or number of leaves fed on for either type of leaf (Dowd and Lagrimini, 1996b). Although the LT_{50} was somewhat longer for the transgenic vs. wild type, respectively, the rate of death was about twofold more rapid for the transgenic vs. wild type leaves.

Newly hatched larvae of the European corn borer (*Ostrinia nubilalis*) and *H. zea* were also caged with smaller leaves. Although this tree would not be a natural host for these insect species, they were used for comparative purposes. Both insect species fed readily on the leaves used. The mortality of *O. nubilalis* larvae caged with the transgenic leaves was significantly higher (39%) than that for larvae caged with wild type leaves (12%) after two days. The amount of damage caused by *O. nubilalis* was significantly less on transgenic leaves compared to wild type after two and four days (about half in both cases). In contrast, the mortality of *H. zea* was significantly lower (8.4%) for larvae caged with transgenic leaves compared to wild type ones (20%) after two days. The amount of damage by *H. zea* was significantly greater for transgenic vs. wild type leaves after two and four days (about twice the amount was consumed in both cases). However, in this case the numbers of insects seen after two days was significantly fewer compared to what was set up, suggesting that the wild type larvae had a greater trend toward cannibalism, confounding the interpretation.

Compared to the studies with the tobacco and tomato, the studies with the sweet gum show some similarities. The transgenic plants were often more resistant compared to the wild type. Again, this appeared to be dependent on the age of the leaves (for *H. cunea*), and the species of insect. There was one case of enhanced susceptibility of the transgenic to an unnatural predator (*H. zea*).

General Conclusions

The studies we have conducted so far suggest enhanced peroxidase activity generally increased resistance to insects whenever any significant effect was seen. However, seeing an effect was dependent on the age of tissue, the type of tissue, and the size and species of insect tested. At this point, it is possible to model the results by considering the net effect of the many possible effects of the peroxidase as a type of

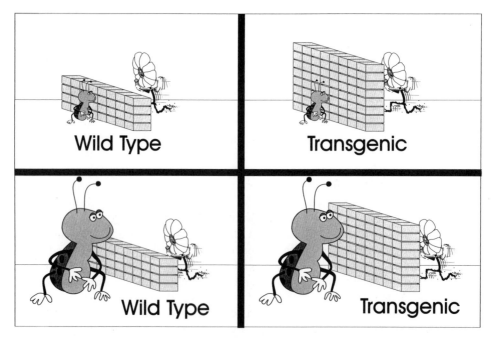

Figure 12.5   Relationship of insect size to barriers set up by transgenic vs. wild type peroxidase

barrier. Depending on the age of the tissue, the barrier may be greater and/or produced more rapidly by the peroxidase. The final size of the barrier may be different for different tissues, and the final size of the barrier may be greater for those plants overexpressing the peroxidase. The size of the insect relative to the barrier may determine how well the insect is able to overcome it. In cases of inappropriate matches of the peroxidase and substrate, the barrier may actually be lowered in the transgenic relative to the wild type.

This relationship is illustrated in a series of cartoons representing different situations. In Figure 12.5, the smaller insect is less able to overcome the barrier set up by the transgenic peroxidase compared to the wild type, while the larger insect can readily overcome both barriers, as was seen with the tomato fruit and the different sized *H. zea* tested. In Figure 12.6, the smaller insect is readily able to overcome the barrier produced by the wild type and underproducer compared to the overproducer, as was seen with leaf disk tests on *N. sylvestris*. In Figure 12.7, the final barrier of the transgenic is greater than the wild type and cannot readily be overcome by the size of insect tested, but the developing barrier can be readily overcome by the same sized insect, as was seen with the *H. zea* larvae caged with the basal and terminal tobacco stems, and as might be predicted with the effects seen with *H. cunea* and the mature *L. styraciflua* leaves. In Figure 12.8, the barrier is altered such that it can be overcome by one insect type, but not by another, as was seen with larvae of *H. zea* and *O. nubilalis*, and immature *L. styraciflua* leaves.

These studies are admittedly limited to closely related dicots (tobacco and tomato) and have extended to another dicot (sweet gum). As already discussed, the chemistry of lignin formation in leaves in herbaceous tissues of herbs and woody plants is similar. The secondary chemistry of the plants is also similar, but of course

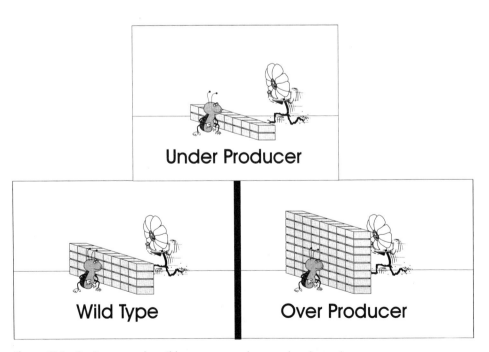

Figure 12.6 Barriers set up by wild type, overproducer and underproducer

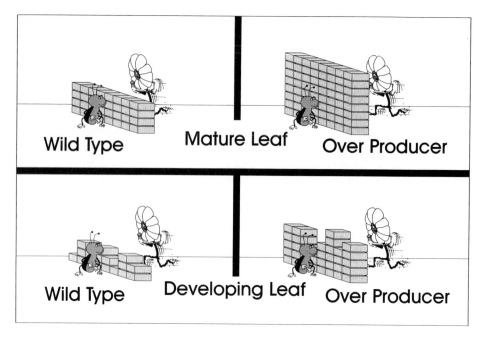

Figure 12.7 Developing vs. final barriers

much more so for the tobacco and tomato. However, the effects on the same insect species (*H. zea*) of the same age on immature leaves were different. Enhanced production of the tobacco anionic peroxidase enhanced resistance of immature *N. sylvestris* leaves to newly hatched *H. zea*, but production of this peroxidase in the immature *L. styraciflua* leaves caused increased susceptibility to this insect. Increased phenolic content, including lignin formed in tobacco stems (Lagrimini, 1991) and tomato stems and fruit (Lagrimini *et al.*, 1993), possibly explains the resistance, although immediate toxic effects are also likely to be involved.

Predictability of the effect in monocots may be even more difficult. Although enhanced peroxidase activity appears to increase resistance to several insects (Dowd,

Figure 12.8 Altered barriers selectively favouring one insect species

1994b; Dowd and Norton, 1995), aberrant lignin production causes increased susceptibility to insects and pathogens in bm mutants of corn (Barriere and Argillier, 1993) (albeit different tissues are involved). As there appears to be some specificity of the different maize peroxidases toward lignin substrates (Dowd and Norton, 1995), introducing a new peroxidase may produce an 'unbalancing' effect on lignin. However, this effect on other processes affected by peroxidases (quinone production, nutrient tie-up, crosslinking of polysaccharides) may offset negative affects on lignin composition. Again it is the net size of the barrier, which may be composed of several factors, that will determine whether enhanced activity of a particular peroxidase introduced through transgenic means will result in increased or decreased resistance to a particular size or species of insect.

Implications and Further Research Needs

Desirable and Undesirable Properties Associated with Transgenic Expression of Peroxidase

As might be expected from the variety of roles in which peroxidases are potentially involved, increasing levels may have undesirable effects on plants. Initially, wilting was noted with transgenic tobacco (Lagrimini *et al.*, 1990) and tomato (Lagrimini *et al.*, 1992) overexpressing tobacco anionic peroxidase. However, plants that received iron supplements and were grown in the greenhouse under temperatures and light conditions in early summer seldom showed wilting compared to wild type plants (even in bolted tobacco plants when soil became dry enough that plants became top heavy and sometimes tipped over) (Dowd and Lagrimini, 1996a; Dowd, personal observation). As indicated, this wilting was apparently a problem with overexpression in roots, which limited root growth (Lagrimini *et al.*, 1996). Plants with tissue selective promoters that avoid roots (such as the chlorophyll a/b binding promoter) have not had this problem (Lagrimini, unpublished data). This problem has also not been noted with the sweet gum. Transgenic tomatoes grown under identical conditions to wild type also had significantly lower numbers of fruit with blossom end rot (Dowd, unpublished observations), a condition that is promoted by a shortage of Ca^{2+} ions. Perhaps the need of the peroxidase for Ca^{2+} ions creates a sink in the fruit that prevents the limited calcium content that leads to this condition. Tobacco plants expressing horseradish prxC1a peroxidase grew more rapidly and flowered earlier than wild type plants, although final plant heights of both types were the same (Kawaoka, 1994).

As would be expected from the wound induced increased browning, bruised flowers of *N. sylvestris* overexpressing tobacco anionic peroxidase browned more severely that those of the corresponding wild type (Dowd, personal observation). This information suggests that general expression would be undesirable for plants where flower quality is important, and in fact it may be desirable to use antisense constructs to reduce browning in flowers, where insects and diseases are adequately controlled by other means. On the other hand, where flowers are not aesthetically important, but are undesirably consumed by insects (such as cucumber beetles), this damage-produced browning of flowers may deter unwanted feeding. Mature leaves of transgenic tobacco, tomato, and sweet gum are visibly darker green than wild type plants (Dowd, personal observation), which may or may not be a desirable horticultural trait.

Greater degrees of lignification would be expected to increase strength of tissues, resulting in less lodging and shredding compared to wild type plants under equivalent environmental conditions. Enhanced cotton fiber strength has been noted in varieties expressing tobacco anionic peroxidase (Maliyakal, 1995). However, increased lignin content or other factors that prevent insect feeding may be undesirable where plant material is used for animal forage (as has been discussed for the bm maize mutants). Similarly, removal of lignin for paper making is a costly process, and increasing lignin content may be undesirable for this use. Again, using antisense constructs may produce more desirable, lower lignin materials, provided that susceptibility to diseases and insects does not increase significantly. This situation may also be of concern to the milling industry, should the grain become much harder than can be handled by the machinery. A higher lignin content may be desirable when the wood is used for burning, due to the higher energy content. On the other hand, more durable grain may be less susceptible to breakage during harvesting and transport, which would be expected to reduce direct product loss and loss as a result of breakage due to more ready accessibility to stored product molds and diseases. Benefits would have to be tempered with digestibility changes and milling machinery requirements. Higher lignin in crop residues may result in more limited decomposition prior to the next growing season, which could cause planting, tillage, or other problems.

Increased levels of phenolics have in several cases caused decreases in the efficacy of insect pathogens, presumably due to binding to the insecticidal protein (Appel, 1993). Simple phenolics can bind the *Bt* crystal protein (e.g. Lord and Undeen, 1990), and viral protein (Appel, 1993). The quinone produced from chlorogenic acid by tomato polyphenol oxidases can covalently bind to viral protein (Felton and Duffey, 1990). Combining *Bt* and peroxidase in transgenic plants may be undesirable or desirable, depending on the net effect on the crystal protein. Combining transgenic proteinase inhibitors and peroxidase may prove synergistic if the peroxidase binds the nutritive proteins (but not the proteinase inhibitors, which would depend on labile groups present), and the peroxidase and proteinase both further inhibit digestive enzymes for the already more limited availability nutritive protein. However, some quinones that may be produced by peroxidases can inhibit activity of some proteinase inhibitors (Felton *et al.*, 1989). Cross-resistance to plant pathogens may also result, although so far this has not been demonstrated in tomato (Lagrimini *et al.*, 1993).

As has been indicated, the net effects of peroxidase overexpression in plants may have a number of modes of action against insects, all taking place concurrently. One strategy of limiting the evolution of resistance, is to use several different agents that act by different mechanisms. One would expect that with the potential multiple modes of action of overexpression of peroxidase against insects, the ability to develop resistance would likely be slower than that of a transgene having a monofunctional effect, such as a proteinase inhibitor. Currently regulatory orientation indicates a greater concern by the United States EPA toward transgene products that affect the insects (such as an insect specific toxin) as opposed to something that primarily affects the plant (*Chemical and Engineering News*, 1994). Although the net effects of the peroxidases are complex, a general toughening does appear to be involved, which would be a plant-oriented trait. As the trait is plant-derived, this is also more desirable from a regulatory standpoint than would be a foreign-introduced trait, such as from an insect pathogen. Consumers are very familiar with the concept of something browning when damaged and being undesirable to them, so a plant trait that causes a similar, understandable response in the plant that makes it undesirable to insects (provided that same part is not to be eaten by the consumer), may make this transgene strategy more acceptable.

Enhanced Predictability of Results – Research Needs and Strategies

Although there are several peroxidase clones now available, enough is not known about their effects and interactions with other plant chemicals to make effects predictable. Availability of isogeneic lines that differ only in peroxidase, and evaluation for insect resistance, may make it more likely to select an appropriate peroxidase(s). More than one peroxidase may be needed, depending on substrate specificities. For example, one would be needed to generate H_2O_2, and another would be needed to lignify. Multiple isoenzymes may be needed for appropriate responses to pathogens (Gaspar *et al.*, 1992), or insects (Dowd, 1994b; Dowd and Norton, 1995). Transposon gene disruption techniques (Bensen *et al.*, 1995) may be useful in delimiting useful peroxidases for insect resistance. However, with the more routine nature of transformation and regeneration for some plant species, it may be more time-effective to 'just do it' and generate transgenic plants expressing peroxidase genes that are available, and study the effects.

References

Allen, R. D. (1995) Dissection of oxidative stress tolerance using transgenic plants, *Plant Physiol.* **107**, 1049–1054.

Angelini, R. and Federico, R. (1989) Histochemical evidence of polyamine oxidation and generation of hydrogen peroxide in the cell wall, *J. Plant Physiol.* **135**, 212–217.

Appel, H. M. (1993) Phenolics in ecological interactions: the importance of oxidation, *J. Chem. Ecol.* **19**, 1521–1553.

Arnason, J. T., Baum, B., Gale, J., Lambert, J. D. H., Bergvinson, D., Philogene, B. J. R., Serratos, J. A., Mihm, J. and Jewell, D. C. (1994) Variation in resistance of Mexican land races of maize to maize weevil *Sitophilus zeamais*, in relation to taxonomic and biochemical parameters, *Euphytica* **74**, 227–236.

Baker, W. L. (1972) *Eastern Forest Insects*, Washington, DC: US Department of Agriculture, Miscellaneous Publication No. 1175.

Barriere, Y. and Argillier, O. (1993) Brown-midrib genes of maize: a review, *Agronomie* **13**, 865–876.

Barz, W. and Hoesel, W. (1978) Metabolism and degradation of phenolic compounds in plants, *Recent Adv. Phytochem.* **12**, 339–369.

Barz, W. and Nicholas, H. A. (1978) Metabolism of phenolics and vitamins in cell cultures, In: Thorpe, T. A. (Ed.), *Frontiers of Plant Tissue Culture*, Calgary: International Association of Plant Tissue Culture, University of Calgary, pp. 345–352.

Bell, A. A. (1981) Biochemical mechanisms of disease resistance, *Annu. Rev. Plant Physiol.* **32**, 21–81.

Bensen, R. J., Johal, G. S., Crane, V. C., Tossberg, J. T., Schnable, P. S., Meeley, R. B. and Briggs, S. P. (1995) Cloning and characterization of the maize An1 gene, *Plant Cell* **7**, 75–84.

Bergvinson, D. J., Hamilton, R. I. and Arnason, J. T. (1995a) Leaf profile of maize resistance factors to European corn borer, *Ostrinia nubilalis, J. Chem. Ecol.* **21**, 343–354.

Bergvinson, D. J., Larsen, J. S. and Arnason, J. T. (1995b) Leaf profile of maize resistance factors to European corn borer, *Ostrinia nubilalis, J. Chem. Ecol.* **21**, 343–354.

Bergvinson, D. J., Arnason, J. T., Hamilton, R. I., Mihm, J. A. and Jewell, D. C. (1994) Determining leaf toughness and its role in maize resistance to the European corn borer (Lepidoptera: Pyralidae), *J. Econ. Entomol.* **87**, 1743–1748.

Berlin, J. and Barz, W. (1975) Oxidative decarboxylation of *para*-hydroxybenzoic acids by peroxidases under *in vivo* and *in vitro* conditions, *Z. Naturforsch.* **30c**, 650–658.

Bernards, M. A. and Lewis, N. G. (1992) Alkyl ferulates in wound healing potato tubers, *Phytochemistry* **31**, 3409–3412.

Brad, I., Terbea, M., Marcu, Z. and Hurduc, N. (1974) Influenta infectarii cu *Fusarium moniliforme* Scheld a unor linii si hibrizi de porumb, a supra spectrului izoenzimelor peroxidazice, *Stud. Cerret. Biochim.* **17**, 11–14.

Bradley, D. J., Kjellbom, P. and Lamb, C. J. (1992) Elicitor- and wound-induced oxidative cross-linking of a proline-rich plant cell wall protein: a novel, rapid defense response, *Cell* **70**, 21–31.

Brewbaker, J. L. and Hasegawa, Y. (1975) Polymorphisms of the major peroxidases of maize, In: Markert, C. L. (Ed.), *Isoenzymes III: Developmental Biology*, New York: Academic, pp. 659–673.

Brewbaker, J. L., Nagai, C. and Liu, E. H. (1985) Genetic polymorphisms of 13 maize peroxidases, *J. Hered.* **76**, 159–167.

Bronner, R., Westphal, E. and Dreger, F. (1991) Enhanced peroxidase activity associated with the hypersensitive response of *Solanum dulcamara* to the gall mite *Aceria cladophthirus* (Acari: Eriophyoidea), *Can. J. Bot.* **69**, 2192–2196.

Buffard, D., Breda, C., van Huystee, R. B., Asemota, O., Pierre, M., Dang Ha, D. B. and Esnault, R. (1990) Molecular cloning of complementary DNAs encoding two cationic peroxidases from cultivated peanut cells, *Proc. Natl. Acad. Sci. USA* **87**, 8874–8878.

Byrne, P. F., Darrah, L. L., Simpson, K. B., Keaster, A. J., Barry, B. D. and Zuber, M. S. (1989) Maize silk pH as an indicator of resistance to the corn earworm (Lepidoptera: Noctuidae), *Environ. Entomol.* **18**, 356–360.

Byrne, P. F., Darrah, L. L., Snook, M. E., Wiseman, B. R., Widstrom, N. W., Mollenbeck, D. J. and Berry, D. B. (1996) Maize silk-browning, maysin content, and antibiosis to the corn earworm, *Helicoverpa zea* (Boddie), *Maydica* **41**, 13–18.

Campa, A. (1991) Biological roles of plant peroxidases: known and potential functions, In: Everse, J., Everse, K. E. and Grisham, M. B. (Eds), *Peroxidases in Chemistry and Biology*, Boca Raton: CRC Press, pp. 25–50.

Cardeña-Gomez, G. and Nicholson, R. L. (1987) Papilla formation and associated peroxidase activity: a non-specific response to attempted fungal penetration of maize, *Physiol. Mol. Plant Pathol.* **31**, 51–67.

Chemical and Engineering News (1994) 'Plant pesticides' attract attention of EPA regulators, *Chem. Eng. News* **72**, 28.

Chen, Z., Silva, H. and Klessig, D. F. (1993) Active oxygen species in the induction of plant systemic acquired resistance by salicylic acid, *Science* **262**, 1883–1886.

Coley, P. D. (1983) Herbivory and defensive characteristics of tree species in lowland tropical forest, *Ecol. Monogr.* **53**, 209–233.

Connell, W. A. (1956) Nitidulidae of Delaware, *Delaware Agric. Exper. Stat. Bull.* **318**, 1–67.

Davin, L. B. and Lewis, N. G. (1992) Phenylpropanoid metabolism: biosynthesis of monolignols, lignans and neolignans, lignins and suberins, In: Stafford, H. A. and Ibrahim, R. K. (Eds), *Phenolic Metabolism in Plants*, New York: Plenum Press, pp. 325–375.

Dowd, P. F. (1992) Corn kernel aryl alcohol oxidases as cross-resistance mechanisms for fungi and insects, In: *Abstracts of the 203rd meeting of the American Chemical Society*, Washington, DC: American Chemical Society, #AGRO-107.

Dowd, P. F. (1994a) Examination of an *Aspergillus flavus* resistant inbred of maize for cross-resistance to sap beetles (Coleoptera: Nitidulidae), *Entomol. Exper. Appl.* **77**, 177–180.

Dowd, P. F. (1994b) Enhanced maize (*Zea mays* L.) pericarp browning: associations with insect resistance and involvement of oxidizing enzymes, *J. Chem. Ecol.* **20**, 2497–2523.

Dowd, P. F. (1995) Sap beetles and mycotoxins in maize, *Food Addit. Contam.* **12**, 497–508.

Dowd, P. F. and Lagrimini, L. M. (1996a) Examination of transgenic tobacco and tomato overexpressing tobacco anionic peroxidase for resistance to insects (submitted).

Dowd, P. F. and Lagrimini, L. M. (1996b) Examination of transgenic sweet gum over-expressing tobacco anionic peroxidase for resistance to insects (submitted).

Dowd, P. F. and Norton, R. A. (1995) Browning-associated mechanisms of resistance to insects in corn callus tissue, *J. Chem. Ecol.* **21**, 583–600.

Dowd, P. F. and Vega, F. E. (1996) Enzymatic oxidation products of allelochemicals as a potential direct resistance mechanism against insects: effects on the corn leafhopper *Dalbulus maidis*, *Nat. Toxins* **4**, 85–91.

Dowd, P. F., Lagrimini, L. M. and Vega, F. E. (1993) Insect resistance of transgenic tobacco and tomato tissues expressing high levels of tobacco anionic peroxidase, presentation at the National Meeting of the Entomological Society of America, Indianapolis, December.

Dowd, P. F., Duvick, J. A., Rood, T. and Norton, R. A. (1996) Enhancement of the toxicity of maize allelochemicals to maize ear pathogens by enzymatic oxidation (in preparation).

Dunford, H. B. (1991) Horseradish peroxidase: structure and kinetic properties, In: Everse, J., Everse, K. E. and Grisham, M. B. (Eds), *Peroxidases in Chemistry and Biology, Vol. II*, Boca Raton: CRC Press, pp. 1–24.

Elstner, E. F. (1982) Oxygen activation and oxygen toxicity, *Ann. Rev. Plant Physiol.* **33**, 73–96.

Elstner, E. F. and Heupel, A. (1976) Formation of hydrogen peroxide by isolated cell walls from horseradish (*Armoracia lapathifolia* Gibb.), *Planta* **130**, 175–180.

Epstein, E., Cohen, J. D. and Bandurski, R. S. (1980) Concentration and metabolic turnover of indoles in germinating kernels of *Zea mays* L. (Maize), *Plant Physiol.* **65**, 415–421.

Espelie, K. E., Franceschi, V. R. and Kolattukudy, P. E. (1986) Immunocytochemical localization and time course appearance of an anionic peroxidase associated with suberization in wound-healing potato tuber tissue, *Plant Physiol.* **81**, 487–492.

Felton, G. W. and Duffey, S. S. (1990) Inactivation of baculovirus by quinones formed in insect-damaged plant tissues, *J. Chem. Ecol.* **16**, 1221–1236.

Felton, G. W. and Duffey, S. S. (1991) Protective action of midgut catalase in lepidopteran larvae against oxidative plant defenses, *J. Chem. Ecol.* **17**, 1715–1732.

Felton, G. W. and Summers, C. B. (1995) Antioxidant systems in insects, *Arch. Insect Biochem. Physiol.* **29**, 187–197.

Felton, G. W., Donato, K., DelVecchio, R. J. and Duffey, S. S. (1989) Activation of plant foliar oxidases by insect feeding reduces nutritive quality of foliage for noctuid herbivores, *J. Chem. Ecol.* **15**, 2667–2694.

Felton, G. W., Donato, K. K., Broadway, R. M. and Duffey, S. S. (1992) Impact of oxidized plant phenolics on the nutritional quality of dietary protein to a noctuid herbivore, *Spodoptera exigua*, *J. Insect Physiol.* **38**, 277–285.

Fincher, G. B. and Stone, B. A. (1986) Cell walls and their components in cereal grain technology, *Adv. Cereal Sci. Technol.* **8**, 207–295.

Fry, S. C. (1979) Phenolic components of the primary cell wall and their possible role in the hormonal regulation of growth, *Planta* **146**, 343–351.

Fry, S. C. (1980) Gibberellin-controlled pectinic acid and protein secretion in growing cells, *Phytochemistry* **19**, 735–740.

Fry, S. C. (1982) Isodityrosine, a new cross-linking amino acid from plant cell-wall glycoprotein, *Biochem. J.* **20**, 449–455.

Fry, S. C. (1986) Polymer-bound phenols as natural substrates of peroxidases, In: Greppin, H., Penel, C. and Gaspar, T. (Eds), *Molecular and Physiological Aspects of Plant Peroxi-*

dases, Geneva: University of Geneva, pp. 169–181.

Fry, S. C. (1989) Gibberellin-controlled pectinic acid and protein secretion in growing cells, *Phytochemistry* **19**, 735–740.

Fujiyama, K., Takemura, H., Shinmyo, A., Okada, H. and Takano, M. (1990) Genomic DNA structure of two new horseradish peroxidase-encoding genes, *Gene* **89**, 163–169.

Fujiyama, K., Takemura, H., Shibayama, S., Kobayashi, K., Choi, J-K., Shinmyo, A. Takano, M. and Okada, H. (1988) Structures of the horseradish peroxidase isoenzyme C. genes, *Eur. J. Biochem.* **173**, 681–687.

Fulcrand, H., Cheminat, A., Brouillard, R. and Cheynier, V. (1994) Characterization of compounds obtained by chemical oxidation of caffeic acid in acidic conditions, *Phytochemistry* **35**, 499–505.

Gaspar, T., Penel, C., Thorpe, T. and Greppin, H. (1982) *Peroxidases: 1970–1980. A Survey of their Biochemical and Physiological Roles in Higher Plants*, Geneva: University of Geneva, Center of Botany.

Gaspar, T., Penel, C., Hagege, D. and Greppin, H. (1992) Peroxidases in plant growth, differentiation and development processes, In: Lobarzewski, J., Greppin, H., Penel, C. and Gaspar, T. (Eds), *Biochemical, Molecular and Physiological Aspects of Plant Peroxidases*, Geneva: University of Geneva, pp. 249–280.

Gazaryan, I. G. and Lagrimini, L. M. (1996a) Tobacco anionic peroxidase overexpressed in transgenic plants. I. Purification and unusual kinetic properties, *Phytochemistry* **41**, 1029–1034.

Gazaryan, I. G. and Lagrimini, L. M. (1996b) Tobacco anionic peroxidase overexpressed in transgenic plants. II. Aerobic oxidation of indole-3-acetic acid, *Phytochemistry* **42**, 1271–1278.

Geiger, W. B. (1946) The mechanism of the antibacterial action of quinones and hydroquinones, *Arch. Biochem.* **11**, 23–31.

Greisbach, H. A. (1981) Lignins, In: Conn, E. E. (Ed.), *The Biochemistry of Plants*, New York: Academic, pp. 457–480.

Gross, G. G. (1978) Recent advances in the chemistry and biochemistry of lignin, *Recent Adv. Phytochem.* **12**, 177–220.

Halliwell, B. (1978) Lignin synthesis: the generation of hydrogen peroxide and superoxide by horseradish peroxidase and its stimulation by manganese (II) and phenols, *Planta* **140**, 81–88.

Hedin, P. A., Davis, F. M., Williams, W. P. and Salin, M. L. (1984) Possible factors of leaf-feeding resistance in corn to the southwestern corn borer, *J. Agric. Food Chem.* **32**, 262–267.

Hertig, C., Rebmann, G., Bull, J., Mauch, F. and Dudler, R. (1991) Sequence and tissue-specific expression of a putative peroxidase gene from wheat (*Triticum aestivum* L.), *Plant Mol. Biol.* **16**, 171–174.

Hoffman-Ostenhof, O. (1947) Mechanisms of the antibiotic action of certain quinones, *Science* **105**, 549–550.

Hoyle, M. C. (1977) High resolution of peroxidase-indoleacetic acid oxidase isoenzymes from horseradish by isoelectrofocusing, *Plant Physiol.* **60**, 787–793.

Hwang, R. H., Kennedy, J. F. and Melo, E. H. M. (1991) A mechanism for lignification in plants, *Carbo. Polymer* **14**, 77–88.

Hwang, S-Y. (1993) ABA induction of corn flooding tolerance through root lignification: physiological study and molecular cloning of a peroxidase gene, unpublished PhD thesis, Ohio State University.

Iiyama, K., Lam, T. B. T. and Stone, B. A. (1994) Covalent cross-links in the cell wall, *Plant Physiol.* **104**, 315–320.

Intapruk, C., Higashimura, N., Yamamoto, K., Okada, N., Shinmyo, A. and Takano, M. (1991) Nucleotide sequences of two genomic DNAs encoding peroxidase of *Arabidopsis thaliana*, *Gene* **98**, 237–241.

International Union of Biochemistry and Molecular Biology on the Nomenclature and Classification of Enzymes, Nomenclature Committee (1992) *Enzyme Nomenclature 1992*, New York: Academic.

Kanofsky, J. (1991) Peroxidase-catalyzed generation of singlet oxygen and of free radicals, In: Everse, J., Everse, K. E. and Grisham, M. B. (Eds), *Peroxidases in Chemistry and Biology, Volume II*, Boca Raton: CRC Press, pp. 219–237.

Kawaoka, A., Kawamoto, T., Moriki, H., Murakami, A., Murakami, K., Yoshida, K., Sekine, M., Takano, M. and Shinmyo, A. (1994) Growth-stimulation of tobacco plant by the introduced horseradish peroxidase gene prxC1a, *J. Fermen. Bioeng.* **78**, 49–53.

Kolattukudy, P. E. and Dean, B. B. (1974) Structure, gas chromatographic measurements and function of suberin synthesized by potato tuber slices, *Plant Physiol.* **54**, 116–121.

Kolattukudy, P. E., Mohan, R., Bajar, M. A. and Sherf, B. A. (1992) Plant peroxidase gene expression and function, *Biochem. Soc. Trans.* **20**, 333–337.

Kostyuk, V. A. (1987) Inhibition of microsomal oxidation by *o*-benzoquinone derivatives, *Biochemistry (USSR)* **52**, 305–308.

Lagrimini, L. M. (1991) Wound-induced deposition of polyphenols in transgenic plants over-expressing peroxidase, *Plant Physiol.* **96**, 577–583.

Lagrimini, L. M. (1992) Plant peroxidases: under- and over-expression in transgenic plants and physiological consequences, In: Penel, C., Gaspar, T. and Greppin, H. (Eds), *Plant Peroxidases 1980–1990*, Geneva: University of Geneva, pp. 59–69.

Lagrimini, L. M. and Rothstein, S. (1987) Tissue specificity of tobacco peroxidase isoenzymes and their induction by wounding and tobacco mosaic virus infection, *Plant Physiol.* **84**, 438–442.

Lagrimini, L. M., Bradford, S. and Rothstein, S. (1990) Peroxidase induced wilting in transgenic tobacco plants, *Plant Cell* **2**, 7–18.

Lagrimini, L. M., Liu, T. Y. and Joly, R. J. (1996) Enhanced peroxidase activity suppresses root growth in transgenic plants, *J. Exper. Bot.* (in press).

Lagrimini, L. M., Burkhart, W., Moyer, M. and Rothstein, S. (1987) Molecular cloning of complementary DNA encoding the lignin forming peroxidase from tobacco: molecular analysis and tissue-specific expression, *Proc. Natl. Acad. Sci.* **84**, 438–442.

Lagrimini, L. M., Vaughn, J., Erb, W. A. and Miller, S. A. (1993) Peroxidase overproduction in transgenic tomato plants: wound induced polyphenol deposition and disease resistance, *Hortscience* **28**, 218–221.

Lagrimini, L. M., Vaughn, J., Finer, J., Klotz, K. and Rubaihayo, P. (1992) Expression of a tobacco peroxidase gene in transformed tomato plants, *J. Am. Soc. Hortic. Sci.* **117**, 1012–1016.

Lamport, D. T. A. (1989) Extensin peroxidase ties the knots in the extensin network, In: Osborne, D. J. and Jackson, M. B. (Eds), *Cell Separation in Plants: Physiology, Biochemistry and Molecular Biology*, Berlin: Springer-Verlag, pp. 101–113.

Levine, A., Tenhaken, R., Dixon, R. and Lamb, C. (1994) Hydrogen peroxide from the oxidative burst orchestrates the plant hypersensitive response, *Cell* **79**, 583–593.

Levings, C. S. and Stuber, C. W. (1971) A maize gene controlling silk browning in response to wounding, *Genetics* **69**, 491–498.

Lord, J. C. and Undeen, A. H. (1990) Inhibition of the *Bacillus thuringiensis* var. *israelensis* toxin by dissolved tannins, *Environ. Entomol.* **19**, 1547–1551.

Macri, F., DiLenna, P. and Vianello, A. (1974) Preliminary research on peroxidase, polyphenol oxidase and phenol content in healthy and infected corn leaves, susceptible and resistant to *Helminthosporium maidis* race, T. *Riv. Pathol. Veg. Ser. 4*, **10**, 109–121.

Mäder, M. and Füssl, R. (1982) Role of peroxidase in lignification of tobacco cells, *Plant Physiol.* **70**, 1132–1134.

Mäder, M., Nessel, A. and Bopp, M. (1977) Uber die physiologie Bedeutung der Peroxidase-Isoenzymgruppen des Tabaks anhand einiger biochemischer Eigenschaften. II. pH Optima, Michaelis Konstanten, Maximale Oxidationsraten, *Zeitschr. Pflanzenphysiol.* **82**,

247–260.

Maliyakal, J. (1995) Transgenic cotton plants producing heterologous peroxidase, International Patent Application No. WO 95/08914.

Marañón, M. J. R. and van Huystee, R. B. (1994) Plant peroxidases: interactions between their prosthetic groups, *Phytochemistry* **37**, 1217–1225.

Markwalder, H.-O. and Neukom, H. (1976) Diferulic acid as a possible crosslink in hemicelluloses from wheat germ, *Phytochemistry* **15**, 836–837.

Mayer, A. M. (1987) Polyphenol oxidases in plants – recent progress, *Phytochemistry* **26**, 11–20.

Mazza, G. and Welinder, K. G. (1980) Covalent structure of turnip peroxidase 7. Cyanogen bromide fragments, complete structure and comparison to horseradish peroxidase C, *Eur. J. Biochem.* **108**, 481–489.

McCune, D. C. and Galston, A. W. (1959) Inverse effects of gibberellin on peroxidase activity and growth in dwarf strains of peas and corn, *Plant Physiol.* **34**, 416–418.

Mehdy, M. C. (1994) Active oxygen species in plant defense against pathogens, *Plant Physiol.* **105**, 467–472.

Metcalf, R. L. and Metcalf, R. A. (1993) *Destructive and Useful Insects*, New York: McGraw-Hill.

Metlintskii, L. V., Ozeretskovskaya, O. L., Savel'eva, O. N., Baliauri, V. D. and Stom, D. I. (1972) Participation of caffeic acid and products of its enzymatic transformation in protective reactions of the potato against *Phytophthora infestans*, *Dokl. Akad. Nauk. SSR. Ser. Biol.* **202**, 228–231.

Meyer, L. (1988) Role of cyclic hydroxamic acids in resistance of maize to microbial diseases and insects, *Zentrabl. Microbiol.* **143**, 39–46.

Miles, P. W. and Oertli, J. J. (1993) The significance of antioxidants in aphid–plant interactions: the redox hypothesis, *Entomol. Exper. Appl.* **67**, 275–283.

Miles, P. W. and Peng, Z. (1989) Studies on the salivary physiology of plant bugs: detoxification of phytochemicals by the salivary peroxidase, *J. Insect Physiol.* **35**, 865–872.

Moerschbacher, B. M. (1992) Plant peroxidases: involvement in response to pathogens, In: Penel, C., Gaspar, T. and Greppin, H. (Eds), *Plant Peroxidases 1980–1990*, Geneva: University of Geneva, pp. 91–99.

Morgens, P. H., Callahan, A. M., Dunn, L. J. and Abeles, F. B. (1990) Isolation and sequencing of cDNA clones encoding ethylene-induced putative peroxidases from cucumber cotyledons, *Plant Mol. Biol.* **14**, 715–725.

Nicholson, R. L., Bauman, L. F. and Warren, H. L. (1976) Association of *Fusarium moniloforme* with brown midrib maize, *Plant Dis. Report.* **60**, 908–910.

O'Connell, R. J., Brown, I. R., Mansfield, J. W., Bailey, J. A., Mazau, D., Rumeau, D. and Esquerre-Tugaye, M. T. (1990) Immunocytochemical localization of hydroxyproline-rich glycoproteins accumulating in melon and bean at sites of resistance to bacteria and fungi, *Mol. Plant Microbe Inter.* **3**, 33–40.

Ogawa, S., Shiro, Y. and Marishima, I. (1979) Calcium binding by horseradish peroxidase C and the heme environmental structure, *Biochem. Biophys. Res. Commun.* **90**, 674–681.

Orr, W. C. and Sohal, R. S. (1992) The effects of catalase gene overexpression on life span and resistance to oxidative stress in transgenic *Drosophila melanogaster*, *Arch. Biochem. Biophys.* **297**, 35–41.

Pang, A., Catesson, A. M., Francesch, C., Rolando, C. and Goldberg, R. (1989) On substrate specificity of peroxidase involved in the lignification process, *J. Plant Physiol.* **135**, 325–329.

Patzlaff, M. and Barz, W. (1978) Peroxidatic degradation of flavonones, *Z. Naturforsch.* **33c**, 675–684.

Penel, C. (1986) The role of calcium in the control of peroxidase activity. In: Greppin, H., Penel, C. and Gaspar, T. (Eds), *Molecular and Physiological Aspects of Plant Peroxidases*,

Geneva: University of Geneva, pp. 155–164.

Penel, C., Gaspar, T. and Greppin, H. (1992) *Plant Peroxidases 1980–1990*, Geneva: University of Geneva.

Peng, Z. and Miles, P. W. (1988) Acceptability of catechin and its oxidative condensation products to the rose aphid, *Macrosiphum rosae*, *Entomol. Exper. Appl.* **47**, 255–265.

Pfanz, H. (1993) Oxidation of IAA by extracellular peroxidases, In: Welinder, K. G., Rasmussen, S. K., Penel, C. and Greppin, H. (Eds), *Plant Peroxidases: Biochemistry and Physiology*, Geneva: University of Geneva, pp. 169–174.

Poulos, T. L., Edwards, S. L., Wariishi, H. and Gold, M. H. (1993) Crystallographic refinement of lignin peroxidase at 2A, *J. Biol. Chem.* **268**, 4429–4440.

Rahbé, Y. and Febvay, G. (1993) Protein toxicity to aphids: an *in vitro* test on *Acyrthosiphon pisum*, *Entomol. Exper. Appl.* **67**, 149–160.

Rasmussen, S. K., Welinder, K. G. and Heijgaard, J. (1991) cDNA cloning, characterization and expression of an endosperm-specific barley peroxidase, *Plant Mol. Biol.* **16**, 317–327.

Raupp, M. J. (1985) Effects of leaf toughness on mandibular wear of the leaf beetle, *Plagiodera versicolora*, *Ecol. Entomol.* **10**, 73–79.

Rebmann, G., Hertig, C., Bull, J., Mauch, F. and Dudler, R. (1991) Cloning and sequencing of cDNAs encoding a pathogen-induced putative peroxidase of wheat (*Triticum aestivum* L.), *Plant Mol. Biol.* **16**, 329–331.

Reigh, D. L., Wender, S. H. and Smith, E. C. (1973) Scopoletin: a substrate for an isoperoxidase from *Nicotiana tabacum* tissue culture W-38, *Phytochemistry* **12**, 1265–1268.

Reimmann, C., Ringli, C. and Dudler, R. (1992) Complementary DNA cloning and sequence analysis of a pathogen-induced putative peroxidase from rice, *Plant Physiol.* **100**, 1611–1612.

Riley, R. G. and Kolattukudy, P. E. (1975) Evidence for covalently attached *p*-coumaric acid and ferulic acid in cutins and suberins, *Plant Physiol.* **56**, 650–654.

Roberts, E. and Kolattukudy, P. E. (1989) Molecular cloning, nucleotide sequence, and abscisic acid induction of a suberization associated highly anionic peroxidase, *Mol. Genes Gene.* **217**, 223–232.

Roberts, E., Kutchan, T. and Kolattukudy, P. E. (1988) Cloning and sequencing of cDNA for a highly anionic peroxidase from potato and the induction of its mRNA in suberizing potato fruits, *Plant Mol. Biol.* **11**, 5–26.

Robinson, T. (1979) The evolutionary ecology of alkaloids. In: Rosenthal, G. A. and Janzen, D. H. (Eds), *Herbivores: Their Interactions with Plant Secondary Metabolites*, New York: Academic Press, pp. 413–448.

Schmeltz, I. (1971) Nicotine and other tobacco alkaloids, In: Jacobson, M. and Crosby, D. G. (Eds), *Naturally Occurring Insecticides*, New York: Marcel Dekker, pp. 99–136.

Scott, G. E. and Zummo, N. (1988) Sources of resistance maize to kernel infection by *Aspergillus flavus* in the field, *Crop Sci.* **28**, 504–507.

Serratos, J. A., Arnason, J. T., Nozzolillo, C., Lambert, J. D. H., Philogene, B. J. R., Fulcher, G., Davidson, K., Peacock, L., Atkinson, J. and Morand, P. (1987) Factors contributing to resistance of exotic maize populations to maize weevil, *Sitophilus zeamais*, *J. Chem. Ecol.* **13**, 751–762.

Shah, D. M., Rommens, C. M. T. and Beachy, R. N. (1995) Resistance to diseases and insects in transgenic plants: progress and applications to agriculture, *Trends Biotechnol.* **13**, 362–368.

Sheen, S. J. (1974) Polyphenol oxidation by leaf peroxidases in *Nicotiana*, *Bot. Gaz.* **135**, 155–161.

Shin, S. Y., Calvisi, E. G., Beaman, T. C., Pankratz, S. H., Gerhardt, P. and Marquis, R. E. (1994) Microscopic and thermal characterization of hydrogen peroxide killing and lysis of spores and protection by transition metal ions, chelators, and antioxidants, *Appl. Environ. Microbiol.* **60**, 3192–3197.

Sissler, H. D. and Cox, C. E. (1960) Physiology of fungitoxicity, In: Horsfall, J. G. and

Dimond, A. E. (Eds), *Plant Pathology*, Vol. 2, New York: Academic Press, pp. 507–552.

Stafford, H. A. (1960) Differences between lignin-like polymers formed by peroxidation of eugenol and ferulic acid in leaf sections of *Phleleum*, *Plant Physiol.* **35**, 108–114.

Stafford, H. A. and Ibrahim, R. K. (1992) Phenolic metabolism in plants, *Recent Advances in Phytochemistry*, Vol. 26, Plenum Press.

Stamopoulos, D. C. (1988) Toxic effect of lignin extracted from the tegument of *Phaseolus vulgaris* seeds on the larvae of *Acanthoscleides obtectus* (Say) (Col., Bruchidae), *J. Appl. Entomol.* **105**, 317–320.

Stipanovic, R. D. (1983) Function and chemistry of plant trichomes and glands in insect resistance, In: Hedin, P. E. (Ed.), *Plant Resistance to Insects*, Washington, DC: American Chemical Society, pp. 69–100.

Stout, M. J., Workman, J. and Duffey, S. S. (1994) Differential induction of tomato foliar proteins by arthropod herbivores, *J. Chem. Ecol.* **20**, 2575–2594.

Styles, E. D. and Ceska, O. (1977) The genetic control of flavonoid synthesis in maize, *Can. J. Genet. Cytol.* **19**, 289–302.

Sullivan, J. and Lagrimini, L. M. (1993) Transformation of *Liquidamber styraciflua* using *Agrobacterium tumefaciens*, *Plant Cell Report* **12**, 303–306.

Summers, C. B. and Felton, G. W. (1994) Prooxidant effects of phenolic acids on the generalist herbivore *Helicoverpa zea* (Lepidoptera: Noctuidae): potential mode of action for phenolic compounds in plant anti-herbivore chemistry, *Insect Biochem. Mol. Biol.* **9**, 943–953.

Swain, T. (1979) Tannins and lignins, In: Rosenthal, G. A. and Janzen, D. H. (Eds), *Interactions of Herbivores with Plant Secondary Metabolites*, New York: Academic Press, pp. 657–682.

Theilade, B. and Rasmussen, S. K. (1992) Structure and chromosomal localization of the gene encoding barley seed peroxidase BP 2A, *Gene* **118**, 261–266.

Thorpe, T. A., Van, M. T. T. and Gaspar, T. (1978) Isoperoxidases in epidermal layers of tobacco and changes during organ formation *in vitro*, *Physiol. Plant.* **44**, 388–394.

Urs, N. V. R. and Dunleavy, J. M. (1975) Enhancement of the bactericidal activity of a peroxidase system by phenolic compounds, *Phytopathology* **65**, 686–690.

Vance, C. P., Kirk, T. K. and Sherwood, R. T. (1980) Lignification as a mechanism of disease resitance, *Annu. Rev. Phytopathol.* **18**, 259–288.

Venere, R. J. (1980) Role of peroxidase in cotton resistant to bacterial blight, *Plant Sci. Lett.* **20**, 47–56.

Wainhouse, D., Cross, D. J. and Howell, R. S. (1990) The role of lignin as a defence against the spruce bark beetle *Dendroctonus micans*: effect on larvae and adults, *Oecologia* **85**, 257–265.

Waiss, A. C., Chan, B. G., Elliger, C. A., Wiseman, B. R., McMillian, W. W., Widstrom, N. W., Zuber, M. S. and Keaster, A. J. (1979) Maysin, a flavone glycoside from corn silks with antibiotic activity toward corn earworm, *J. Econ. Entomol.* **72**, 256–258.

Welinder, K. G. (1991) The plant peroxidase superfamily, In: Lobarzewski, J., Greppin, H., Penel, C. and Gaspar, T. (Eds), *Biochemical, Physiological and Molecular Aspects of Plant Peroxidases*, Geneva: University of Geneva Press, pp. 3–14.

Welinder, K. G. and Gajhede, M. (1993) Structure and evolution of peroxidases, In: Welinder, K. G., Rasmussen, S. K., Penel, C. and Greppin, H. (Eds), *Plant Peroxidases: Biochemistry and Physiology*, Geneva: University of Geneva Press, pp. 35–42.

Williams, W. G., Kennedy, G. G., Yamamota, R. T., Thacker, J. D. and Bordner, J. (1980) 2-tridecanone: a naturally occurring insecticide from the wild tomato *Lycopersicon hirsutum f. glabratum*, *Science* **207**, 888–889.

Wiseman, B. R. and Isenhour, D. J. (1993) Interaction of diet ingredients with levels of silk of a corn genotype resistant to corn earworm (Lepidoptera: Noctuidae), *J. Econ. Entomol.* **86**, 1291–1296.

Xu, Y. and van Huystee, R. B. (1993) Association of calcium and calmodulin to peroxidase secretion and activation, *J. Plant Physiol.* **141**, 141–146.

Young, S. A., Guo, A., Guikema, J. A., White, F. F. and Leach, J. E. (1995) Rice cationic peroxidase accumulates in xylem vessels during incompatible interactions with *Xanthomonas oryzae* pv *oryzae*, *Plant Physiol.* **107**, 1333–1341.

Acknowledgments

We thank the many students and collaborators who have been involved in these studies over the years, without whom little progress would have been made.

Disclaimer: This article reports the results of research only. Mention of a proprietary product does not constitute an endorsement or a recommendation for its use by USDA.

13

Insecticidal Compounds Induced by Regulated Overproduction of Cytokinins in Transgenic Plants

ANN SMIGOCKI, SUNGGI HEU, IRIS MCCANNA, CHRIS WOZNIAK and GEORGE BUTA

Introduction

Cytokinins are a major group of plant growth regulators that modulate a number of physiological and biochemical processes (Mok and Mok, 1994). They markedly affect flowering, fruit set, ripening, leaf senescence, seed germination, and stomatal function (Davies, 1987; Skoog and Miller, 1957) and, as a result, are being used as bioregulators in commercial applications. The combination of endogenous and exogenously applied growth hormones is likely to induce physiological changes that also affect plant responses to pathogens and pests (Bailiss, 1977; Balazs and Kiraly, 1981; Faccioli *et al.*, 1984; Hallahan *et al.*, 1992; Hedin and McCarty, 1994; Hedin *et al.*, 1988; Mills *et al.*, 1986; Nicholson, 1992; Orr and Lynn, 1992; Plich, 1976; Thomas and Blakesley, 1987). On several occasions, cytokinin applications have been shown to suppress virus- and fungus-induced hypersensitive responses that are generally considered to be part of the disease resistance response in plants (Balazs and Kiraly, 1981; Beckman and Ingram, 1994). However, an overwhelmingly larger number of other reports correlate cytokinin applications with enhanced pathogen resistance. In one of the first reports, tissue culture-derived *Solanum tuberosum* plants became more resistant to the fungal pathogen *Phytophthora infestans* when certain combinations of cytokinins and auxins were added to the medium (Ingram, 1967). Foliar applications of cytokinins have also been shown to be effective against viral infection. Symptom development was reduced in tobacco mosaic virus (TMV) infected *Nicotiana tabacum*, and cytokinin analysis of TMV-resistant and -susceptible tomato cultivars revealed a strong correlation between resistance and higher cytokinin concentrations (Balazs and Kiraly, 1981; Li and Qiu, 1986). Although cytokinin influence on insect infestations has only been addressed indirectly, results from field trials using commercial formulations of natural cytokinins suggest that the overall increase in yields may in part have been due to reduced insect populations (Blanco-Montero and Ward, 1995; Hedin and McCarty, 1994).

The mechanism of cytokinin-mediated enhanced pest and pathogen resistance is not known, but a need to exploit natural, environmentally compatible pesticide alternatives exists. Therefore, defining the role of cytokinins in plant defense responses may potentially be of use in an integrated pest management program for reducing the usage of environmentally damaging synthetic pesticides.

Secondary Metabolism and Cytokinins

Protection against insects without the aid of externally applied chemicals has been achieved most recently with biological insecticides and genetically engineered insect-resistant plants. The least exploited approaches have been the natural defense mechanisms of plants. This in part is due to the lack of a thorough understanding of the rather complex and integrated defense systems that have evolved in plants. What is known is that the defense response is initially comprised of physical and chemical barriers. The second stage of the response is induced by the pathogen, resulting in the strengthening of the cell wall and production of toxic compounds (Graham and Graham, 1991). The toxic compounds are mainly products of second-ary metabolic pathways that are often sequestered in specialized structures such as trichomes or vacuoles, away from many primary functions. The biosynthetic pathway for each secondary metabolite consists of numerous enzymatic steps, and on occasion more than one pathway is involved in the biosynthesis of a particular compound. The complexity of the pathways presents a difficult task for effective genetic engineering to enhance the production of these toxic compounds for use in plant protection. However, since these pathways appear to be common to all crop plants, appropriate strategies that will only require limited modifications for increased production of a potentially useful secondary metabolite can probably be devised. Currently there are a number of reports on a combination of biochemical and molecular techniques that are being used to explore the role played by individ-ual secondary metabolites in plant physiological processes including defense (Rhodes, 1994).

Cytokinin effects on secondary metabolism have been studied less than that of other plant hormones, but several reports clearly document the correlation between cytokinin application and accumulation of secondary metabolites (Binns *et al.*, 1987; Orr and Lynn, 1992; Teutonico *et al.*, 1991). In tissue-cultured plant cells increased concentrations of numerous secondary products were observed following cytokinin application. These compounds included anthocyanins (Ozeki and Koma-mine, 1986), betacyanins (Mothes *et al.*, 1985), tannins (Lees, 1986), coumarins, sco-poletin and scopolin (Hino *et al.*, 1982), rhodoxanthin (Kayser and Gemmrich, 1984), berberine (Nakagawa *et al.*, 1984) and indole alkaloids (Decendit *et al.*, 1992; Merillon *et al.*, 1991). A similar increase in the concentration of four allelochemicals which included gossypol, condensed tannins, flavonoids and anthocyanins, was observed in field-grown cotton sprayed with cytokinin or commercial formulations of cytokinin (Hedin and McCarty, 1994). All of these secondary compounds have previously been shown to be toxic to a major pest of cotton, the tobacco budworm.

Endogenous Modulation of Cytokinin Concentrations

The effect of exogenously applied cytokinins on disease development and pest damage is most likely profoundly affected by the known differences in endogenous concentrations of free phytohormones. Interpretation of effects is further complicated by problems of uptake, compartmentalization and metabolism of exogenously added compounds. As is often the case these days, application of molecular genetics is used as an alternate approach. Manipulation of endogenous concentrations of cytokinins in plant cells has been made possible by the discovery of cytokinin bio-synthesis genes carried by bacteria (Akiyoshi *et al.*, 1983; Barry *et al.*, 1984). These genes have been well characterized and shown to increase cytokinin concentrations in transformed plant cells. One particular gene, the isopentenyl transferase gene (*ipt*) from *Agrobacterium tumefaciens*, codes for a key enzyme in the cytokinin biosyn-thetic pathway and has been used extensively to analyze the mechanisms of cyto-kinin action in regulation of plant growth and development in transgenic plant systems (Akiyoshi *et al.*, 1984; Gan and Amasino, 1995; Li *et al.*, 1992; Martineau *et al.*, 1994; Medford *et al.*, 1989; Smart *et al.*, 1991; Smigocki, 1991, 1995; Smigocki and Owens, 1988). Continuous or unregulated expression of this gene has been shown to have a negative impact on normal plant development (Smigocki and Owens, 1989). In most cases, despite prolific shoot development and delay of the onset of senescence, root development is completely inhibited, prohibiting regener-ation of whole plants.

It was in shooty teratomas induced by the constitutive expression of the *ipt* gene that extremely high cytokinin levels were first shown to be associated with the up-regulation of expression of pathogenesis-related (PR) genes (Memelink *et al.*, 1987). Proteins encoded by PR genes are part of the defense response in plants and their increased concentrations in other transgenic plants have been shown to suppress fungal infections (Alexander *et al.*, 1993; Broglie *et al.*, 1991; Liu *et al.*, 1994). Whether these PR proteins are induced in shooty teratomas by elevated cytokinin levels or simply by the stress associated with growth in the absence of roots remains an open issue. Most recently, Storti *et al.* (1994) introduced the *ipt* gene into a *Fusarium oxysporum* susceptible tomato cultivar and observed that the regenerated teratomas were more resistant to this fungal pathogen. In all of these studies, however, other hormone biosynthetic genes normally carried on the same plasmid in *A. tumefaciens* were also introduced along with the *ipt* gene; therefore, their involvement in the disease process cannot be completely ruled out despite the fact that some of these genes were inactivated by transposons. Cloning and reconstruc-tion of the *ipt* gene for temporal and tissue-specific expression provides for a better means of assessing the participation of cytokinins in disease development.

Developmental Effects in Transgenic Plants

Since wounding is usually the first insect-induced damage incurred by the plant, fusion of the *ipt* gene with a promoter from a known wound-inducible gene would allow for cytokinin gene up-regulation upon insect attack. A developmentally regu-lated gene has been isolated from the potato tuber that has also been demonstrated to be induced by mechanical wounding and insect feeding in leaves of transgenic plants (Johnson *et al.*, 1989; Lorberth *et al.*, 1992; Thornburg *et al.*, 1987). There-fore, we fused the promoter from this proteinase inhibitor II (PI-IIK) gene to the *ipt* gene and introduced it into tobacco and tomato (Figure 13.1A and B). A number of independent transformants of *Nicotiana plumbaginifolia*, *N. tabacum*, and *Lycopersi-con esculentum* carrying the chimeric gene (PI-II-*ipt*) were regenerated (Smigocki, 1995; Smigocki *et al.*, 1993; Smigocki, unpublished). Transcripts of the *ipt* gene were

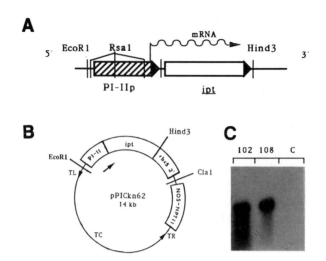

Figure 13.1 Construction of a chimeric wound-inducible *ipt* gene. A: the potato proteinase inhibitor IIK gene (PI-II) promoter was fused at its 5′-untranslated region to the coding region of the isopentenyl transferase gene (*ipt*); B: a plant transformation vector, pPICkn62, was used for *Agrobacterium*-mediated gene transfer; C: Northern blot of total leaf RNA from wounded transgenic PI-II-*ipt* plants (l02, l08) and a control (C) plant

wound-inducible in the leaves at all developmental stages of growth (Figure 13.1C). The transcripts were also detected in leaves directly above or below the damaged leaves as has been previously reported when heterologous genes fused to the PI-IIK gene promoter were analyzed in transgenic plants (Keil *et al.*, 1989; Sanchez-Serrano *et al.*, 1987; Siebertz *et al.*, 1989; Thornburg *et al.*, 1990). Interestingly, typical cytokinin effects were not apparent in transgenic *N. plumbaginifolia* plants until after the plants flowered, but appeared much earlier in two cultivars of *N. tabacum* and tomato transformants. *N. plumbaginifolia* plants exhibited pronounced apical dominance, an underdeveloped root system and much larger leaves throughout growth. Their total chlorophyll content was reduced by about 40%. As the plants matured and flowered, the fully expanded basal leaves developed a darker green color that corresponded to a threefold higher chlorophyll content than that of the controls. This was followed by the emergence of numerous lateral shoots at the base of the stem. Leaf concentrations of the cytokinins zeatin and zeatinriboside reached a maximum increase of 70-fold over those in control plants.

The *N. tabacum* and tomato transformants exhibited more typical cytokinin effects early in development. Plants were generally shorter, had thicker stems, increased node number, and smaller leaves (Smigocki, 1995; Smigocki, unpublished). Chlorophyll levels were similar to the controls until the plants started to bolt or branch and then the leaves turned a dark green color.

The variability in the appearance of the cytokinin effects among the transformants seems to be partly influenced by the expression pattern of the PI-II-*ipt* gene. Presence of the *ipt* transcripts in roots and stems was governed by the developmental stage of the plant. In seedlings and pre-bolted *N. plumbaginifolia* transformants, root and stem *ipt* messages were not detected before or following leaf wounding and only direct wounding of roots induced a low level of stem expression. Upon bolting, however, elevated levels of *ipt* gene transcripts were detected in the roots even prior to leaf wounding, but in stems they were generally absent and wound-inducible to only relatively low levels. The emergence of numerous lateral shoots from the base of the stem and increase in chlorophyll concentrations in the fully expanded basal leaves of unwounded plants followed the developmental induction of relatively high levels of root *ipt* gene transcripts.

Transgenic *N. tabacum* plants exhibited the opposite developmental regulation of root and stem expression in that no *ipt* transcripts were detected in bolted, pre-flowering and flowering plants but a slight increase was detected in rosette plants. The appearance of cytokinin effects, therefore, seems to be correlated with root and, to a lesser extent, stem expression of the *ipt* gene. The earlier expression in rosette *N. tabacum* plants induced a number of characteristic cytokinin effects early in plant development that persisted throughout the life of the plant, a phenomenon often observed following single or multiple treatment of tissues with cytokinin.

Insect Resistance

To relate the cytokinin effects and wound-inducible expression of the *ipt* gene to the degree of insect infestation, transgenic PI-II-*ipt* plants were exposed to the herbivorous pest, tomato (tobacco) hornworm (*Manduca sexta*), and a virus-transmitting pest, the green peach aphid (*Myzus persicae*). All transformants exhibited enhanced insect resistance that was correlated with the appearance of cytokinin effects (Smigocki *et al.*, 1993; Smigocki, unpublished). When the *M. sexta* larvae were fed either leaf material from flowering, transgenic *N. plumbaginifolia* plants or were allowed to feed on the whole plants, they consumed significantly less than larvae feeding on leaves from control plants (Figure 13.2). On average, second and third instar larvae consumed 60% less leaf material than those feeding on control leaves and their weight gain was reduced by approximately 20 to 60%. The effect on green peach aphid nymphs was more significant in that only about half as many (30–40%) of the newly hatched nymphs developed into adult females when feeding on transgenic plants, and of those that reached adulthood, approximately 75% were able to reproduce.

The combination of induction of the cytokinin gene by wounding and normal developmental regulation appears to raise the *ipt* concentrations to levels that are needed for the increased resistance. Highest concentrations of zeatin and zeatinriboside cytokinins were found in leaves remaining on the transgenic *N. plumbaginifolia* plants after tomato hornworm feeding (Smigocki *et al.*, 1993). Cytokinin levels were elevated by about 70-fold (500 pmol/g fresh weight) in comparison to controls (7 pmol/g). Dipping leaf petioles in solutions of zeatinriboside further decreased insect feeding on transgenic leaves. This response was not observed when leaves from normal, untransformed plants were placed in cytokinin solutions and may reflect problems associated with exogenous applications referred to earlier. Most likely in transgenic plants even low levels of exogenous cytokinins are sufficient to increase further the endogenous cytokinin concentrations up to levels that are effective in retarding insect feedings, unlike those in normal plants. The effects of zeatinriboside applications on delaying the green peach aphid development were more dramatic in that most of the nymphs did not reach maturity.

Enhanced resistance to tomato hornworm larvae was also demonstrated with two cultivars of *N. tabacum* and one cultivar of tomato transgenic plants (Smigocki,

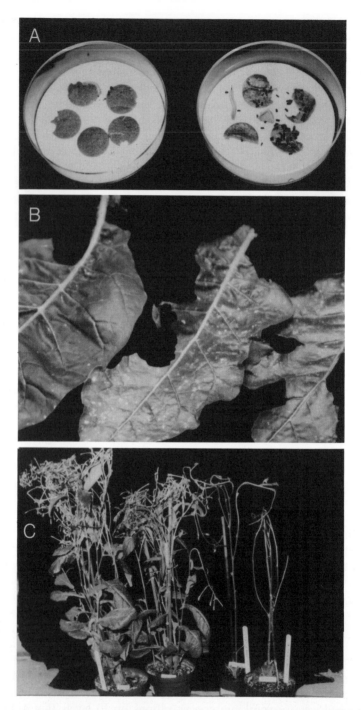

Figure 13.2   A: leaf disk bioassay using transgenic (left) and normal (right) *N. plumbaginifolia* plant material at 24 hours after infestation with third instar *M. sexta* larvae; B: whole leaf bioassay using detached leaves from transgenic PI-II-*ipt* plants (left) and two negative controls, transgenic, promoterless *ipt* (center) and normal, untransformed (right); C: enhanced resistance of transgenic PI-II-*ipt* plants (two on the left) compared to control plants (two on the right) after 19 days' infestation with tomato hornworm larvae

unpublished). In leaf disk assays, feeding was reduced by about 90% in a 48-hour test. Unlike the transgenic *N. plumbaginifolia* plants, resistance was observed much earlier in the development of *N. tabacum*, prior to bolting and flowering, and was correlated with the appearance of typical cytokinin effects.

Cytokinin-induced Insecticidal Compound(s)

From these studies, it appears that the beneficial effects of cytokinin on enhancing the level of resistance to some insects may be applicable to other plant species and may contribute to a more universal effect on defense properties of plants. Cytokinins themselves are unlikely to affect insect pests negatively, but rather, as has already been demonstrated, modify some secondary metabolic pathway that leads to enhanced production or secretion of an insecticidal secondary metabolic compound (Binns *et al.*, 1987; Orr and Lynn, 1992; Teutonico *et al.*, 1991). To explore this possibility, we are in the process of analyzing extracts from transgenic plants for insecticidal activity. After selective solvent extraction and partitioning, insecticidal activity was found to be associated with surface extracts. This same activity was minor and inconsistent in normal plants, probably due to a low concentration of the active compound(s). The active fractions were found not to contain insecticidal sucrose esters (Buta *et al.*, 1993; Neal *et al.*, 1994). A dilute solution (0.1%) of the fractionated extracts was lethal to second instar tomato hornworm larvae (Figure 13.3) Larvae were killed within two hours after being placed on normal leaf disks dipped in the extracts. Similar levels of activity were retained for greater than two months when the extracts were stored in the cold.

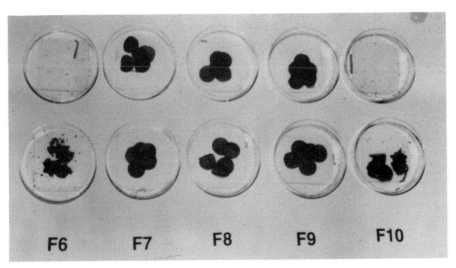

Figure 13.3 Analysis of insecticidal activity in fractionated leaf extracts from transgenic PI-II-*ipt* plants. Leaf disks from normal plants were dipped in fractions 6 through I0 (F6–FI0). Second instar tomato hornworm larvae exposed to F7, F8 and F9 were killed within two hours. Consumption of leaf disks at 24 (bottom) and 48 hours (top) after infestation

231

Specificity of Insecticidal Compound(s)

The specificity of the unfractionated extract from the *ipt* transformed plants was tested against a number of plant pests and pathogens. In general, no activity was detected against pathogenic fungi, bacteria or nematodes although in some cases, crude extracts had antibacterial activity, promoted spore germination of *F. oxysporum* and killed *C. elegans* (Heu and Smigocki, unpublished). We were able to determine that the compounds responsible for this activity were in the sugar ester fraction that previously had been shown to have bactericidal activity (Cutler *et al.*, 1990).

So far it appears that the activity is only specific to insects. Preliminary analysis of four insects, each representing a different insect order, indicates that only certain orders may be targeted. Three of the orders tested were negatively affected by the active compound(s). In a leaf disk bioassay, using either transgenic leaves or normal leaves dipped in the extracts, lepidopteran and homopteran but not the coleopteran (alfalfa weevil) insects were either killed or their feeding was significantly inhibited which, in some cases, affected normal development and reproduction. In an agar bioassay, addition of the extract to sugar beet cell cultures infested with first instar root maggot larvae (*Tetanops myopaeformis*, Diptera: Otitidae) induced a wild thrashing and twitching behavior during the initial four-hour observation period that was reduced, but still identifiable, after four days of incubation (Figure 13.4; Wozniak and Smigocki, unpublished). Similar behavior was not observed with non-transgenic plant extracts. Based on these preliminary results, we speculate that the active insecticidal compound may have a similar mechanism of action on species of these susceptible insect orders.

Future Perspectives

Presently, little is known about the mode of action of the *ipt* gene on enhanced disease and pest resistance in transgenic plants, but involvement of secondary metabolites is strongly suspected based on our results. Currently, we are focusing on

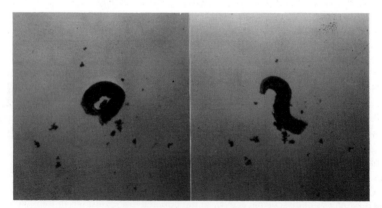

Figure 13.4 Twitching and thrashing behavior of the sugar beet root maggot exposed to the extract from transgenic PI-II-*ipt* plants (left) compared to the extract from non-transformed plants (right)

the purification of the active compound and to date, our efforts have yielded a highly active two-component fraction. Once identified, the effect of elevated cytokinin levels on its accumulation or increased secretion will be determined. The potential for direct genetic manipulation of the corresponding biosynthetic pathway will be explored as a possible means of enhancing natural plant defense mechanisms.

References

Akiyoshi, D. E., Klee, H., Amasino, R., Nester, E. W. and Gordon, M. P. (1984) T-DNA of *Agrobacterium tumefaciens* encodes an enzyme of cytokinin biosynthesis, *Proc. Natl. Acad. Sci. USA* **81**, 5994–5998.

Akiyoshi, D. E., Morris, R. O., Hinz, R., Mischke, B. S., Kosuge, T., Garfinkel, D., Gordon, M. P. and Nester, E. W. (1983) Cytokinin/auxin balance in crown gall tumors is regulated by specific loci in the T-DNA, *Proc. Natl. Acad. Sci. USA* **80**, 407–411.

Alexander, D., Goodman, R. M., Gut-Rella, M., Glascock, C., Weymann, K., Friedrich, L., Maddox, D., Ahl Goy, P., Luntz, T., Ward, E. and Ryals, J. (1993) Increased tolerance to two Oomycete pathogens in transgenic tobacco expressing pathogenesis-related protein la, *Proc. Natl. Acad. Sci. USA* **90**, 7327–7331.

Bailiss, K. W. (1977) The effect of benlate and cytokinins on the content of tobacco mosaic virus in tomato leaf disks and cucumber mosaic virus in cucumber cotyledon disks and seedlings, *Annu. Appl. Biol.* **87**, 383–392.

Balazs, E. and Kiraly, Z. (1981) Virus content and symptom expression in Samsun tobacco treated with kinetin and a benzimidazole derivative, *Phytopath. Z.* **100**, 356–360.

Barry, G. F., Rogers, S. G., Fraley, R. T. and Brand, L. (1984) Identification of a cloned cytokinin biosynthetic gene, *Proc. Natl. Acad. Sci. USA* **81**, 4776–4780.

Beckman, K. B. and Ingram, D. S. (1994) The inhibition of the hypersensitive response of potato tuber tissues by cytokinins: similarities between senescence and plant defence responses, *Physiol. Mol. Plant Path.* **44**, 33–50.

Binns, A. N., Chen, R. H., Wood, H. N. and Lynn, D. G. (1987) Cell division promoting activity of naturally occurring dehydrodiconiferyl glucosides: do cell wall components control cell division?, *Proc. Natl. Acad. Sci. USA* **84**, 980–984.

Blanco-Montero, C. A. and Ward, C. R. (1995) Mitigation effects of cytokinin plant growth regulator on turfgrass root-biomass loss by white grubs, *SW Entomol.* **20**, 11–15.

Broglie, K., Chet, I., Holliday, M., Cressman, R., Biddle, P., Knowlton, C., Mauvais, C. J. and Broglie, R. (1991) Transgenic plants with enhanced resistance to the fungal pathogen *Rhizoctonia solani*, *Science* **254**, 1194–1197.

Buta, J. G., Lusby, W. R., Neal, J. W., Waters, R. M. and Pittarelli, G. W. (1993) Sucrose esters from *Nicotiana gossei* active against the greenhouse whitefly *Trialeurodes vaporariorum*, *Phytochemistry* **32**, 859–864.

Cutler, H. G., Severson, R. F., Montemurro, N., Cole, P. D., Sisson, V. A. and Stephenson, M. G. (1990) *Plant Growth Inhibitory and Antimicrobial Properties of Sucrose Esters from Nicotiana tabacum Cultivars and Species*, Plant Growth Regulation Society of America.

Davies, P. J. (1987) *Plant Hormones and Their Role in Plant Growth and Development*, Boston: Kluwer Academic Publishers.

Decendit, A., Liu, D., Ouelhazi, L., Doireau, P., Merillon, J-M. and Rideau, M. (1992) Cytokinin-enhanced accumulation of indole alkaloids in *Catharanthus roseus* cell cultures – the factors affecting the cytokinin response, *Plant Cell Report* **11**, 400–403.

Faccioli, G., Rubies-Autonell, C. and Albertini, R. (1984) Role of cytokinins in the acquired resistance of *Chenopodium amaranticolor* towards an infection of tobacco necrosis virus, *Phytopathol. Medit.* **23**, 15–33.

Gan, S. and Amasino, R. M. (1995) Inhibition of leaf senescence by autoregulated production of cytokinin, *Science* **270**, 1986–1988.

Graham, T. L. and Graham, M. Y. (1991) Cellular coordination of molecular responses in plant defense, *Mol. Plant Microbe Inter.* **4**, 415–422.

Hallahan, D. L., Pickett, J. A., Wadhams, L. J., Wallsgrove, R. M. and Woodcock, C. (1992) Potential of secondary metabolites in genetic engineering of crops for resistance, In: Gatehouse, A. M. R., Hilder V. A. and Boulter, D. (Eds), *Plant Genetic Manipulation for Crop Protection*, Melksham: Redwood Press, pp. 212–248.

Hedin, P. A. and McCarty, J. C. (1994) Multiyear study of the effects of kinetin and other plant growth hormones on yield, agronomic traits, and allelochemicals of cotton, *J. Agric. Food Chem.* **42**, 2305–2307.

Hedin, P. A., Williams, W. P., Davis, F. M. and Thompson, A. C. (1988) Effects of bio-regulators on nutrients, insect resistance, and yield of corn (*Zea mays* L.), *J. Agric. Food Chem.* **36**, 746–748.

Hino, F., Okazaki, M. and Miura, Y. (1982) Effects of kinetin on formation of scopoletin and scopolin in tobacco tissue cultures, *Agric. Biol. Chem.* **46**, 2195–2202.

Ingram, D. S. (1967) The expression of R-gene resistance to *Phytophthora infestans* in tissue cultures of *Solanum tuberosum*, *J. Gen. Microbiol.* **57**, S-91.

Johnson, R., Narvaez, J., An, G. and Ryan, C. (1989) Expression of proteinase inhibitors I and II in transgenic tobacco plants: effects on natural defense against *Manduca sexta* larvae, *Proc. Natl. Acad. Sci. USA* **86**, 9871–9875.

Kayser, H. and Gemmrich, A. R. (1984) Hormone induced changes in carotenoid composition in *Ricinus* cell cultures: I. Identification of rhodoxanthin, *Z. Naturforsch.* **39**, 50–54.

Keil, M., Sanchez-Serrano, J. J. and Willmitzer, L. (1989) Both wound-inducible and tuber-specific expression are mediated by the promoter of a single member of the potato proteinase inhibitor II gene family, *EMBO J.* **8**, 1323–1330.

Lees, G. L. (1986) Condensed tannins in the tissue culture of sainfoin (*Onobrychis viciifolia* Scop.) and birdsfoot trefoil (*Lotus corniculatus* L.), *Plant Cell Report* **5**, 247–251.

Li, H. and Qiu, W. (1986) Effect of cytokinins on the disease resistance of tomato cultivars to the infection of tobacco mosaic virus (TMV-T), *Sci. Sinica* **24**, 626–633.

Li, Y., Hagen, G. and Guilfoyle, T. J. (1992) Altered morphology in transgenic tobacco plants that overproduce cytokinins in specific tissues and organs, *Dev. Biol.* **153**, 386–395.

Liu, D., Raghothama, K. G., Hasegawa, P. M. and Bressan, R. A. (1994) Osmotin over-expression in potato delays development of disease symptoms, *Proc. Natl. Acad. Sci. USA* **91**, 1888–1892.

Lorberth, R., Dammann, C., Ebneth, M., Amati, S. and Sanchez-Serrano, J. (1992) Promoter elements involved in environmental and developmental control of potato proteinase inhibitor II expression, *Plant J.* **2**, 477–486.

Martineau, B., Houck, C. M., Sheehy, R. E. and Hiatt, W. R. (1994) Fruit-specific expression of the *A. tumefaciens* isopentenyl transferase gene in tomato: effects on fruit ripening and defense-related expression in leaves, *Plant J.* **5**, 11–19.

Medford, J. I., Horgan, R., El-Sawi, Z. and Klee, H. J. (1989) Alterations of endogenous cytokinins in transgenic plants using a chimeric isopentenyl transferase gene, *Plant Cell* **1**, 403–413.

Memelink, J., Hoge, J. H. C. and Schilperoort, R. A. (1987) Cytokinin stress changes the developmental regulation of several defence-related genes in tobacco, *EMBO J.* **6**, 3579–3583.

Merillon, J-M., Liu, D., Huguet, F., Chenieux, J-C. and Rideau, M. (1991) Effects of calcium entry blockers and calmodulin inhibitors on cytokinin-enhanced alkaloid accumulation in *Catharanthus roseus* cell cultures, *Plant Physiol. Biochem.* **29**, 289–296.

Mills, P. R., Gussine, J. and Wood, R. S. K. (1986) Induction of resistance in cucumber to *Colletotrichum lagenarium* by 6-benzylaminopurine, *J. Phytopathol.* **116**, 11–17.

Mok, D. W. and Mok, M. C. (1994) *Cytokinins: Chemistry, Activity and Function*, Boca Raton: CRC Press.

Mothes, K., Schutte, H. R. and Luckner, M. (1985) *Biochemistry of Alkaloids*, Berlin: VEB

Deutscher Verlag der Wissenschaften.

Nakagawa, K., Konagai, A, Fukui, H. and Tabata, M. (1984) Release and crystallization of berberine in the liquid medium of *Thalictrum minus* cell suspension cultures, *Plant Cell Rep.* **3**, 254–257.

Neal, J. W., Buta, J. G., Pittarelli, G. W., Lusby, W. R. and Bentz, J-A. (1994) Novel sucrose esters from *Nicotiana gossei*: effective biorationals against selected horticultural insect pests, *Hortic. Entomol.* **87**, 1600–1607.

Nicholson, R. L (1992) Phenolic compounds and their role in disease resistance, *Annu. Rev. Phytopathol.* **30**, 369–389.

Orr, J. D. and Lynn, D. G. (1992) Biosynthesis of dehydrodiconiferyl alcohol glucosides: implications for the control of tobacco cell growth, *Plant Physiol.* **98**, 343–352.

Ozeki, Y. and Komamine, A. (1986) Effects of growth regulators on the induction of anthocyanin synthesis in carrot suspension cultures, *Plant Cell Physiol.* **27**, 1361–1368.

Plich, M. (1976) Influence of growth regulators on the development of collar knot disease caused by the fungus *Phytophthora cactorum* in apple trees, *Fruit Sci. Rep.* **3**, 33–42.

Rhodes, M. J. C. (1994) Physiological roles for secondary metabolites in plants: some progress, many outstanding problems, *Plant Mol. Biol.* **24**, 1–20.

Sanchez-Serrano, J. J., Keil, M., Connor, O., Schell, J. and Willmitzer, L. (1987) Wound-induced expression of a potato proteinase inhibitor II gene in transgenic tobacco plants, *EMBO J.* **6**, 303–306.

Siebertz, B., Logemann, J., Willmitzer, L. and Schell, J. (1989) *cis*-analysis of the wound-inducible promoter *wun1* in transgenic tobacco plants and histochemical localization of its expression, *Plant Cell* **1**, 961–968.

Skoog, F. and Miller, C. O. (1957) Chemical regulation of growth and organ formation in plant tissues cultured *in vitro*, *Symp. Soc. Exp. Biol.*, **11**, 118–130.

Smart, C. M., Scofield, S. R., Bevan, M. W. and Dyer, T. A. (1991) Delayed leaf senescence in tobacco plants transformed with *tmr*, a gene for cytokinin production in *Agrobacterium*, *Plant Cell* **3**, 647–656.

Smigocki, A. C. (1991) Cytokinin content and tissue distribution in plants transformed by a reconstructed isopentenyl transferase gene, *Plant Mol. Biol.* **16**, 106–115.

Smigocki, A. C. (1995) Expression of a wound-inducible cytokinin biosynthesis gene in transgenic tobacco: correlation of root expression with induction of cytokinin effects, *Plant Sci.* **109**, 153–163.

Smigocki, A. C. and Owens, L. D. (1988) Cytokinin gene fused with a strong promoter enhances shoot organogenesis and zeatin levels in transformed plant cells, *Proc. Natl. Acad. Sci. USA* **85**, 5131–5135.

Smigocki, A. C. and Owens, L. D. (1989) Cytokinin-to-auxin ratios and morphology of shoots and tissues transformed by a chimeric isopentenyl transferase gene, *Plant Physiol.* **91**, 808–811.

Smigocki, A. C., Neal, J. W., McCanna, I. and Douglass, L. (1993) Cytokinin-mediated insect resistance in *Nicotiana* plants transformed with the *ipt* gene, *Plant Mol. Biol.* **23**, 325–335.

Storti, E., Bogani, P., Bettini, P., Bittini, P., Guardiola, M. L., Pellegrini, M. G., Inze, D. and Buiatti M. (1994) Modification of competence for *in vitro* response to *Fusarium oxysporum* in tomato cells. II. Effect of the integration of *Agrobacterium tumefaciens* genes for auxin and cytokinin synthesis, *Theor. Appl. Genet.* **88**, 89–96.

Teutonico, R. A., Dudley, M. W., Orr, J. D., Lynn, D. G. and Binns, A. N. (1991) Activity and accumulation of cell division-promoting phenolics in tobacco tissue cultures, *Plant Physiol.* **97**, 288–297.

Thomas, T. H. and Blakesley, D. (1987) Cytokinins – plant hormones in search of a role – practical and potential uses of cytokinins in agriculture and horticulture, *Monogr. Br. Growth Reg. Group* **14**, 69–83.

Thornburg, R. W., An, G., Cleveland, T. E., Johnson, R. and Ryan, C. W. (1987) Wound-inducible expression of a potato inhibitor II-CAT gene fusion in transgenic tobacco plants, *Proc. Natl. Acad. Sci. USA* **84**, 744–748.

Thornburg, R. W., Kernana, A. and Molin, L. (1990) Chloramphenicol acetyl transferase (CAT) protein is expressed in transgenic tobacco in field tests following attack by insects, *Plant Physiol.* **92**, 500–505.

14

Genetic Engineering of Plant Secondary Metabolism for Insect Protection

SCOTT CHILTON

Secondary metabolites are generally produced by multigene pathways despite their low molecular weight and apparent structural simplicity. At present genetic engineering of plants is limited to the introduction and expression of single genes, although recovery of transformants does depend on introduction of at least a second gene providing a scorable marker or a selectable trait. Some vectors are capable of transferring as many as four genes, but in no case have even two genes of a pathway been transferred and coordinately expressed as yet. Research directed toward the introduction of coordinately expressed genes is in progress. Plants expressing single gene products that have sufficient economic value for commercial development have been confined to antisense mRNA or to protein products such as *Bt* toxin. There are many secondary metabolites of low molecular weight that have very high economic value in medicine, and others of proven efficacy in crop protection that will eventually be economically attractive targets for genetic engineering once coordinate expression of multigene pathways becomes feasible.

Many natural products have been pointed out as having potential value in protecting crops against insects (Hallahan *et al.*, 1992). Some have actually been used for this purpose prior to the explosion of synthetic chemical pesticides after the Second World War. Indeed, a few of the natural products (pyrethrum, derris root, sabadilla, nicotine) were the main source of crop protection between 1900 and 1950. At the end of the twentieth century a few of these pesticidal natural products continue to be used in specialized niches, and a few new ones, such as ryania and neem tree extracts (Margosan O), have been introduced as commercial plant protectants in more recent times.

Several examples of the pathway complexity of agronomically valuable secondary metabolites are discussed in this chapter. Cyanogenic glucoside synthesis is probably accomplished by four enzymes (genes) beyond the amino acid starting point, glucosinolates by about six structural genes, DIBOA glucoside by four genes and DIMBOA glucoside by six genes beyond the tryptophan pathway; and rotenone by about seven genes beyond the flavonoid pathway. Pyrethrin probably requires several genes beyond the phytodienoic acid pathway, and at least two genes beyond isopentenyl-pyrophosphate. By contrast, nature's own genetic engineer, *Agrobacterium tumefaciens*, needs to transfer only one gene to cause the plant tumor to synthesize the secondary metabolites nopaline or octopine.

Knowledge of biosynthetic pathways, enzymes, site of synthesis, translocation and storage are in a very rudimentary state or non-existent for most agronomically valuable secondary products. Much groundwork needs to be laid in understanding the biosynthesis of the medically and agronomically important secondary metabolites. Typically even less (nothing) is known about genetics and molecular biology of the vast majority of useful secondary metabolites. The potential for crop protection by genetic engineering of secondary metabolism is great; the need for more information is pressing.

Selected examples of progress in genetic engineering of plant secondary metabolism have been reviewed recently (Ellis *et al.*, 1994). More specifically, a number of secondary metabolites have been suggested as candidates for crop protection by gene transfer (Hallahan *et al.*, 1992). This review concentrates on a few examples of high commercial potential specifically as insecticides or insect deterrents. The examples presented here are atypical in that they have been better studied than most of the secondary metabolites of economic interest. The first four secondary metabolites (nicotine, cyanogenic glucosides, glucosinolates and DIMBOA) were selected for discussion because of their proven anti-insect value and the considerable progress in understanding their biosynthesis, enzymology and molecular biology. Two other natural products (pyrethrin and rotenone) were chosen because of their proven and continuing importance as insecticides, and a final example is discussed because the class of unsaturated isobutylamides is recognized as having insect knockdown potential equal to that of pyrethrin and has attracted attention as a model (lead chemical) for preparation of new synthetic insecticides (Elliott *et al.*, 1987a,b; Masakazu *et al.*, 1989).

Nicotine

Powdered tobacco leaf (*Nicotiana tabacum*) or aqueous extracts have been used as an insecticide in Europe and America since the beginning of the eighteenth century. Commercial preparation of extracts containing a standardized amount of nicotine sulfate was introduced at the end of the nineteenth century (Shepherd, 1951). Standardized dusts also became available in the early twentieth century. Tobacco processing wastes and the tree tobacco, *N. rustica*, originally from South America, have been used for commercial extraction of nicotine. Nicotine constitutes as much as

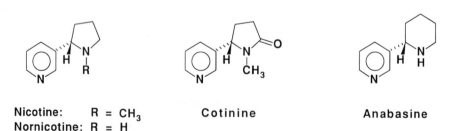

Nicotine: R = CH₃   Cotinine   Anabasine
Nornicotine: R = H

Figure 14.1   Structures of nicotine and related metabolites

15% of the dry weight of the leaf of *N. rustica* (Collison *et al.*, 1929). Nicotine free base is slightly volatile (b.p. 247°C). This property has been widely used for fumigating greenhouses by heating either tobacco or an extract. Sunlight and air destroy nicotine relatively rapidly, so it is eminently environmentally degradable; but it is highly toxic to humans.

Nicotine is accompanied by a number of minor related alkaloids including nornicotine and anabasine (Figure 14.1). Anabasine, originally isolated from *Anabasis aphylla* (Chenopodaceae) in Russia, is the major alkaloid in *N. glauca*. Anabasine has similar insecticidal properties to nicotine and has been used as a commercial insecticide in both Russia and the United States.

Nicotine Toxicity

Nicotine, an analog of the insect and mammalian neurotransmitter acetylcholine, is a stimulant of the autonomic nervous system, and at higher doses, is a depresssant. Nicotine is toxic to insects, mammals and birds by ingestion, inhalation and contact. Death can be very rapid (5–30 min), about as rapid as with hydrogen cyanide (Shepherd, 1951). Oral LD_{50} values in the range 3–18 mg/kg have been reported for many laboratory animals and birds (RTECS, 1994a). Direct uptake of nicotine free base into the bloodstream through the eyes and tongue is very rapid, and skin absorption is also a danger. A cat was killed within minutes after 40% nicotine free base was rubbed on its abdomen (Shepherd, 1951). The hazards associated with use of nicotine as a pesticide have led to its replacement with safer synthetic insecticides.

Insect Resistance Mechanisms

A single tobacco hornworm larva (*Manduca sexta*) can consume an entire tobacco leaf containing 100–150 mg nicotine in one day (Guthrie and Apple, 1961). The larvae can tolerate such large doses because most of the nicotine is actively excreted unmetabolized (Mandrell and Gardiner 1976; Self *et al.*, 1964a); also, although there is no detectable gross metabolism of nicotine, there is selective detoxification and active export of any nicotine which does penetrate the CNS (Morris, 1983a,b). Aphids are able to feed on tobacco leaves containing more than enough nicotine to kill them, because they suck the photosynthate-containing phloem and not the xylem which carries nicotine from the site of synthesis in the root to the leaf. Other tobacco-feeding insects extensively degrade nicotine. In the tobacco wireworm (*Conoderus vespertinus*), cigarette beetle (*Lasioderma serricorne*) and differential grasshopper (*Melanoplus differentialis*), the main product is cotinine (Figure 14.1), as in humans (Self *et al.*, 1964b). The nicotine-sensitive housefly (*Musca domestica*) also rapidly oxidizes nicotine to cotinine, suggesting that metabolism alone may not be a sufficient resistance mechanism.

Nicotine Biosynthesis, Intermediates and Enzymes

Nicotine is the only insecticidal secondary metabolite which has been subjected to attempts to alter level of accumulation by gene transfer methods (Robins *et al.*,

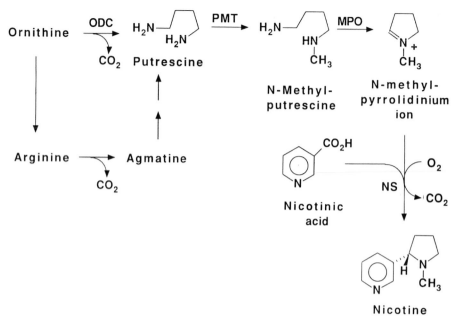

Figure 14.2   Nicotine biosynthesis

1994). All of the intermediates between ornithine or arginine and nicotine are well-known (Figure 14.2). Putrescine is an early branch point intermediate, serving as precursor to the polyamines spermine and spermidine, the hydroxycinnamoyl-putrescines and nicotine-related alkaloids. Nicotine-producing species also produce anabasine from a related pathway involving lysine in place of ornithine. Anabasine is a minor alkaloid of most *Nicotiana* species, but is the major component of *N. glauca*. *Duboisia nicotiana* produces nicotine and tropane alkaloids, both of which are derived from the second branch point intermediate N-methylpyrrolidinium ion. The enzymes ornithine decarboxylase (ODC), arginine decarboxylase (ADC), putrescine methyl transferase (PMT) and N-methylputrescine oxidase (PMO) have been purified and characterized (McLauchlan *et al.*, 1993; Walton *et al.*, 1994). The enzyme of the last step, nicotine synthase, has not been purified, but activity in coupling nicotinic acid to methylpyrrolidinium ion has been demonstrated with a crude extract of *N. tabacum* and *N. glauca* (Friesen and Leete, 1990). The coupling reaction requires oxygen. A mechanism involving NADPH as cofactor has been proposed.

Nicotine is synthesized in the roots only, but transported to the leaves (Dawson, 1948). Low nicotine tobacco cultivars contain as little as 0.02% d.w. nicotine, while the highest nicotine varieties contain 8% d.w., a concentration far too high for unblended use in smoking tobacco. *N. rustica*, with a leaf content of 15% d.w., has been grown for commercial preparation of nicotine insecticide (Collison *et al.*, 1929). It is not known what factors regulate nicotine levels. Exogenously administered ornithine and arginine have no effect on nicotine concentration; putrescine and agmatine have a small effect and nicotinic acid has a slightly larger effect (Robins *et al.*, 1987). Nicotine synthase could be limiting in nicotine synthesis, but this enzyme has not yet been purified, nor has the gene, unique to nicotine-producing species, been cloned.

Transgenic Modification of Nicotine Biosynthesis

Although administration of exogenous intermediates has only a small effect on nicotine level, transgenic tobacco roots have been prepared to test the effect of providing a second enzyme producing an intermediate (Robins *et al.*, 1994). Introduction of genes via the root-inducing (Ri) plasmid of *Agrobacterium rhizogenes* is particularly attractive when studying genes affecting nicotine biosynthesis because the synthesis is root-specific. A yeast *odc* gene driven by the CaMV 35S promoter was introduced into *N. rustica* hairy roots (Hamill *et al.*, 1990). The presence of *odc* mRNA was confirmed and increased and ODC enzyme activity was detected. The level of putrescine and N-methylputrescine was increased in some clones and the average nicotine content of the cultured roots increased twofold at day 14.

Genes for the remaining enzymes of the pathway have not been cloned yet, although an attempt is being made to screen a DNA expression library using an antibody to N-methylputrescine oxidase (McLauchlan *et al.*, 1993). Preliminary evidence suggests that this enzyme may be a glycoprotein. Glycosylation of the transgenic protein may be an additional obstacle to introducing a functional PMO enzyme.

Other experiments aimed at diverting the pathway to higher production of the insecticidal anabasine have been carried out on *N. tabacum* using the leaf *rbcS* promoter with a bacterial lysine decarboxylase (*ldc*) gene (Hemminghaus *et al.*, 1991). Leaf expression of LDC enzyme had no effect on nicotine and anabasine production in the root. The same *ldc* gene introduced into *N. glauca* hairy root culture with a root-expressing CaMV 35S promoter did raise the anabasine concentration while lowering nicotine in one transformed root line (Fecker *et al.*, 1992). However only one of 54 kanamycin resistant root lines contained detectable bacterial LDC enzyme activity under assay conditions designed to discriminate against plant LDC. Although the bacterial *ldc* gene was correctly integrated in many LDC-negative roots, they did not produce detectable *ldc* mRNA. This may be an example of gene silencing which may often defeat the strategy of addition of a gene to increase production of a limiting intermediate.

Nicotine is degraded in tobacco leaf by demethylation to nornicotine. If the level is to be raised in plants which already produce nicotine, it might be desirable to introduce antisense to the mRNA encoding the demethylating enzyme. Raising the level of nicotine or anabasine in plants by genetic engineering will eventually also require consideration of the problem of increasing the supply of nicotinic acid available for the synthesis. This is particularly true should it be desirable to introduce nicotine biosynthesis in a plant which does not already produce it and in which only enough nicotinic acid is synthesized to provide for the low level of the essential cofactor NADH.

Cyanogenic Glucosides and Glucosinolates

Cyanogenic Glucosides

Hydrogen cyanide-releasing cyanogenic glucosides are found sporadically, but widely distributed in ferns, monocots and dicots. Over 3000 species have been identified as cyanogenic. Despite this wide phylogenetic distribution of cyanogenesis, the

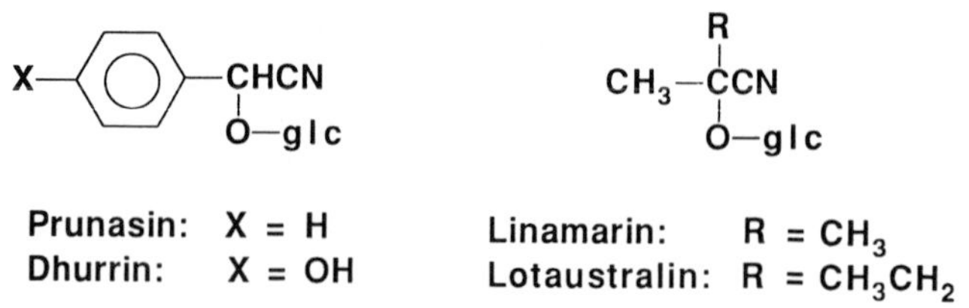

Prunasin: X = H   Linamarin: R = CH$_3$
Dhurrin: X = OH   Lotaustralin: R = CH$_3$CH$_2$

Figure 14.3   Structures of some cyanogenic glucosides

great majority of plant species are acyanogenic. The most intensively studied cyanogenic glucosides are dhurrin in sorghum (Figure 14.3) and linamarin and lotaustralin in white clover and cassava.

Release of Hydrogen Cyanide at the Wound Site

Toxic, volatile and oxidatively unstable hydrogen cyanide is released by the action of a glucosidase on the cyanogenic glucoside (Figure 14.4). The initial aglucone released is a cyanohydrin. A second release enzyme, hydroxynitrile lyase, may be present to accelerate the spontaneous decompostion of the cyanohydrin into an aldehyde (or acetone) and hydrogen cyanide. The hydroxynitrile lyase is not present in all cyanogenic species. In the case of sorghum the glucoside dhurrin is stored in the vacuole of leaf epidermal cells, the releasing glucosidase is in the cytosol of mesophyll cells, and hydroxynitrile lyase is in the chloroplast of mesophyll cells. In white clover linamarin is stored in the vacuole of leaf epidermal cells, the glucosidase is extracellular, and there is no hydroxynitrile lyase. Hydrogen cyanide is very toxic, but also very volatile (b.p. 26°C); consequently after an initial burst the toxic defense chemical is cleared very rapidly from the wound site.

Toxicity of Hydrogen Cyanide

There is no question that hydrogen cyanide is lethal to all organisms, and that very young plants of sorghum and bamboo can release a sufficient amount of hydrogen cyanide (up to 2 mg/g fresh tissue) sufficiently rapidly to kill both insects and man. Hydrogen cyanide has been used for years as an insect fumigant in sealed spaces. The extreme volatility of hydrogen cyanide presents problems in designing experiments that mimic natural conditions. Hydroxybenzaldehyde is released in addition to hydrogen cyanide on damaging sorghum. Many studies have overlooked the toxicity of p-hydroxybenzaldehyde, which is less volatile and more persistent than hydrogen cyanide. Other carbonyl components released on hydrolysis of a cyanogenic glucoside may have an effect as great or even greater than that of hydrogen

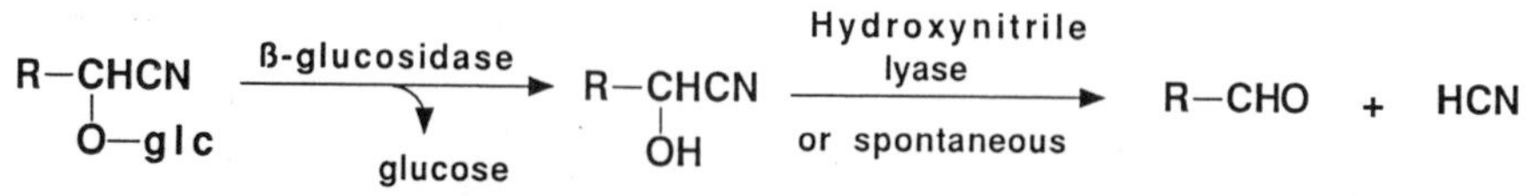

Figure 14.4   Release of HCN from cyanogenic glucosides

cyanide. Much attention has been paid to dimorphic populations of *Lotus cor-niculata* and *Trifolium repens* which have stable, interbreeding populations of cyano-genic and acyanogenic individuals (reviewed by Jones, 1988). Dimorphism provides a control for study of the role of cyanogenesis on leaf herbivory in natural popu-lations.

Glucosinolates

Alkylisothiocyanates (mustard oils) are released from thioglucosides called glucosin-olates (reviewed by Fenwick and Heaney, 1983). Glucosinolates (Figure 14.5) are universal in the family Brassicaceae (over 1000 species tested). They are also a hall-mark of a few other related families in the Capparidales. The rare, scattered individ-ual occurrences in only a few species outside of the Capparidales have been the subject of chemotaxonomic study (Rodman, 1981). The pungent and spicy alkyl-isothiocyanates are responsible for much of the taste of the many Brassicaceae vegetables in cultivation (e.g. cabbage, kale, cauliflower, Brussels sprouts, turnips, cress, mustard). The release of isothiocyanates requires a thioglucosidase (myrosinase) and is accompanied by a rearrangement of the carbon skeleton of the glucosinolate (Figure 14.6). The enzymatic release may be modified in the presence of epithiospecifier protein or heavy metals to give a range of other products. The taste and odor of the more volatile glucosinolates deter many insect herbivores from feeding on Brassicaceae (Chew, 1988). However, some insects, such as the cabbage moth, have become tolerant of mustard oils and even use them as clues to finding suitable plants. Glucosinolates have frequently been named for the plant from which they were first isolated; more systematic nomenclature names the substituent on the thiocarboximate carbon. Thus sinigrin (Figure 14.5), the major glucosinolate of black mustard and horseradish, is also known as allyl glucosinolate or allyl-GS and releases allylisothiocyanate, while sinalbin, the major glucosinolate of white mustard, is also known as *p*-hydroxybenzyl-GS and releases *p*-hydroxy-benzylisothiocyanate.

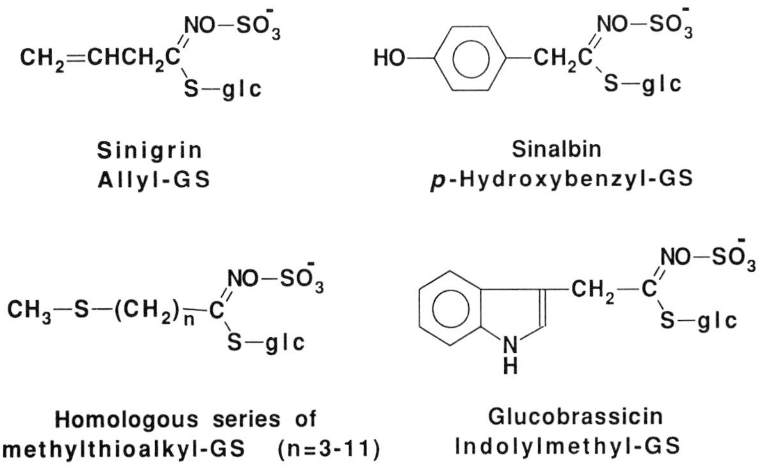

Sinigrin
Allyl-GS

Sinalbin
***p*-Hydroxybenzyl-GS**

Homologous series of
methylthioalkyl-GS   (n=3-11)

Glucobrassicin
Indolylmethyl-GS

Figure 14.5   Structure of some glucosinolates

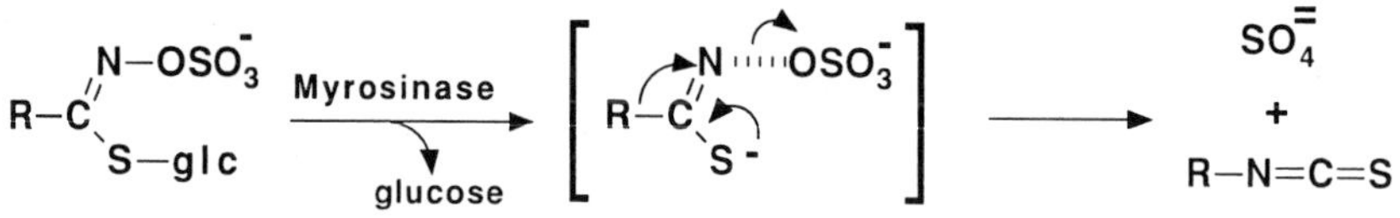

Figure 14.6 Relase of mustard oils from glucosinolates

Alkylisothiocyanate Release at the Wound Site

The glucosinolate is stored in the vacuole of specialized cells called myrosin cells and a thioglucosidase (myrosinase) is located in the cytoplasm of the specialized myrosin cells. The myrosin cells, dispersed in a variety of tissues, have been called mustard bombs because they release the pungent mustard oils when myrosin cells are damaged (Lüthy and Mattile, 1984). Myrosin cells are found in the second-outermost cell layer of the peripheral cortex of the radicle, and in cotyledons of *Brassica napus* (Thangstad *et al.*, 1990). Myrosinase is also found in guard cells and in xylem, which are prime sites for fungal or bacterial invasion (Höglund *et al.*, 1991). On wounding and breakdown of the compartmentalization in myrosin cells, the glucosinolate is hydrolyzed to an extremely unstable thiohydroximate sulfate ester (Figure 14.6). Non-enzymatic, spontaneous cleavage of the sulfate ester and migration of an alkyl group from C to S produces the volatile isothiocyanate responsible for the typical taste and pungency of Brassica vegetables and is also responsible for defense against fungi and insects. Recent studies on the effect of isothiocyanates on insects are listed in Table 14.1.

Cyanogenic Glucoside Biosynthesis

The enzyme carrying out the first, committed step of the cyanogenic glucoside pathway, Cyt P-450$_{TYR}$, has been purified and shown to be a multifunctional enzyme responsible for the first four steps of the pathway (Sibbeson *et al.*, 1994, 1995). Cyt P-450$_{TYR}$ converts tyrosine to *p*-hydroxyphenylacetaldoxime without

Table 14.1 Recent studies of glucosinolates and isothiocyanates in insect resistance

Insect	Scientific name	References
Crucifer flea beetle	*Phyllotetra cruciferae*	Bodnaryk (1991)
Wireworm	*Limonius infuscatus*	Brown *et al.* (1991)
	L. californicus	Williams *et al.* (1993)
Desert locust	*Schistocerca gregaria*	Ghaout *et al.* (1991)
Bertha armyworm	*Mamestra configurata*	Bodnaryk (1991); McCloskey and Isman (1993)
Stored grain insects	*Prostephanus truncatus*	Seck *et al.* (1993)
	Sitotroga cerealella	Seck *et al.* (1993)
	Tribolium castaneum	Seck *et al.* (1993)
	Sitophilus zeamais	Seck *et al.* (1993)
	Callosobruchus maculatus	Seck *et al.* (1993)
Sugar beet cyst nematode	*Heterodera schactii*	Lazzeri *et al.* (1993)

detectable intermediates (Figure 14.7). Tyrosine is first oxidized to N-hydroxytyrosine which is further oxygenated, possibly via the N,N-dihydroxylamine and subsequent dehydration to 3-(*p*-hydroxyphenyl)-2-nitrosopropionic acid. Based on chemical models, decarboxylation of the 2-nitroso acid is expected to be extremely facile and would lead directly to 2-*p*-hydroxyphenylacetaldoxime. The decarboxylated 2-(*p*-hydroxyphenyl)-nitroethane, or its *aci*-tautomer has also been proposed as an intermediate (Halkier and Møller, 1990; Halkier *et al.*, 1991); however this nitro compound is at a higher oxidation state than the aldoxime, and there is no evidence of involvement of a reduction step. Chemical dehydration of nitroalkanes gives the higher oxidation state nitrile oxide (=cyanate). Chemical conversion of nitroalkanes to nitriles requires a reductant as well as a dehydrating agent (Wherli and Schaer, 1977), while oximes are easily dehydrated chemically to give nitriles. For these reasons the pathway is shown via the nitroso intermediate in Figure 14.7. The initial enzymatic dehydration product in the microsomal preparation is the (*E*)-isomer which is isomerized, presumably by an enzyme, to the (*Z*)-isomer.

Decarboxylation of 2-nitrosocarboxylic acids occurs spontaneously and does not necessarily require an enzyme; however the multifunctional Cyt P-450$_{TYR}$ enzyme catalyzes at least two oxygenations and a dehydration between tyrosine and the aldoxime. The evidence that each of the catalytic activities resides in the same protein, that tyrosine and N-hydroxytyrosine are substrates for the enzyme, and that both oxygenations take place at the same active site without substrate dissociation has been reviewed (Sibbesen *et al.*, 1995). Antibody to Cyt P-450$_{TYR}$ and degenerate oligonucleotide probes based on the amino acid sequence have been used to screen a cDNA library and obtain a full-length clone encoding the committing enzyme of the cyanogenic glucoside pathway (Sibbeson *et al.*, 1995). The antibody to sorghum Cyt P-450$_{TYR}$ has been shown to cross-react with a major polypeptide of similar molecular weight in cassava microsomes, which convert valine into acetone cyanohydrin, the aglycone of linamarin (Du *et al.*, 1995).

Intermediacy of *p*-hydroxyphenylacetaldoxime has been demonstrated by its conversion by the sorghum microsomal system into mandelonitrile in a two-step process: dehydration of the aldoxime to phenylacetonitrile, followed by C-2 hydroxylation to mandelonitrile. However, if carbon monoxide is introduced to block the second cytochrome P-450 step, then both the dehydration and hydroxylation steps are blocked, suggesting that the dehydration step is linked in some way to the subsequent cytochrome P-450 C-hydroxylation. The enzyme(s) affecting the dehydration and C-hydroxylation have not been purified. The soluble mandelonitrile : UDP glucose transferase has been purified (Reay and Conn, 1974). The last step of glucosylation is necessary to stabilize the cyanohydrin glucoside for long-term storage in the vacuole of the leaf epidermis.

The Glucosinolate Biosynthetic Pathway

Progress on the glucosinolate pathway, which is not as advanced as that of the cyanogenic glucoside pathway, has been reviewed (Wallsgrove and Bennett, 1995). The aldoxime-type intermediate of the cyanogenic glucoside pathway also serves as a good intermediate to glucosinolates when administered to whole plants: [^{14}C]phenylacetaldoxime labels benzyl glucosinolate in *Lepidium sativum* (garden

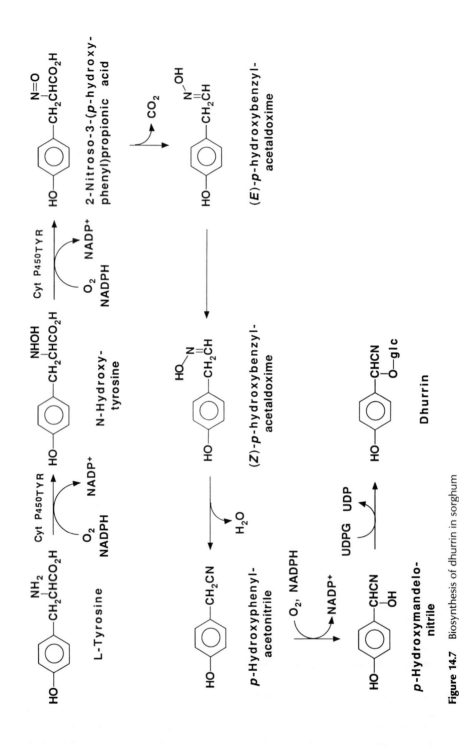

Figure 14.7 Biosynthesis of dhurrin in sorghum

cress) and in *Tropaeolum majus*; and isobutylaldoxime labels isopropyl-GS in *Cochlearia officinalis*. A microsomal preparation from *Brassica campestris* converted tryptophan into indoleacetaldoxime (Ludwig-Muller and Hilgenberg, 1988). The aldoxime could be an intermediate in the pathway to indolmethyl-GS of Brassica. Conversion of tryptophan into the aldoxime has also been shown in maize sunflower, tobacco and pea which do not produce glucosinolates. Indoleacetaldoxime may be a precursor to indoleacetonitrile and indoleacetic acid (IAA) in nonglucosinolate plants.

The conversion of the aldoxime into thiohydroximic acid is presumed to occur via sulfhydryl addition of cysteine or glutathione to the aldoxime, followed by cleavage of the addition product by a cysteine lyase to produce a thiohydroximic acid, which is then glucosylated (Figure 14.8). The presence of an appropriate glutathione transferase and CS lyase has not yet been proven. A soluble UDPG : thiohydroximate glucosyl transferase has been purified from rape seedling (Reed *et al.*, 1993) and *Arabidopsis thaliana* (Guo and Poulton, 1994). The purified enzyme selectively glucosylated thiohydroximates, but did not show selectivity for the side chain moiety. The sulfotransferase purified from *Lepidium sativum* was also tolerant of variation in the sidechain of its substrates (Glendening and Poulton, 1988, 1990). Glucosyl transferase and sulfotransferase activity of *B. juncea* copurified through extensive ion exchange, metal chelate and affinity chromatographies, but glucotransferase activity could be differentially degraded by heating (Jain *et al.*, 1990). Substrate promiscuity of the glucotransferase and sulfotransferase of *B. juncea* permitted synthesis of unnatural *o*-nitrophenyl glucosinolate from synthetic *o*-nitrobenzaldoxime in cell suspension culture (Grootwassink *et al.*, 1990). The isothiocyanate releasing enzyme myrosinase has been purified from oilseed rape (Bones and Slupphaug, 1989) and white mustard (Björkman and Janson, 1972), and clones for two gene families of myrosinases have been isolated from white mustard (Xue *et al.*, 1992).

Canola (*Brassica napus*) plants have been transformed with a gene that encodes tryptophan decarboxylase from *Catharanthus roseus* (Chavadej *et al.*, 1994). In transgenic plants that express this gene, tryptophan is diverted away from production of indolylmethylglucosinolates by decarboxylation to tryptamine which accumulates. Indolylmethylglucosinolate concentration is lowered in all plant parts, but particularly strikingly in the seed where its concentration is only 3% of that in seed of untransformed plants.

Mechanistic Heterogeneity of Glucosinolate Pathway Committing Enzymes

Considerable note has been taken of the common intermediates in the first four steps of the cyanogenic glucoside and glucosinolate pathways up to the aldoxime (Mahadevan, 1975). Numerous authors have speculated on the possible evolutionary significance and some have suggested that the glucosinolate pathway evolved from the cyanogenic glucoside pathway (Kjaer and Larsen, 1976; Rodman, 1981). Although the glucosinolate and cyanogenic glucoside pathways probably proceed through common intermediates, evidence has been recently presented that there may be two mechanistically distinct amino acid N-hydroxylating enzymes within the group of glucosinolate-producing species. In glucosinolate-producing *B. napus* the first enzyme of the pathway appears to be a flavin-containing peroxidase, while

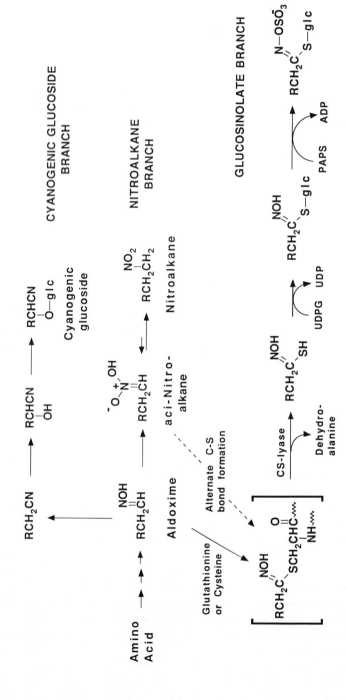

Figure 14.8 Proposed biosynthetic relationship of cyanogenic glucosides, glucosinolates and nitroalkanes

in *Tropaeolum majus* the enzyme accomplishing the same step may be a cytochrome P-450 (Wallsgrove and Bennett, 1995). The existence of mechanistically different enzymes in glucosinolate-producing plants suggests that the glucosinolate pathway may have evolved more than once.

An Engineered Glucosinolate-Cyanogenic Glucoside Hybrid Pathway

The fact that an aldoxime is a common intermediate to the pathways of three classes of plant defensive chemicals, cyanogenic glucosides, glucosinolates and nitro-alkanes (Figure 14.8), means that addition of late genes of the glucosinolate pathway to a cyanogenic plant might lead to the production of both cyanogenic glucosides and glucosinolates. In fact, nature may already have produced such a pathway hybrid in papaya which is unique in that it produces both the phenylalanine-derived benzyl glucosinolate and a small amount of the phenylalanine-derived cyanogenic glucoside, prunasin, along with the novel cyclopentenylglycine-derived cyanogenic glucoside tetraphyllin B (Spencer and Siegler, 1984). Cyanogenic glucosides are produced from just five amino acids in addition to cyclopentenylglycine: valine, leucine, isoleucine, phenylalanine, and tyrosine. Glucosinolate-producing plants however are able to oxidize a wider range of amino acids via the aldoxime to glucosinolates: alanine, valine, leucine, isoleucine, glutamic acid, phenylalanine, tyrosine, tryptophan, and methionine. In addition, both phenylalanine and methionine can be homologated before oxidation to the aldoxime (Mithen and Toros, 1995), and a number of the intermediates can be hydroxylated in the side chain. Thus introduction of early glucosinolate pathway genes into a cyanogenic plant could lead to a profusion of new cyanogenic glucosides not previously seen in nature. The conversion of a glucosinolate plant to cyanogenesis would provide it with a new insect defense against those Brassica specialists that home in on mustard oil odor for oviposition or use glucosinolates as feeding stimulants.

Cyclic Hydroxamic Acids (DIBOA and DIMBOA)

The cyclic hydroxamic acids are a series of progressively methoxylated 2,4-dihydroxy-1,4-benzoxazin-3-ones: DIBOA, DIMBOA and DIM_2BOA (Figure 14.9). These metabolites are major chemical defenses in young maize, wheat and rye against bacteria, fungi and insects (Niemeyer, 1988). There is no DIMBOA present in endosperm or embryo, but it can be readily detected in the germinating plant within ten hours after excision of the embryo from immature kernels (Chilton, unpublished). The concentration of DIMBOA-glucoside in the seedling peaks at 1–10 mmol/kg f.w. during the first week after germination and then slowly declines. Maize and wheat generally contain much more DIMBOA than DIBOA. Some maize varieties also contain a high level of DIM_2BOA. Only the unmethoxylated DIBOA is found in rye. The benzoxazolones (BOA, MBOA and M_2BOA) are less toxic, abiotic decomposition products of the hydroxamic acids. DIMBOA is the easiest hydroxamic acid to isolate because of its abundance and its crystallinity, and is therefore the best-studied hydroxamic acid.

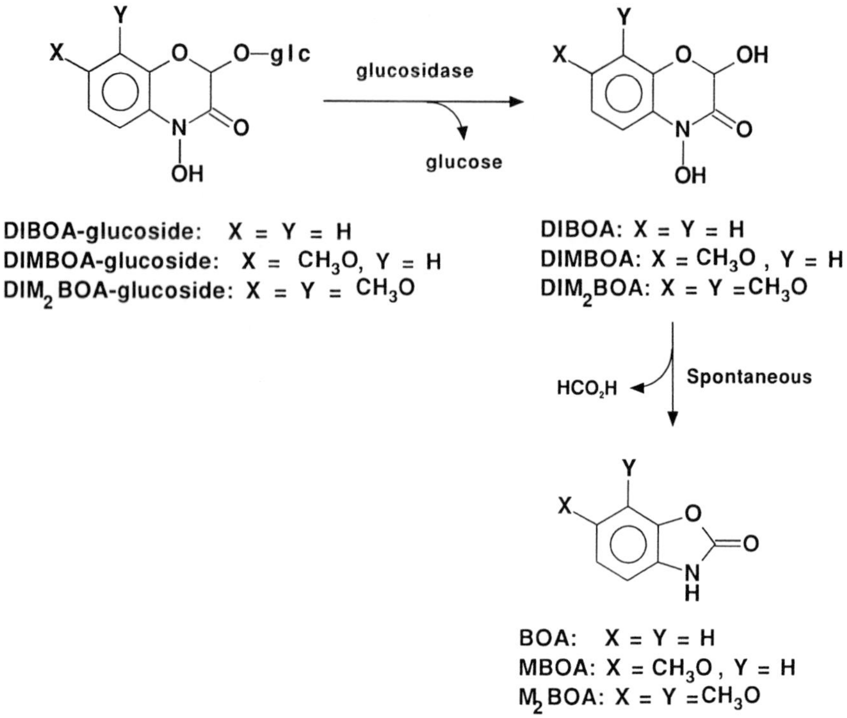
Y
X
O
O—glc
N
O
OH
glucosidase
glucose
Y
X
O
OH
N
O
OH
DIBOA-glucoside: X = Y = H
DIMBOA-glucoside: X = CH3O, Y = H
DIM2BOA-glucoside: X = Y = CH3O
DIBOA: X = Y = H
DIMBOA: X = CH3O, Y = H
DIM2BOA: X = Y = CH3O
HCO2H
Spontaneous
Y
X
O
O
N
H
BOA: X = Y = H
MBOA: X = CH3O, Y = H
M2BOA: X = Y = CH3O

Figure 14.9 Wound release of cyclic hydroxamic acids and their spontaneous decomposition

Storage, Release and Decomposition of DIMBOA

The cyclic hydroxamates are unstable, wound-released, defensive metabolites. They are stored as stable, low-toxicity glucosides separated from a β-glucosidase located in plastids (Esen, 1992). Enzymatic release (Figure 14.9) of the toxic aglycone DIMBOA is complete within 30 minutes of wounding. DIMBOA, which is generally toxic at 0.1–1 mM to bacteria, fungi, insects and plants, including maize, has a half-life of about 24 hours in maize wound juice (pH 5.6) (Woodward *et al.*, 1978). The decomposition product MBOA (70% yield) has much lower toxicity. Thus the defensive system stores a stable, non-toxic glucoside, liberates a very toxic, unstable aglycone in a matter of minutes after wounding, and subsequently destroys the toxic compound in the course of a few days.

Defensive Role against Insects, Fungi and Bacteria

Hydroxamic acids have been correlated with resistance of cereals to *Fusarium nivale*, northern leaf blight, *Helminthosporium turcicum*, *Septoria nodorum*, and wheat stem rust, *Puccinia graminis* var. *tritici* (reviewed by Niemeyer, 1988). DIMBOA is the major active metabolite of maize against *Agrobacterium tumefaciens* (Sahi *et al.*, 1990) and soft rotting *Erwinia* bacteria (Corcuera *et al.*, 1978). The role of DIMBOA in resistance to a number of insect pests of maize has been an active area of study in recent years (Table 14.2) and is the subject of several reviews (Meyer, 1988; Niemey-

Table 14.2 Recent studies of the role of DIMBOA in insect resistance

Insect	Scientific name	References
European corn borer	*Ostrinia nubilalis*	Barry *et al.* (1994); Campos *et al.* (1990); Reid *et al.* (1991)
Asian corn borer	*Ostrinia funacalis*	Tseng (1989)
Southwestern corn borer	*Diatraea grandiosella*	Hedin *et al.* (1993)
Western corn rootworm	*Diabrotica virgifera virgifera*	Xie *et al.* (1992)
African armyworm	*Spodoptera exempta*	Okelloekochu and Wilkins (1994)
Aphids on maize	*Rhopalosiphum padi*	Givovich *et al.* (1994); Morse *et al.* (1991)
Aphids on wheat	*Rhopalosiphum, Metopolophium, Schizaphis*	Cuevas and Niemeyer (1993); Nicol *et al.* (1993)

er, 1988). DIMBOA also exerts more indirect effects through parasitoids of maize-destroying insects (Campos *et al.*, 1990; Martos *et al.*, 1992), while its breakdown product MBOA acts as an attractant for the western corn rootworm (Bjostad and Hibbard, 1992).

Biosynthesis of DIMBOA

Synthesis of DIMBOA takes place in both roots and in shoots (Peng and Chilton, 1994). DIMBOA biosynthesis diverges formally from the tryptophan pathway at indole (Figure 14.10). The latter is converted by a membrane-bound enzyme into 1,4-2H-benzoxazin-3(4H)-one (Desai *et al.*, 1996). No intermediates are detectable between indole and benzoxazinone. Two further steps are catalyzed by NADPH-dependent microsomal enzymes inhibitable by carbon monoxide and by other cytochrome P-450 monooxygenase inhibitors. Benzoxazinone is hydroxylated to give 2-hydroxybenzoxazinone (N-deoxyDIBOA) (Kumar *et al.*, 1994) and the latter is further hydroxylated to DIBOA (Bailey and Larson, 1991). Each intermediate, indole, benzoxazinone, or 2-hydroxybenzoxazinone, administered separately to maize shoots is converted into DIBOA. Thus there is no metabolic channeling between indole and DIBOA with the possible exception of the conversion of indole into benzoxazinone. This reaction involves at least two steps: insertion of one oxygen atom into the indole ring and addition of a second oxygen atom at C-3 of the benzoxazine. Experiments with ^{12}O label show that both of these oxygens are derived from air O_2. A dioxygenase may produce an enzyme-bound, unstable indolenine hydroperoxide which rearranges to benzoxazinone before leaving the enzyme. The oxygenations as far as DIBOA occur before glucosylation which is effected by two UDPG : DIBOA glucosyltransferases of maize (Bailey and Larson, 1989). The maize glucosyltransferase also catalyzes glucosylation of DIMBOA by UDPG but not N-deoxyDIBOA nor N-deoxyDIMBOA (Leighton *et al.*, 1994).

Although maize shoots convert indole, benzoxazinone and 2-hydroxybenzoxazinone into DIMBOA in high yield, further oxygenation of DIBOA to 7-hydroxyDIBOA (TRIBOA) was not observed with microsomal preparations in any of the three laboratories. If 7-hydroxylation is carried out on the glucoside of

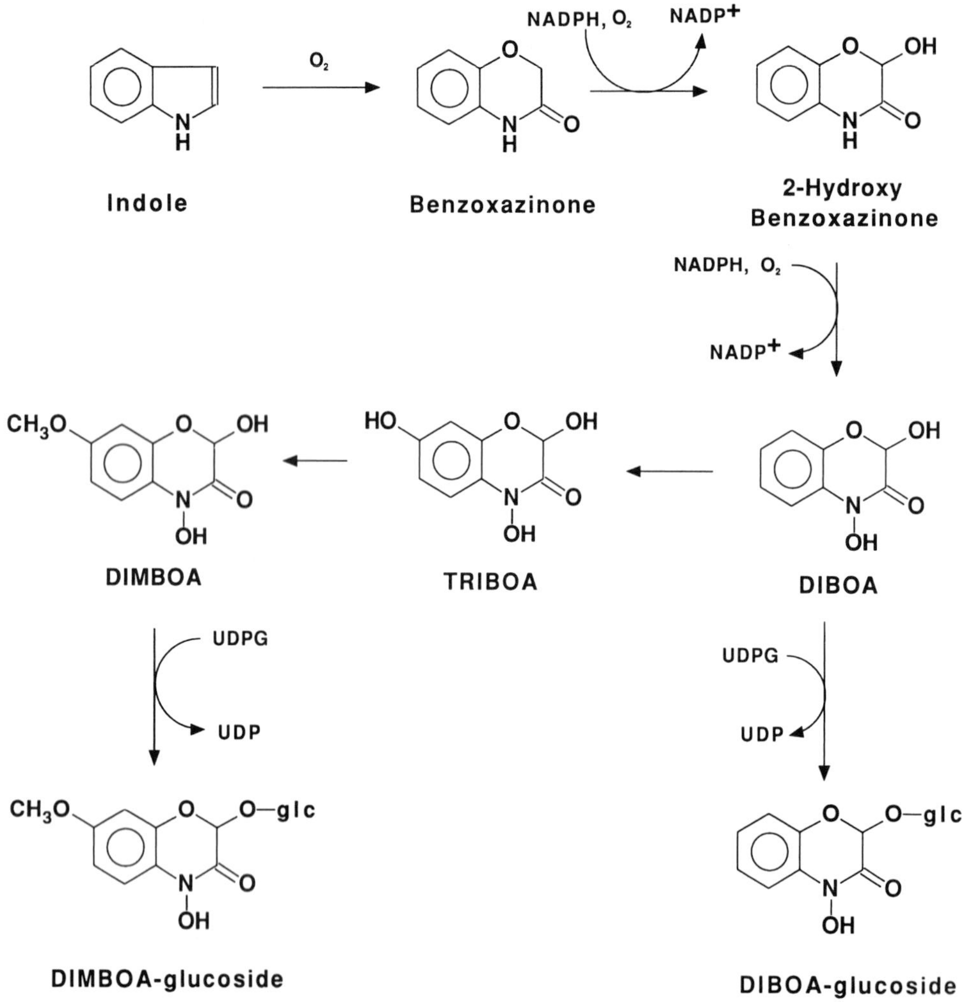

Figure 14.10   Biosynthesis of the cyclic hydroxamic acids

DIBOA, the microsome fraction, which lacks the soluble glucosyl transferase and UDPG, would be unable to perform this last oxygenation.

DIMBOA Pathway Genes

Four cytochrome P-450 genes have been cloned from a maize seedling-specific library (Frey *et al.*, 1995). Whole seedling activities of DIBOA : UDPG glucosyl-transferase and of the probable cytochrome P-450 which N-hydroxylates N-deoxyDIBOA reach a maximum at three days after germination (Bailey and Larson, 1991). This is the same time at which the steady state levels of mRNA of the four cytochrome P-450s reach a maximum in the root; mRNA for these four cytochrome P-450s peaks in the shoot at day 7. The highest mRNA levels are in the coleoptile, the first leaflets, the ground tissue of the nodular complex and in the cortex and pith of the region of cell division in the root. Some of these cytochrome P-450 enzymes may be involved in DIMBOA biosynthesis which also peaks at an early stage. The level of DIMBOA-releasing β-glucosidase is also correlated with the time of maximum concentration of DIMBOA. The β-glucosidase has been purified and characterized and the gene has been isolated and sequenced (Babcock and Esen, 1994). A DIMBOA-deficient maize mutant, containing only 5 μM DIMBOA, may be a pathway regulatory mutant (Chilton, unpublished). Transposon tagging of the mutated gene is in progress.

Pyrethrin

Use of a powder of dried flowers of *Chrysanthemum* (*Pyrethrum*) *cinerariaefolium* as an insecticide was introduced into Europe from Middle Eastern folk usage in 1828 (Casida and Quistad, 1995; Matsui and Yamamoto, 1971). A pyrethrum concentrate containing 20% active ingredient is prepared by alkane extraction. The major insecticidal metabolites in the extract are pyrethrin I and II, esters of chrysanthemic and pyrethric acids (Figure 14.11). Pyrethrins and the closely related insecticidal jasmolins, and cinerins, can be extracted from hydrocarbon solvents into nitromethane and separated by chromatography. The content of pyrethrins peaks in the flower at

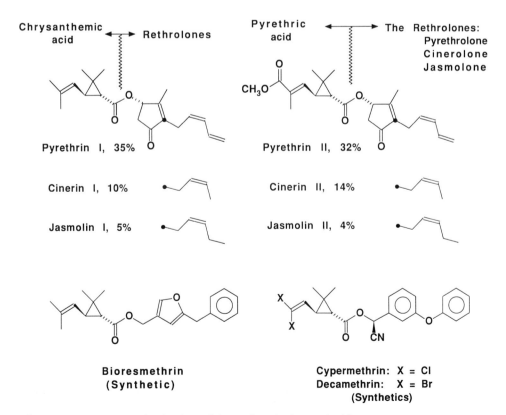

Figure 14.11 Structure of natural pyrethrins and synthetic pyrethroids

about 1% dry weight at the time of full bloom. The pyrethrins are highly localized in the seed wall of disk florets (83%) (Brewer, 1973). The ovule in the achene contains about 10%; the corolla, style and pollen contain lesser amounts.

Pyrethrin is particularly valuable for its quick 'knockdown' of flying insects, low toxicity to man and other mammals, and short lifetime due to photochemical oxidation. These properties, combined with its cost, restrict the use of pyethrin to the house and garden and in the control of mosquitoes and flies. Pyrethrin has served as a hugely successful lead chemical in the development of more than 40 synthetic pyrethroids with higher specific activity than pyrethrin and greater stability to sunlight (Elliott, 1983, 1995; Naumann, 1990). Despite the advances in synthetic pyrethroids, natural pyrethrin continues to hold its special niche in the insecticide market.

Insect Toxicity and Resistance

Pyrethrin sprayed in the air knocks flying insects to the ground almost instantaneously. At a low dose insects may recover from paralysis. The dose required to kill is decreased by lignan-related synergists such as sesamin and synthetic piperonyl butoxide. The methylenedioxy group of synergists is the active site for mechanism-based inhibition of the cytochrome P-450 monoxygenase involved in detoxification of pyrethrin in most insects (Hodgson et al., 1995). Pyrethroids act as neurotoxins by delaying closure of voltage-sensitive sodium channels following depolarization of the nerve membrane potential (Soderlund, 1995). DDT knockdown resistance (kdr) in the housefly also confers cross-resistance to pyrethrin (Knipple et al., 1994). The kdr mechanism of pyrethrin resistance involves lower sensitivity of sodium channels and presynaptic terminals to pyrethroids (Pepper and Osborne, 1993). A kdr-like reduced neuronal sensitivity to DDT and pyrethroids may exist in other crop pests (Soderlund and Bloomquist, 1990). Esterase mechanisms of resistance to synthetic pyrethroids are suggested in Heliothis virescens (Dowd et al., 1987) and Spodoptera exigua (Delorme et al., 1988). Detoxification of pyrethroids by cytochrome P-450 monooxygenases is implied by synergism of pyrethroid toxicity by methylenedioxy-phenyl inhibitors. The cytochrome P-450 responsible for pyrethroid resistance in the housefly has been characterized (Scott and Wheelock, 1992). An increase in cytochrome P-450 reductase may be involved in resistance of the housefly to permethrin (Scott and Georghiou, 1986).

Biosynthesis of Chrysanthemic and Pyrethric Acids

Chrysanthemic acid is a monoterpene derived from two units of iso-pentenylpyrophosphate (Crombie et al., 1971; Crowley et al., 1961, 1962; Godin et al., 1963; Pattenden et al., 1975). The first identifiable intermediate is chrysanthemyl alcohol, which occurs in several other Asteraceae species. The fungus Aspergillus ochraceus can oxidize a mixture of isomeric chrysanthemyl alcohols to chrysanthemic acid with high optical purity (Miski and Davis, 1988). Pyrethrin I, an ester of chrysanthemic acid, is oxidized by P. cinarariaefolium to pyrethrin II, an ester of pyrethric acid (Abou Donia et al., 1973).

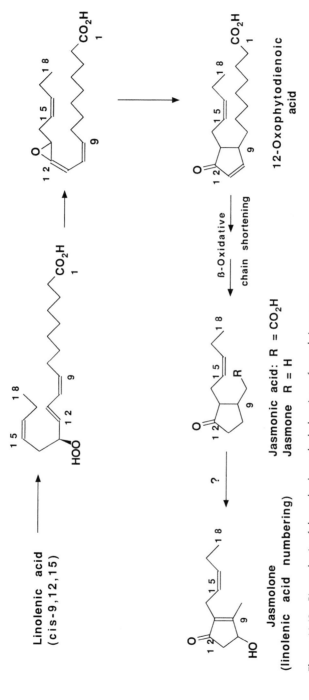

Linolenic acid
(cis-9,12,15)

12-Oxophytodienoic
acid

β-Oxidative
chain shortening

Jasmonic acid: R = CO₂H
Jasmone   R = H

Jasmolone
(linolenic acid numbering)

Figure 14.12 Biosynthesis of the rethrolone alcohol moiety of pyrethrin esters

Biosynthesis of Rethrolones and Pyrethrins

Pyrethrin I, cinerin I and jasmolin I are esters of chrysanthemic acid and the 'rethrolones' pyrethrolone, cinerolone or jasmolone respectively (Figure 14.11). The relationship of the rethrolones to jasmone (Figure 14.12), a component of the essential oils of many plant families, and to the plant hormone jasmonic acid, is evident from their structures. These cyclopentenones are biosynthesized from the 18 : 3 fatty acid, linolenic acid, via phytodienoic acid, an intermediate which bears resemblance to mammalian prostaglandin produced from the 20 : 4 fatty acid, arachidonic acid; however the mechanism of formation of the cyclopentyl ring is different, involving a C-13 peroxide and a very unstable allene epoxide intermediate (Crombie, 1995; Crombie and Morgan, 1991). A cascade of chain shortening then produces jasmonic acid, jasmone and the related alcohols (Crombie and Holloway, 1985). Jasmonic acid may be a universal plant hormone. The ubiquity of jasmone suggests that a major part of the pathway to jasmolone and other rethrolones necessary for biosynthesis of the pyrethrins may already exist in most plants. Cinerolone and jasmolone have been identified by gas chromatography–mass spectrometry in oil of *Micromeria congesta* (Kirimer *et al.*, 1991) and jasmolone in oil of *Ophiopogon japonicus* (Zhu *et al.*, 1991). Cinerolone has also been reported in blueberry (*Vaccinium ashei*) (Horvat *et al.*, 1983).

The esterification of chrysanthemic acid to the rethrolones has not been studied. Some enzymology has been done on the phytodienoic acid–jasmonic acid pathway, but none on the rethrolone extensions of that pathway. Any attempt to use the phytodienoic acid pathway as a starting point for transgenic production of pyrethrin would clearly require increasing metabolic flux through that pathway with unforseen consequences to the level of jasmonic acid and its hormonal action.

Rotenone

Rotenone (Figure 14.13) is a highly modified isoflavonoid of the Leguminosae. Rotenone-containing *Derris* spp. added to water have been used for centuries in the Malay Peninsula to paralyze fish for harvesting and to kill insects (Fukami and Nakajima, 1971). Tribes in South America and Africa use other rotenone-containing genera (*Lonchocarpus*, *Tephrosia* and *Mundulea*) of the pea family for the same purpose. These folk uses were brought to the attention of Europeans in the mid-nineteenth century. A root dust preparation was commercialized for control of insects in the United States in 1911. A crude solvent extract was introduced later. During the Second World War the commercial source of rotenone preparations shifted from Sumatran *Derris* to Amazonian *Lonchocarpus*. Plant-produced rotenone is still competitive with modern synthetic insecticides principally because of its low toxicity to humans. The low mammalian toxicity and rapid degradation of rotenone in the field are major factors enabling rotenone to maintain its special market niche.

About 30 minor insecticidal rotenoids (e.g. amorphin, deguelin and elliptone) have been isolated from rotenone-containing species. Structurally and biosynthetically rotenone is closely related to dehydrokievitone (Figure 14.13), the phytoalexin of French bean (*Phaseolus vulgaris*), another member of the pea family. Further insecticidal rotenoids continue to be isolated (Kole *et al.*, 1992; Lin *et al.*, 1993;

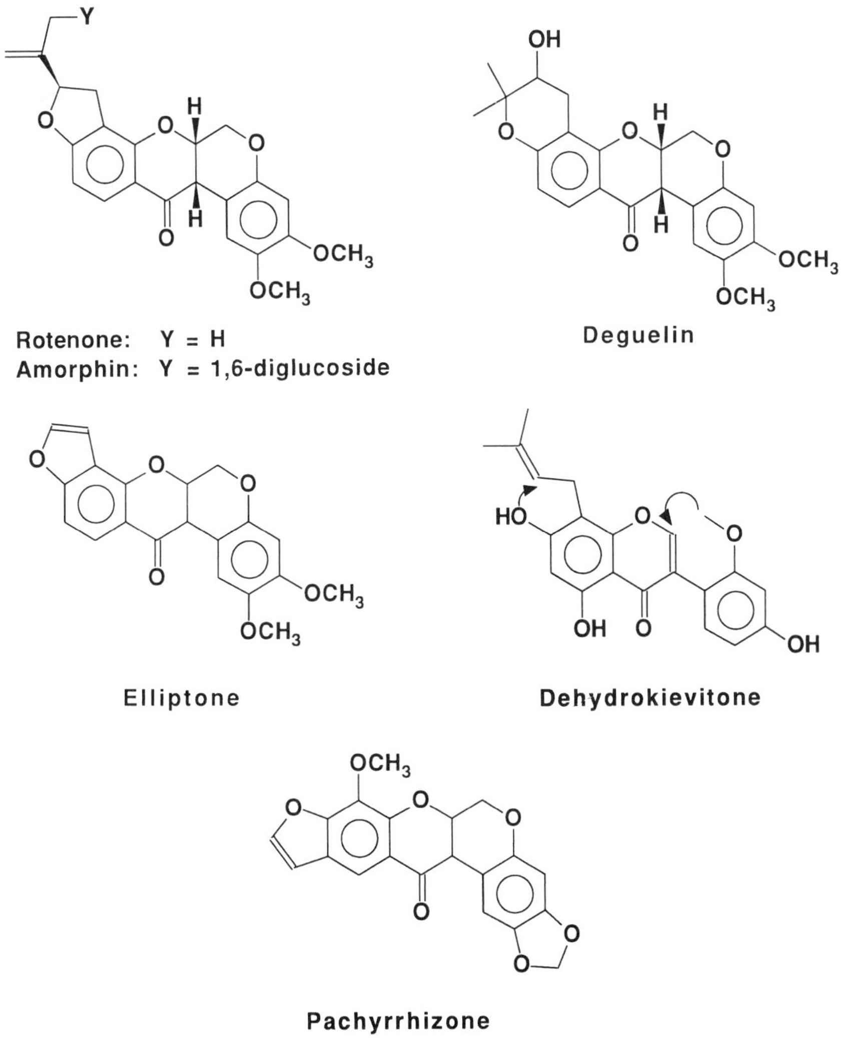
Y
H
O
O
O
H
O
OCH3
OCH3
Rotenone: Y = H
Amorphin: Y = 1,6-diglucoside
OH
O
H
O
O
H
O
OCH3
OCH3
Deguelin
O
O
O
O
OCH3
OCH3
Elliptone
HO
O
O
OH
O
OH
Dehydrokievitone
OCH3
O
O
O
O
O
O
Pachyrrhizone

Figure 14.13 Structures of some rotenoids and related metabolites

Phrasant and Krupadanam, 1993). The plant jicama (*Pachyrrhizus erosus* Urban) has long been cultivated in Mexico for its edible tuber which has recently appeared in North American supermarkets as well. In the areas where jicama is grown, a contact insecticide is prepared by crushing the large, non-edible seeds (Crosby, 1971). The major active component is another rotenoid, pachyrrhizone (Figure 14.13), a contact poison for the striped flea beetle, Mexican bean beetle, silkworm, cabbage worm and aphids.

Toxicity of Rotenone

Rotenone has an oral LD_{50} in laboratory animals in the range 60–300 mg/kg. The lowest observed lethal dose in a human was 143 mg/kg, approximately 10 g rotenone (RTECS, 1994b). Fish and insects are much more sensitive. The lethal toxic

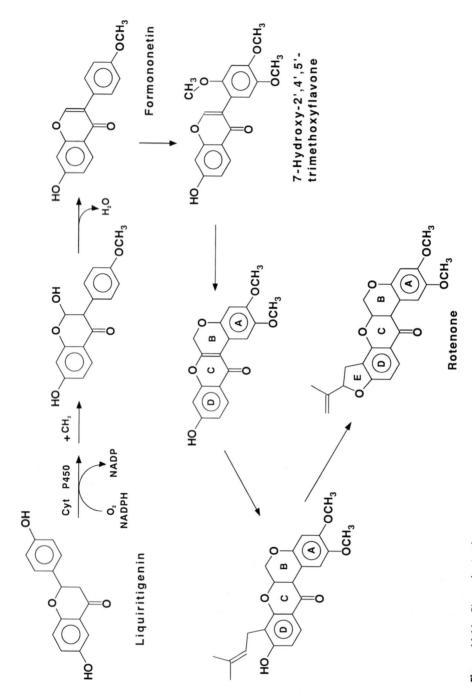

Figure 14.14 Biosynthesis of rotenone

concentration for fish is between 8 and 500 ppb in water and 1–3 ppm for insect larvae in water. Crude rotenone extract is also very effective against insects as a dust on plants. Rotenone acts as an inhibitor of mitochondrial oxidation of NADH (Oberg, 1961). Mammals are less sensitive than insects and fish to rotenone because of malabsorption and more active detoxification. There is some concern that rotenone may be carcinogenic (Greenway *et al.*, 1993; reviewed by Gosalvez, 1983).

Biosynthesis of Rotenone

Extensive use has been made of stable isotopes in the study of the biosynthesis of rotenone (Figure 14.14). Enzymes of the main pathway to rotenone have not yet been characterized, although the enzyme forming the chromene E ring of deguelin has been recently characterized (Crombie *et al.*, 1992). Early intermediates are common to the chalcone-flavonoid pathway. Migration of the *p*-hydroxyphenyl ring A of liquiritigenin (Figure 14.14) from C-2 to C-3 gives the isoflavone formononetin (Crombie and Whiting, 1992), also known as a phytoalexin in lentil, alfalfa and vetch (Bailey and Mansfield, 1982). Hydroxylation and *O*-methylation in the A ring of formononetin leads to 7-hydroxy-2′,4′,5′-trimethoxyisoflavone. Oxidative cyclization involving the 2′-methoxyl group produces the B ring and completes the A/B/C/D rings systems of rotenoids (Bhandari *et al.*, 1992a). The furan E ring is added via prenylation (Bhandari *et al.*, 1992b).

Although all of the research remains to be done on the enzymology of rotenoid biosynthesis, it is clear that the pathway up to formononetin is already present in a large number of economically important legume vegetables. The existence of cell cultures which produce rotenoids should be a useful tool in studying the molecular biology of this pathway (Lambert *et al.*, 1993).

Unsaturated Amides

The commercial success of rotenone and pyrethrin, derived from folk usage of derris root and pyrethrum flowers, stimulated the search in the mid-twentieth century for other plant metabolites with high, selective lethality toward insects (Jacobson, 1971). A number of unsaturated amides in plants were found to possess physiological activity including lethality to insects. Naturally occurring unsaturated isobutyl amides (Figure 14.15) and unsaturated amides of the heterocycle piperidine taste peppery or spicy hot to humans, and in fact piperine and capsaicin are the major active metabolites of black pepper and chili pepper respectively. This class of metabolites is found particularly in Asteraceae, Rutaceae, Piperaceae and Solanaceae. These neutral, non-polar metabolites are generally extracted from dried plant material with a hydrocarbon solvent and then partitioned into nitromethane and purified by chromatography.

Pellitorine

Pellitory root (*Anacyclus pyrethrum* DC, Asteraceae) is used in North Africa as a local anesthetic for treatment of toothache. The major insecticidal amide pellitorine

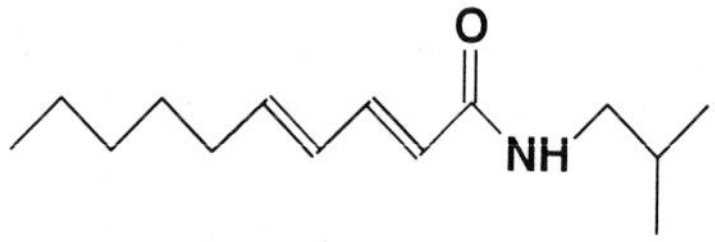

Pellitorine

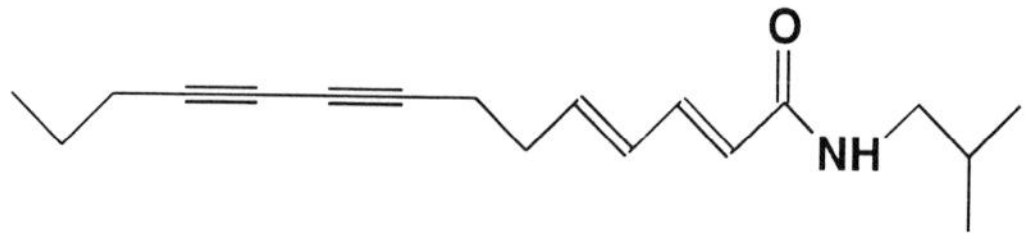

Anacyclin

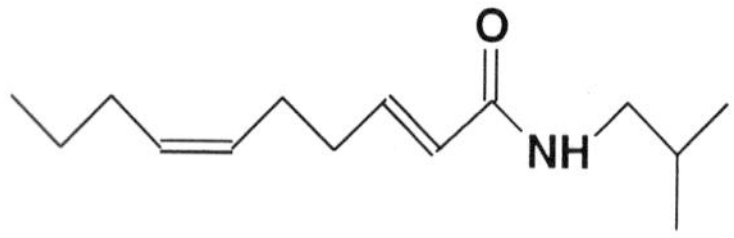

Affinin/Spilanthol

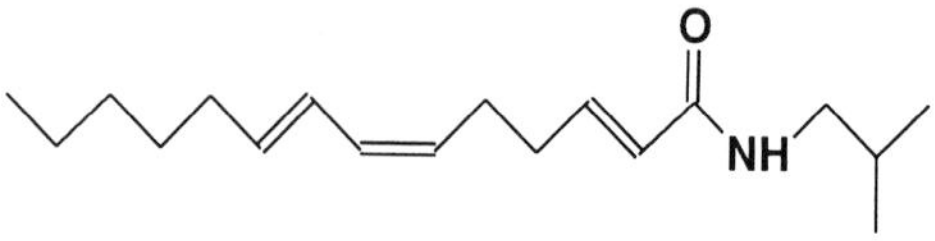

Neoherculin

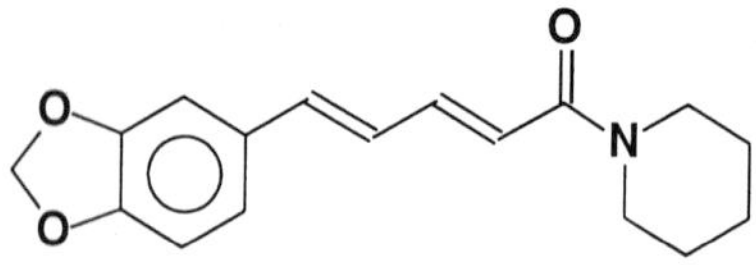

Piperine

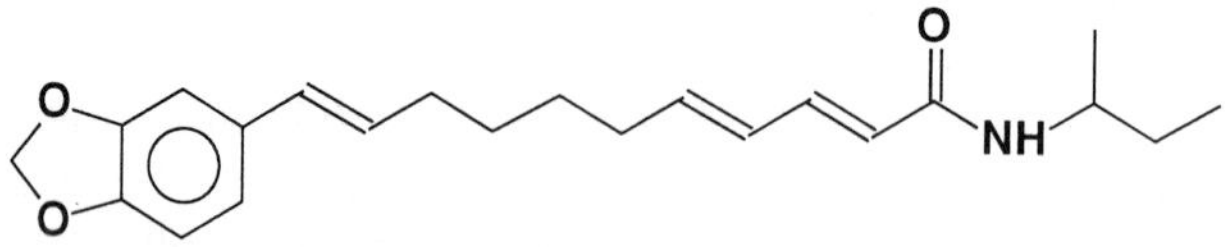

Pipercide

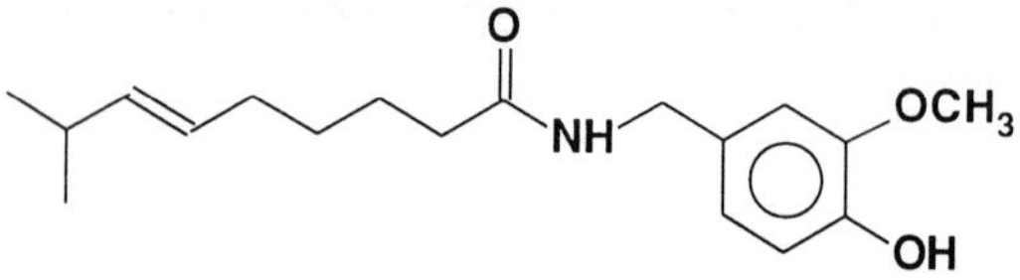

Capsaicin

Figure 14.15   Insecticidal unsaturated amides

can be extracted from pellitory root with hydrocarbon solvent in 0.1% yield (Jacobson, 1949). A co-occurring analog, anacylin, is insecticidally inactive and lacks pungency. Pellitorine is pungent on the tongue and possesses about one half the knockdown and lethal activities of pyrethrins tested at the same concentration. Crude petroleum extract of pellitory root is lethal to adult yellow mealworm (*Tenebrio molitor* L.) following application of 47 µg topically (88% kill) (Crombie, 1955).

Spilanthol

Chewing aerial parts of paracress (*Spilanthes oleraceae*, Asteraceae) causes numbing of the lips and tongue. The plant has been used for alleviation of toothache in Europe (Jacobson, 1971). Dried flower heads contain greater than 1% spilanthol (Crombie *et al.*, 1963). Spilanthol was rediscovered (as affinin) by investigation of a folk use of roots of *peritre del pais* ('native pyrethrum', *Erigeron affinis*, Asteraceae) as an insecticide (Acree *et al.*, 1945). The active component, affinin, constitutes greater than 1% of the dried root of *E. affinis* and is identical to spilanthol. *E. affinis* extract has been used as an anesthetic in tooth extraction. Affinin is found in the root of another Asteraceae, *Heliopsis longipes*, also used as a folk insecticide in Mexico (Jacobson, 1971). Affinin/spilanthol has the same order of paralyzing action and lethality to houseflies as pyrethrin.

Neoherculin

The active principle of toothache tree (*Zanthoxylum clava-herculis*, Rutaceae) and of American coneflower (*Echinacea angustifolia*, Asteraceae) is N-isobutyl-trans, cis, trans, trans-2,6,8,10-dodecatetraenamide (neoherculin = echinacein). Neoherculin has also been isolated in 1.8% yield from fruit of Japanese *Z. piperitum*. Bark of the toothache tree has been available commercially where the tree grows along the southern US coast. A hydrocarbon extract of the bark has been patented as an insecticide (LaForge and Haller, 1943). The toxicity of neoherculin toward houseflies is comparable to that of pyrethrum (Crombie, 1952), and it is highly toxic to yellow mealworm (Crombie, 1955).

A number of other insecticidal, unsaturated isobutyl amides have been isolated from other species and genera of Asteraceae, Rutaceae and Piperaceae (Jacobson, 1971; Parmar *et al.*, 1993). Both the polyunsaturated fatty acid isobutyl amides and the pungent, insecticidal amides of phenolic acids and piperidine are characterisic of the family Piperaceae (Ahn *et al.*, 1992: Gbewonyo *et al.*, 1993). The methylenedioxyphenyl unsaturated amide pipercide (Figure 14.15) has LD_{50} and knockdown activity comparable to pyrethrin I (Masakuzu *et al.*, 1989).

Toxicity

There is little mammalian toxicity information available on the unsaturated amides. However, some experience with human use of these amides as local anesthetics or flavoring agents has been mentioned. Affinin is similar to other insecticidally active unsaturated amides in that it is very pungent to human taste. This is consistent with the use of *H. longipes* in Mexico as a spice in beans and alcoholic drinks; however overdosing has caused death by choking (Jacobson, 1971). *Z. clava-herculis* (source of neoherculin) adminstered orally to rats at 500 mg/kg caused salivation but no other symptoms (Jacobson, 1948).

Prospects for Crop Protection by Unsaturated Amides

Biosynthesis of the unsaturated fatty acid precursors of the amides has not been studied. The steps of chain extension of the cinnamates, involved in formation of amides of Piperaceae and Solanaceae, also remain to be investigated. The amine component comes from decarboxylation of valine. The amide-forming enzyme of piperine biosynthesis, piperoyl-CoA : piperidine N-piperoyl transferase, has been purified (Geisler and Gross, 1990). The unsaturated amides have been demonstrated to have insecticidal activity comparable to that of pyrethrin, and several laboratories have used them as starting points for chemical synthesis programs aimed at duplicating the success obtained with synthetic pyrethroids (Crombie and Horsham, 1987; Elliott *et al.*, 1987a,b; Masakuzu *et al.*, 1989; Miyakado *et al.*, 1989). The relatively simple unsaturated isobutyl amides are probably only a few enzymatic steps removed from the fatty acids of primary metabolism. Their insecticidal activity and biosynthetic simplicity make them attractive candidates for crop protection by genetic engineering.

References

Abou Donia, S. A., Doherty, C. F. and Pattenden, G. (1973) Biosynthesis of chrysanthemum-dicarboxylic acid, and the origin of the Pyrethrin IIs, *Tetrahedron Lett.*, 3477–3481.

Acree, F., Jacobson, M. and Haller, H. L. (1945) An amide possessing insecticidal properties from the roots of *Erigeron affinis* DC, *J. Org. Chem.* **10**, 236–242.

Ahn, J-W., Ahn, M-J., Zee, O-P., Kim, E-J., Lee, S. G., Kim, H. J. and Kubo, I. (1992) Piperidine alkaloids from *Piper retrofractum* fruits, *Phytochemistry*, **31**, 3609–3612.

Babcock, G. D. and Esen, A. (1994) Substrate specificity of maize β-glucosidase, *Plant Sci.* **101**, 31–39.

Bailey, B. A. and Larson, R. L. (1989) Hydroxamic acid glucosyltransferases from maize seedlings, *Plant Physiol.* **90**, 1071–1076.

Bailey, B. A. and Larson, R. L. (1991) Maize microsomal benzoxazinone N-monooxygenase, *Plant Physiol.* **95**, 792–796.

Bailey, J. A. and Mansfield, J. W. (1982) *Phytoalexins*, New York: John Wiley and Sons, pp. 150–152.

Barry, D., Alfaro, D. and Darrah, L. L. (1994) Relation of European corn borer (Lepidoptera: Pyralidae) leaf feeding resistance and DIMBOA content in maize, *Environ. Entomol.* **23**, 177–182.

Bhandari, P., Crombie, L., Daniels, P., Holden, I., van Bruggen, N. and Whiting, D. A. (1992a) Biosynthesis of the A/B/C/D ring system of the rotenoid amorphigenin by *Amorpha fruticosa* seedlings, *J. Chem. Soc. Perkin Trans.* 1, 839–849.

Bhandari, P., Crombie, L., Kilbee, G. W., Pegg, S. J., Proudfoot, G., Rossiter, J., Sanders, M. and Whiting, D. A. (1992b) Biosynthesis of rotenone and amorphigenin. Study of the origins of isopropenyl-substituted dihydrofuran E-rings using isotopically labelled late precursors, *J. Chem. Soc. Perkin Trans.* 1, 851–863.

Bjostad, L. B. and Hibbard, B. E. (1992) 6-Methoxy-2-benzoxazolinone: a semiochemical for host location by western corn rootworm larvae, *J. Chem. Ecol.* **18**, 931–943.

Björkman, R. and Janson, J-C. (1972) Studies on myrosinase I. Purification and characterization of a myrosinase from white mustard seed (*Sinapis alba* L.), *Biochim. Biophys. Acta* **276**, 508–518.

Bodnaryk, R. P. (1991) Developmental profile of sinalbin in mustard seedlings, *Sinapis alba* L. and its relationship to insect resistance, *J. Chem. Ecol.* **17**, 1543–1556.

Bones, A. M. and Slupphaug, G. (1989) Purification, characterization and partial amino acid sequencing of β-thioglucosidase from *Brassica napus* L, *J. Plant Physiol.* **134**, 722–729.

Brewer, J. G. (1973) Microhistological examination of the secretory tissue in pyrethrum florets, *Pyrethrum Post* **12**, 17–22.

Brown, P. D., Morra, M. J., McCaffrey, J. P., Auld, D. L. and Williams, L. (1991) Allelochemicals produced during glucosinolate degradation in soil, *J. Chem. Ecol.* **17**, 2021–2034.

Campos, F., Atkinson, J., Arnason, J. T., Philogène, B. J. R., Atkinson, J., Morand, P. and Werstiuk, N. H. (1990) Biological effects and toxicokinetics of DIMBOA in *Diadegma terebrans* (hymenoptera: Ichneumonidae), an endoparasite of *Ostrinia nubilalis* (Lepidoptera: Pyralidae), *J. Econ. Entomol.* **83**, 356–360.

Casida, J. E. and Quistad, G. B. (1995) *Pyrethrum Flowers: Production, Chemistry, Toxicology, and Uses*, New York: Oxford University Press.

Chavadej, S., Brisson, N., McNeil, J. N. and De Luca, V. (1994) Redirection of tryptophan leads to production of low indole glucosinolate canola, *Proc. Natl. Acad. Sci. USA* **91**, 2166–2170.

Chew, F. S. (1988) Biological effects of glucosinolates, In: Cutler, H. C. (Ed.), *Biologically Active Natural Products Potential Use in Agriculture*, Washington, DC: American Chemical Society, pp. 155–181.

Collison, R. C., Harlan, J. D. and Streeter, L. R. (1929) High nicotine tobacco, *NY* (*Geneva*) *Agr. Exp. Sta. Bull.*, p. 562 (cited by Shephard, 1951).

Corcuera, L. J., Woodward, M. D., Helgeson, J. P., Kelman, A. and Upper, C. D. (1978) 2,4-Dihydroxy-7-methoxy-2H-1,4-benzoxazin-3(4H)-one. An inhibitor from *Zea mays* with differential activity against soft rotting *Erwinia* species, *Plant Physiol.* **61**, 791–795.

Crombie, L. (1952) Amides of vegetable origin. I. Stereisomeric N-isobutyl undeca-1 : 7-diene-1-carboxamides and the structure of herculin, *J. Chem. Soc.*, 2997.

Crombie, L. (1955) Amides of vegetable origin. IV. The nature of pellitorine and anacyclin, *J. Chem. Soc.*, 999.

Crombie, L. (1995) Chemistry of pyrethrins, In: Casida, J. E. and Quistad, G. B. (Eds), *Pyrethrum Flowers: Production, Chemistry, Toxicology, and Uses*, New York: Oxford University Press, pp. 168–177.

Crombie, L. and Holloway, S. J. (1985) Biosynthesis of the pyrethrins: unsaturated fatty acids and the origins of the rethrolone segment, *J. Chem. Soc. Perkin Trans.* 1, 1393–1400.

Crombie, L. and Horsham, M. A. (1987) Synthetic approaches to isobutylamides of insecticidal interest, *Tetrahedron Lett.*, 4879–4882.

Crombie, L. and Morgan, D. O. (1991) Synthesis of [14,14-^{2}H$_2$]-linolenic acid and its use to confirm the pathway to 12-oxophytodienoic acid in plants: a conspectus of the epoxy carbonium ion derived metabolites from linoleic and linolenic acid hydroperoxides, *J. Chem. Soc.*, 581–587.

Crombie, L. and Whiting, D. A. (1992) The mechanism of the enzymic induced flavanone–isoflavone change, *Tetrahedron Lett.*, **33**, 3663–3666.

Crombie, L., Krasinki, A. H. A. and Mazoor-i-Khuda, M. (1963) Amides of vegetable origin. Part X. The stereochemistry and synthesis of affinin, *J. Chem. Soc.*, 4970–4976.

Crombie, L., Doherty, C. F., Pattendon, G. and Woods, D. K. (1971) The acid thermal decomposition products of natural chrysanthemum dicarboxylic acid, *J. Chem. Soc. C.*, 2739–2743.

Crombie, L., Rossiter, J. T., Van Bruggen, N. and Whiting, D. A. (1992) Deguelin cyclase, a prenyl to chromene transforming enzyme from *Tephrosia vogelii*, *Phytochemistry* **31**, 451–461.

Crosby, D. G. (1971) Minor insecticides of plant origin, In: Jacobson, M. and Crosby, D. G. (Eds), *Naturally-Occurring Insecticides*, New York: Marcel Dekker, pp. 205–213.

Crowley, M. P., Inglis, H. S., Snarey, M. and Thain, E. M. (1961) Biosynthesis of the pyrethrins, *Nature* **191**, 281–282.

Crowley, M. P., Godin, P. J., Inglis, H. S., Snarey, M. and Thain, E. M. (1962) The biosynthesis of the 'pyrethrins'. The incorporation of ^{14}C-labelled compounds into the flowers of *Chrysanthemum cinerariaefolium* and the biosynthesis of chrysanthemum mono-carboxylic acid, *Biochem. Biophys. Acta* **60**, 312–319.

Cuevas, L. and Niemeyer, H. M. (1993) Effect of hydroxamic acids from cereals on aphid cholinesterase, *Phytochemistry* **34**, 983–985.

Dawson, R. F. (1948) Aldaloid biogenesis, *Adv. Enzymol.* **8**, 203– 251.

Delorme, R., Fournier, D., Chaufaux, J., Cuany, A., Bride, J. M., Auge, D. and Berge, J. B. (1988) Esterase metabolism and reduced penetration are causes of resistance to deltamethrin in *Spodoptera exigua* HUB (Noctuidae: Lepidoptera), *Pestic. Biochem. Physiol.* **32**, 240–246.

Desai, S. R., Kumar, P. and Chilton, W. S. (1996) Indole is an intermediate in the biosynthesis of cyclic hydroxamic acids in maize, *J. Chem. Soc. Chem. Commun.* 1321.

Dowd, P. C. F., Gagne, C. C. and Sparks, T. C. (1987) Enhanced pyrethroid resistant larvae of the tobacco budworm *Heliothis zea* (F.), *Pestic. Biochem. Physiol.* **28**, 9–16.

Du, L., Bokanga, M., Møller, B. L. and Halkier, B. S. (1995) The biosynthesis of cyanogenic glucosides in roots of cassava, *Phytochemistry* **39**, 323–326.

Elliott, M. (1983) Development in the chemistry and action of pyrethroids, In: Whitehead, D. L. and Bowers, W. S. (Eds), *Natural Products for Innovative Pest Management*, New York: Pergamon Press, chapter 8.

Elliott, M. (1995) Chemical in insect control, In: Casida, J. E. and Quistad, G. B. (Eds), *Pyrethrum Flowers: Production, Chemistry, Toxicology, and Uses*, New York: Oxford Press, pp. 15–23.

Elliott, M., Farnham, A. W., Janes, N. F., Johnson, D. M. and Pulman, D. A. (1987a) Synthesis and insecticidal activity of lipophilic amides. Part 6: 6-(disubstituted-phenyl)hexa-2,4-dienamides, *Pestic. Sci.* **18**, 239–244.

Elliott, M., Farnham, A. W., Janes, N. F., Johnson, D.M. and Pulman, D. A. (1987b) Synthesis and insecticidal activity of lipophilic amides. Part 4: the effect of substituents on the phenyl group of 6-phenylhexa-2,4-dieneamides, *Pestic. Sci.* **18**, 223–228.

Ellis, B. E., Kuroki, G. W. and Stafford, H. A. (1994) *Genetic Engineering of Plant Secondary Metabolism*, New York: Plenum Press.

Esen, A. (1992) Purification and partial characterization of maize (*Zea mays* L.) β-glucosidase, *Plant Physiol.* **98**, 174–182.

Fecker, L. F., Hillebrandt, S., Rügenhagen, C., Herminghaus, S., Landsmann, J. and Berlin, J. (1992) Metabolic effect of a bacterial lysine decarboxylase gene expressed in a hairy root culture of *Nicotiana glauca*, *Biotech. Lett.* **14**, 1035–1040.

Fenwick, G. R. and Heaney, R. K. (1983) Glucosinolates and their breakdown products in cruciferous crops, foods and feedingstuffs, *Food Chem.* **11**, 249–271.

Frey, M., Kliem, R., Saedler, H. and Gierl, A. (1995) Expression of a cytochrome P450 gene family in maize, *Mol. Gen. Genet.* **246**, 100–109.

Friesen, J. B. and Leete, E. (1990) Nicotine synthase – an enzyme from *Nicotiana* species which catalyzes the formation of (*S*)-nicotine from nicotinic acid and 1-methyl-Δ^1-pyrrolinium chloride, *Tetrahedron Lett.* **31**, 6295–6298.

Fukami, H. and Nakajima, M. (1971) Rotenone and the rotenoids, In: Jacobson, M. and Crosby, D. G. (Eds), *Naturally-Occurring Insecticides*, New York: Marcel Dekker, pp. 71–97.

Gbewonyo, W. S. K., Candy, D. J. and Anderson, M. (1993) Structure–activity relationships of insecticidal amides from *Piper guineense* root, *Pestic. Sci.* **37**, 57–66.

Geisler, J. G. and Gross, G. G. (1990) The biosynthesis of piperine in *Piper nigrum*, *Phytochemistry*, **29**, 489–492.

Ghaout, S., Louveaux, A., Mainguet, A. M., Deschamps, M. and Rahal, Y. (1991) What defense does *Schouwia purpurea* (Cruciferae) have against the desert locust?, *J. Chem. Ecol.* **17**, 1499–1515.

Givovich, A., Sandström, J., Niemeyer, H. M. and Pettersson, J. (1994) Presence of a hydroxamic acid glucoside in wheat phloem sap, and its consequences for performance of *Rhopalosiphum padi*, *J. Chem. Ecol.* **20**, 1923–1930.

Glendening, T. M. and Poulton, J. E. (1988) Glucosinolate biosynthesis, *Plant Physiol.* **86**, 319–321.

Glendening, T. M. and Poulton, J. E. (1990) Partial purification and characterization of a 3′-phosphoadenosine- 5′-phosphosulfate : desulphoglucosinolate sulphotransferase from cress, *Plant Physiol.* **94**, 811–818.

Godin, P. J., Inglis, H. S., Snarey, M. and Thain, E. M. (1963) Biosynthesis of the pyrethrins. Part II. Pyrethric acid and the origin of the ester methyl group, *J. Chem. Soc.*, 5878–5880.

Gosálvez, M. (1983) Carcinogenicity with the insecticide rotenone, *Life Sci.* **32**, 809–816.

Greenway, D. L., Allben, M. T., Burger, G. T. and Kodell, R. L. (1993) Bioassay for carcinogenicity of rotenone in female Wistar rats, *Fundam. Appl. Toxic. Off. J. Soc. Toxicol.* **20**, 383–390.

Grootwassink, J. W. D., Balsevich, J. J. and Kolenovsky, A. D. (1990) Formation of sulfatoglucosides from exogenous aldoximes in plant cell cultures and organs, *Plant Sci.*, 11–20.

Guo, L. and Poulton, J. E. (1994) Partial purification and characterization of *Arabidopsis thaliana* UDPG : thiohydroximate glucosyltransferase, *Phytochemistry* **36**, 1133–1138.

Guthrie, F. E. and Apple, J. L. (1961) The role of microorganisms in the detoxification of nicotine by insects, *J. Insect Pathol.* **3**, 426–438.

Halkier, B. A. and Møller, B. L. (1990) The biosynthesis of cyanogenic glucosides in higher plants: identification of three hydroxylation steps in the biosynthesis of dhurrin in *Sorghum bicolor* (L.) Moench and the involvement of 1-*aci*-nitro-2-(p-hydroxyphenyl)-ethane as an intermediate, *J. Biol. Chem.* **265**, 21114–21121.

Halkier, B. A., Lykkesfeldt, J. and Møller, B. L. (1991) 2-Nitro-3-(p-hydroxyphenyl)propionate and 1-*aci*-nitro-2-(p-hydroxyphenyl)ethane, two intermediates in the biosynthesis of the cyanogenic glucoside dhurrin in *Sorghum bicolor* (L.) Moench, *Proc. Natl. Acad. Sci. USA* **88**, 487–491.

Hallahan, D. L., Pickett, J. A., Wadhams, L. J., Wallsgrove, R. M. and Woodcock, C. M. (1992) Potential of secondary metabolites in genetic engineering of crops for resistance, In: Gatehouse, A. M. R., Hilder, V. A. and Boulter, D. (Eds), *Plant Genetic Manipulation for Crop Protection*, Wallingford, UK: CAB International, pp. 215–248.

Hamill, J. D., Robins, R. J., Parr, A. J., Evans, D. M., Furze, J. M. and Rhodes, M. J. C. (1990) Over-expressing a yeast ornithine decarboxylase gene in transgenic roots of *Nicotiana rustica* can lead to enhanced nicotine accumulation, *Plant Mol. Biol.* **15**, 27–38.

Hedin, P. A., Davis, F. M. and Williams, W. P. (1993) 2-Hydroxy-4,7-dimethoxy-1,4-benzoxazin-3-one (N-O-Me-DIMBOA), a possible toxic factor in corn to the southwestern corn borer, *J. Chem. Ecol.* **19**, 531–542.

Hemminghaus, S., Schreier, P. H., McCarthy, J. E. G., Landsmann, J., Botterman, J. and Berlin, J. (1991) Expression of a bacterial lysine decarboxylase gene and transport of the protein into chloroplasts of transgenic tobacco, *Plant Mol. Biol.* **17**, 475–486.

Hodgson, E., Rose, R. L., Adams, N. H., Deamer, N. J., Genter, M. B., Venkatesh, K. and Levi, P. (1995) Role of oxidative enzymes in metabolism and toxicity of pesticides, In: Arinc, E., Schenkmann, J. B. and Hodgson E. (Eds), *Molecular Aspects of Oxidative Drug Metabolizing Enzymes*, Berlin: Springer-Verlag.

Höglund, A. S., Lenman, M., Falk, A. and Rask, L. (1991) Distribution of myrosinase in rapeseed tissues, *Plant Physiol.* **95**, 213–221.

Horvat, R. J., Senter, S. D. and Dekazos, E. D. (1983) GLC–MS analysis of volatile constituents in rabbiteye blueberries, *J. Food. Sci.* **48**, 278–279.

Jacobson, M. (1948) Herculin, a pungent insecticidal constituent of southern prickly ash bark, *J. Amer. Chem. Soc.* **70**, 4234–4237.

Jacobson, M. (1949) The structure of pellitorine, *J. Am. Chem. Soc.* **71**, 366–367.

Jacobson, M. (1971) The unsaturated isobutylamides, In: Jacobson, M. and Crosby, D. G. (Eds), *Naturally Occurring Insecticides*, New York: Marcel Dekker, pp. 137–176.

Jain, J. C., Grootwassink, J. W. D., Reed, D. W. and Underhill, E. W. (1990) Persistent co-purification of enzymes catalyzing the sequential glucosylation and sulfation steps in glucosinolate biosynthesis, *J. Plant Physiol.* **136**, 356–361.

Jones, D. A. (1988) Cyanogenesis in animal–plant interactions, In: *Cyanide Compounds in Biology*, Ciba Foundation Symposium 140, New York: John Wiley and Sons, pp. 151–170.

Kirimer, N., Ozek, T. and Baser, K. H. C. (1991) Composition of the essential oil of *Micromeria congesta*, *J. Essent. Oil Res.* **3**, 387–393.

Kjaer, A. and Larsen, P. O. (1976) Non-protein amino acids, cyanogenic glycosides and glucosinolates, *Biosynthesis* **4**, 179–203.

Knipple, D. C., Doyle, K. E., Marsella-Herrick, P. A. and Soderlund, D. M. (1994) Tight genetic linkage between the *kdr* insecticide resistance trait and a voltage-sensitive sodium channel gene in the house fly, *Proc. Natl. Acad. Sci. USA* **91**, 2483–2487.

Koch, B., Nielsen, V. S., Halkier, B. A., Olsen, C. E. and Møller, B. L. (1992) The biosynthesis of cyanogenic glucosides in seedlings of cassava (*Manihot esculenta* Crantz), *Arch. Biochem. Biophys.* **292**, 141–150.

Kole, R. K., Satpahti, C., Chowdhury, A., Ghosh, M. R. and Adityachaudhury, N. (1992) Isolation of amorpholone, a potent rotenoid insecticide from *Tephrosia candida*, *J. Agric. Food Chem.* **40**, 1208–1210.

Kumar, P., Moreland, D. E. and Chilton, W. S. (1994) 2H-1,4-benzoxazin-3(4H)-one, an intermediate in the biosynthesis of cyclic hydroxamic acids in maize, *Phytochemistry* **36**, 893–898.

LaForge, F. B. and Haller, H. L. J. (1943) Insecticides suitable for use on plants, trees or animals, US Patent No. 2 328 726 (7 September 1943) *Chem. Abstr.* **38**, 10689.

Lambert, N., Trouslot, M. F., Nef-Campa, D. and Crestin, H. (1993) Production of rotenoids by heterotrophic and photomixotrophic cell cultures of *Tephrosia vogelii*, *Phytochemistry* **34**, 1515–1520.

Lazzeri, L., Tacconi, R. and Palmieri, S. (1993) *In vitro* activity of some glucosinolates and their reaction products toward a population of the nematode *Heterodera schachtii*, *J. Agric. Food Chem.* **41**, 825–829.

Leighton, V., Niemeyer, H. M. and Jonsson L. M. V. (1994) Substrate specificity of a glucosyltransferase and an N-hydroxylase involved in the biosynthesis of cyclic hydroxamic acids in Gramineae, *Phytochemistry* **36**, 887–892.

Lin, Y. L., Chen, Y. L. and Kuo, Y. H. (1993) A novel 12-deoxyrotenone, 12-deoxo-12-α-acetoxyelliptone from the roots of *Derris oblonga*, *J. Nat. Prod.* **56**, 1187–1189.

Ludwig-Muller, J. and Hilgenberg W. (1988) A plasma membrane bound enzyme oxidises L-tryptophan to indole-3-acetaldoxime, *Physiol. Plantarum* **74**, 240–250.

Lüthy, B. and Mattile, P. (1984) The mustard oil bomb. Rectified analysis of the subcellular organization of the myrosinase system, *Biochem. Physiol. Pflanzen* **179**, 5–12.

Lykkesfeldt, J. and Møller, B. L. (1995) On the absence of 1-(2′-cyclopentenyl)-glycine-derived cyanogenic glucosides in cassava, *Manihot esculenta* Crantz, *Acta Chem. Scand.* **49**, 540–542.

Mahadevan, S. (1975) The roles of oximes in nitrogen metabolism in plants, *Annu. Rev. Plant Physiol.* **24**, 69–88.

Mandrell, S. H. P. and Gardiner, B. O. C. (1976) Excretion of alkaloids by Malphigian tubules of insects, *J. Exp. Biol.* **64**, 267–281.

Martos, A., Givovich, A. and Niemeyer, H. M. (1992) Effect of DIMBOA, an aphid resistance factor in wheat, on the aphid predator *Eriopsis connexa* Germar (Coleoptera: coccinellidae) *J. Chem. Ecol.* **18**, 469–479.

Masakazu, M., Nakayama, I. and Ohno, N. (1989) Insecticidal unsaturated isobutylamides: from natural products to agrochemical leads, In: Arnason, J. T., Philogène, B. J. R. and Morand, P. (Eds), *Insecticides of Plant Origin*, Washington, DC: ACS Symposium Series 387, pp. 173–187.

Matsui, M. and Yamamoto, I. (1971) Pyrethroids, In: Jacobson, M. and Crosby, D. G. (Eds), *Naturally Occurring Insecticides*, New York: Marcel Dekker, pp. 4–70.

McCloskey, C. and Isman, M. B. (1993) Influence of foliar glucosinolates in oilseed rape and mustard on feeding and growth of the Bertha armyworm *Mamestra configurata* Walker, *J. Chem. Ecol.* **19**, 249–266.

McLauchlan, W. R., McKee, R. A. and Evan, D. M. (1993) The purification and immuno-characterization of N-methylputrescine oxidase from transformed root cultures of *Nicotiana tabacum* L. cv SC58, *Planta* **191**, 440–445.

Meyer, L. (1988) Zur Bedeutung cyklischer Hydroxamsäuren bei der Resistenz des Maises gegenüber mikrobiellen Schadenerregern und Insekten, *Zentralbl. Mikrobiol.* **143**, 39–46.

Miski, M. and Davis, P. J. (1988) Microbiologically catalysed enantio- and diastereo-selective oxidation of chrysanthemol stereoisomers to chrysanthemic acids, *Appl. Environ. Microbiol.* **54**, 2268–2272.

Mithen, R. and Toros, D. (1995) Biochemical genetics of aliphatic glucosinolates in Brassica and Arabidopsis, In: Wallsgrove, R. M. (Ed.), *Amino Acids and Their Derivatives in Higher Plants*, Society for Experimental Biology Seminar Series 56, Cambridge, UK: Cambridge University Press, pp. 261–275.

Miyakado, M., Nakayuma, I. and Ohno, N. (1989) Insecticidal unsaturated isobutylamides, In: Arnason, J. T., Philogène, B. J. R. and Morand, P. (Eds), *Insecticides of Plant Origin*, American Chemical Society Symposium Series No. 387, Washington, DC: American Chemical Society, pp. 173–187.

Morris, C. E. (1983a) Uptake and metabolism of nicotine by the CNS of a nicotine-resistant insect, the tobacco hornworm (*Manduca sexta*), *J. Insect Physiol.* **29**, 807–817.

Morris, C. E. (1983b) Efflux of nicotine and its CNS metabolites from the nerve cord of the tobacco hornworm *Manduca sexta*, *J. Insect Physiol.* **29**, 953–959.

Morris, C. E. (1983c) Efflux patterns for organic molecues from the CNS of the tobacco hornworm, *Manduca sexta*, *J. Insect Physiol.* **29**, 961–966.

Morse, S., Wratten, S. D., Edwards, P. J. and Niemeyer, H. M. (1991) Changes in the hydrox-amic acid content of maize leaves with time and after artificial damage: implications for insect attack, *Ann. Appl. Biol.* **119**, 239–249.

Naumann, K. (1990) *Chemistry of Plant Protection: Vol. 4. Synthetic Pyrethroid Insecticides: Structures and Properties*, and *Vol. 5. Synthetic Pyrethroid Insecticides: Chemistry and Patents*, Berlin: Springer-Verlag.

Nicol, D., Wratten, S. D., Eaton, N. and Copaja, S. V. (1993) Effects of DIMBOA levels in wheat on the susceptibility of the grain aphid (*Sitobion avenae*) to deltamethrin, *Ann. Appl. Biol.* **122**, 427–433.

Niemeyer, H. M. (1988) Hydroxamic acids (4-hydroxy-1,4-benzoxazin-3-ones), defence chemi-cals in the Graminae, *Phytochemistry* **27**, 3349–3358.

Oberg, K. E. (1961) Site of action of rotenone in the respiratory chain, *Exptl. Cell. Res.* **24**, 163–164.

Okelloekochu, E. J. and Wilkins, R. M. (1994) Biological activity of maize leaf extracts against the African armyworm, *Spodoptera exempta*, *Entomol. Exp. Appl.* **72**, 17–23.

Parmar, V. S., Sinha R., Shakil, N. A., Tyagi, O. D., Boll, P. M. and Wengel, A. (1993) An insecticidal amide from *Piper falconeri*, *Ind. J. Chem.* **32B**, 392–393.

Pattenden, G., Popplestone, C. R. and Storer, R. (1975) Investigation of the role of chrysanthemyl, lavandulyl and artemisyl alcohols in the biosynthesis of chrysanthemic acid, *J. Chem. Soc. Chem. Commun.* 290–291.

Peng, S. and Chilton, W. S. (1994) Biosynthesis of DIMBOA in maize using deuterium oxide as a tracer, *Phytochemistry* **37**, 167–171.

Pepper, D. R. and Osborne, M. P. (1993) Electrophysiological identification of site-insensitve mechanisms in knockdown resistant strains (*kdr, super-kdr*) of the housefly larva (*Musca domestica*), *Pestic. Sci.* **39**, 279–286.

Phrasant, A. and Krupadanam, G. L. D. (1993) Dehydro-6-hydroxyrotenoid and lupenone from *Tephrosia villosa*, *Phytochemistry* **32**, 484–486.

Reay, P. F. and Conn, E. E. (1974) The purification and properties of a uridine diphosphate glucose:aldehyde cyanohydrin β-glucosyltransferase from sorghum seedlings, *J. Biol. Chem.* **249**, 5826–5830.

Reed, D. W., Davin, L., Jain, J. C., Deluca, V., Nelson, L. and Underhill, E. W. (1993) Purfication and properties of UDP-glucose : thiohydroximate glucosyl transferase from *Brassica napus* L. seedlings, *Arch. Biochem. Biophys.* **205**, 526–532.

Reid, L. M., Arnason, J. T., Nozzolillo, C. and Hamilton, R. I. (1991) Laboratory and field resistance to the European corn borer in maize germplasm, *Crop Sci.* **31**, 1496–1502.

Robins, R. J., Hammill, J. D., Parr, A. J., Smith, K., Walton, N. J. and Rhodes, M. J. C. (1987) Potential for use of nicotinic acid as a selective agent for isolation of high nicotine-producing lines of *Nicotiana rustica* hairy root cultures, *Plant Cell Rep.* **6**, 122–126.

Robins, R. J., Walton, N. J., Parr, A. J., Aird, E. L. H., Rhodes, M. J. C. and Hamill, J. D. (1994) Progress in the genetic engineering of the pyridine and tropane alkaloid biosynthetic pathways of solanaceous plants, In: Ellis, B. E., Kuroki, G. W. and Stafford, H. A. (Eds), *Recent Advances in Phytochemistry 28, Genetic Engineering of Plant Secondary Metabolism*, New York: Plenum Press, pp. 1–34.

Rodman, J. E. (1981) Divergence, convergence and parallelism in phytochemical characters: the glucosinolate-myrosinase system, In: Young, D. A. and Seigler, D. S. (Eds), *Phytochemistry and Angiosperm Phylogeny*, New York: Praeger Publishers, pp. 43–79.

RTECS (1994a) Nicotine, In: *Registry of Toxic Effects of Chemical Substances* (microfiche) Washington, DC: National Institute of Occupational Safety and Health, Registry Number 67816.

RTECS (1994b) Rotenone, In: *Registry of Toxic Effects of Chemical Substances* (microfiche), Washington, DC: National Institute of Occupational Safety and Health, Registry Number 23961.

Sahi, S. V., Chilton, M. D. and Chilton, W. S. (1990) Corn metabolites affect growth and virulence of *Agrobacterium tumefaciens*, *Proc. Natl. Acad. Sci. USA* **87**, 3879–3883.

Scott, J. G. and Georghiou, G. P. (1986) Mechanisms responsible for high levels of permethrin resistance in the house fly, *Pestic. Sci.* **17**, 195–206.

Scott, J. G. and Wheelock, G. D. (1992) In: *Molecular Mechanisms of Insecticide Resistance*, American Chemical Society Symposium Series No. 505, Washington, DC: American Chemical Society, pp. 16–30.

Seck, D., Lognay, G., Haubruge, E., Wathelet J-P., Marlier, M., Gaspar, C. and Severein, M. (1993) Biological activity of the shrub *Boscia senegalensis* (Pers.) Lam. ex Poir. (Capparaceae) on stored grain insects, *J. Chem. Ecol.* **19**, 377–389.

Self, L. S., Guthrie, F. E. and Hodgson, E. (1964a) Adaptation of tobacco hornworms to the ingestion of nicotine, *J. Ins. Physiol.* **10**, 907–914.

Self, L. S., Guthrie, F. E. and Hodgson, E. (1964b) Metabolism of nicotine by tobacco-feeding insects, *Nature* **204**, 300–301.

Shepherd, H. H. (1951) *The Chemistry and Action of Insecticides*, New York: McGraw-Hill, pp. 117–132.

Sibbeson, O., Koch, B., Halkier, B. A. and Møller, B. L. (1994) Isolation of the heme-thiolate enzyme cytochrome P-450TYR, which catalyzes the committed step in the biosynthesis of the cyanogenic glucoside dhurrin in *Sorghum bicolor* (L.) Moench, *Proc. Natl. Acad. Sci. USA* **91**, 9740–9744.

Sibbeson, O., Koch, B. M., Rouzé, P., Møller, B. L. and Halkier, B. A. (1995) Biosynthesis of cyanogenic glucosides. Elucidation of the pathway and characterization of the cyto-chromes P-450 involved, In: Wallsgrove, R. M. (Ed.), *Amino acids and their derivatives in higher plants*, Society for Experimental Biology Seminar Series 56, New York: Cambridge University Press, pp. 227–241.

Soderlund, D. M. (1995) Mode of action of pyrethrins and pyrethroids, In: Casida, J. E. and Quistad, G. B. (Eds), *Pyrethrum Flowers: Production, Chemistry, Toxicology, and Uses*, New York: Oxford University Press, pp. 217–233.

Soderlund, D. M. and Bloomquist, J. R. (1990) Molecular mechanisms of insecticide resist-ance. In: Roush, R. T. and Tabashniks, B. E. (Eds), *Pesticide Resistance in Arthropods*, New York: Chapman and Hall, pp. 74–84.

Spencer, K. C. and Siegler, D. S. (1984) Cyanogenic glycosides of *Carica papaya* and its phylogenetic position with respect to the Violales and Capparales, *Am. J. Bot.* **71**, 1444–1447.

Thangstad, O. P., Iversen, T. H., Slupphaug, G. and Bones, A. (1990) Imunocytochemical localization of myrosinase in *Brassica napus* L., *Planta* **180**, 245–248.

Tseng, C. T. (1989) Concentration of DIMBOA in leaf tissues and various stages of corn plant in relation to resistance of host plant to the Asian corn borer, *Chi Wu Pao Hu Hsueh Hui K'an* **31**, 34–43 (*Chem. Abstr.* **111**, 74932).

Wallsgrove, R. M. and Bennett, R. N. (1995) The biosynthesis of glucosinolates in Brassicas. In: Wallsgrove, R. M. (Ed.), *Amino Acids and their Derivatives in Higher Plants*, Society for Experimental Biology Seminar Series 56, New York: Cambridge University Press, pp. 243–275.

Walton, N. J., Peerless, A. C. J., Robins, R. J., Rhodes, M. J. C., Boswell, H. D. and Robins, D. J. (1994) Purification and properties of putrescine N-methyltransferase from trans-formed roots of *Datura stramonium* L., *Planta* **193**, 9–15.

Wherli, P. A. and Schaer, B. (1977) Direct transformation of nitro compounds into nitriles, *J. Org. Chem.* **42**, 3956–3958.

Williams, L., Morra, M. J., Brown, P. D. and McCaffrey, J. P. (1993) Toxicity of allyl isothiocyanate-amended soil to *Limonius californicus* (mann.) (Coleoptera: Elateridae) wireworms, *J. Chem. Ecol.* **19**, 1033–1046.

Woodward, M. D., Corcuera, L. J., Helgeson, J. P. and Upper, C. D. (1978) Decomposition of 2,4-dihydroxy-7-methoxy-1,4-benzoxazin-3(4H)-one in aqueous solutions, *Plant Physiol.* **61**, 796–802.

Xie, Y. S., Arnason, J. T., Philogène, B. J. R., Olechowski, H. T. and Hamilton, R. I. (1992) Variation of hydroxamic acid content in maize roots in relation to geographic origin of maize germ plasm and resistance to western corn rootworm (Coleoptera: Chrysomelidae), *J. Econ. Entomol.* **85**, 2478–2485.

Xue, J., Lenman, M., Falk, A. and Rask, L. (1992) The glucosinolate-degrading enzyme myro-sinase in Brassicaceae is encoded by a gene family, *Plant Mol. Biol.* **18**, 387–398.

Zhu, Y., Liu, L., Wang, W., Ling, D. and Sun, Y. Z. (1991) Constituents of oil of *Ophiopogon japonicus*, *Fenxi Zazhi* **11**, 21–23 (*Chem. Abstr.* **115**, 99055).

15

Managing Resistance to Transgenic Crops

RICK ROUSH

Introduction

As demonstrated throughout the previous chapters, the development of transgenic plants offers a powerful and attractive new method for managing insect pests. From an environmental standpoint, transgenic crops have clear advantages over the pesticides they could replace (Roush, 1994). However, transgenic crops seem now in 1996 to be at a stage of development rather similar to that of synthetic insecticides in the late 1940s, shortly after the first uses of DDT on crops. Throughout the late 1940s, there was great enthusiasm for the new and stunningly effective insecticides, but there were already a few cases where insects had evolved resistance to DDT after only a few years of use (Metcalf, 1980). Insecticide resistance had been raised as a concern more than 30 years earlier (Melander, 1914), but even through the mid-1950s, relatively few people considered resistance to be a serious threat to control of any pest. In those heydays of pesticide discovery, the general attitude seemed to be that resistance could be overcome by ever newer pesticides. However, by the 1970s, it had become very clear that at least some major pests were evolving resistance faster than new and environmentally acceptable insecticides could be discovered and brought to market (Georghiou, 1986; Metcalf, 1980). It is noteworthy that two of the first three *Bacillus thuringiensis* transgenic crops registered in the USA, cotton and potatoes, are targeted at markets essentially created by the recurrent evolution of resistance to insecticides in the pink bollworm (*Pectinophora gossypiella*), two cotton bollworms (*Heliothis virescens* and *Helicoverpa armigera*), and the Colorado potato beetle (*Leptinotarsa decemlineata*).

When it became clear that the strategy of controlling resistant pests with new pesticides was failing, the concept of 'resistance management' began to evolve in the late 1950s. The aim of resistance management programs is to slow the evolution of resistance and thereby extend the useful life of valuable toxicants. Resistance was most effectively delayed by reduction in pesticide use and rotation of pesticides, but not by use of pesticide mixtures (Metcalf, 1980; Roush, 1989; Roush and Tabashnik, 1990; Tabashnik, 1989). Resistance management practices might have easily been called 'susceptibility conservation'. As will be further discussed in this chapter, susceptibility to toxicants is effectively a non-renewable resource, comparable to soil lost via erosion. Even when insecticides have been removed from use for extended periods, it is rare that they can ever again be used extensively in the same cropping system (Metcalf, 1980). The substitution of new toxicants to replace those that have failed is not resistance management, as it does nothing to preserve susceptibility. The availability of several insecticides with different modes of action and metabolism has certainly helped in the management of resistance by reducing dependence on any one insecticide, but only when used with other non-chemical practices which also reduced selection pressure.

The historical lesson from insecticides that remains most relevant to transgenic crops is that resistance will most effectively be delayed by a combination of clever molecular biology (the modern equivalent of insecticide chemistry and screening) *and* careful management practices in the field. Excessive reliance on a toxin-based resistance management strategy will surely doom transgenics to repeat the history of insecticides, where the legacy of clever chemistry is resistance to a dozen distinct classes of toxicants scattered among more than 500 species of arthropods (Georghiou and Lagunes-Tejeda, 1991). It is the thesis of this chapter that the future of transgenics rests on the use of transgenic plants not as just another pest control, but as an extremely valuable tool for the management of insect population dynamics in concert with other control tactics. In many cases, the greatest advantages offered by transgenics are that they will simplify pest management and facilitate the use of non-transgenic controls for both the targeted pest and other pests in the cropping system.

Potential for Resistance to Transgenic Plants

Of all of the types of transgenic technologies under investigation for insect control, we have data only for the risks for crops based on *Bacillus thuringiensis* (*Bt*) Cry toxins. For the other technologies described in this book, we can only speculate in the broadest terms that resistance is possible, based on the wide range of resistance mechanisms shown by insects to insecticides. The most important of these mechanisms are increased metabolism and reduced binding of synthetic pesticides, but other mechanisms include sequestration by specific enzymes, decreased penetration, altered behavior, and reduced activation (Roush and Tabashnik, 1990; Siegfried *et al.*, 1990).

Resistance to transgenic plants expressing *Bt* Cry toxins is already documented. Several studies have shown that resistance to *Bt* can be selected in the laboratory (as summarized in Chapter 2). However, because laboratory selection does not always give results that are comparable to the field (Roush, 1994; Roush and McKenzie, 1987), the most compelling evidence for *Bt* resistance comes from field populations of the diamondback moth, *Plutella xylostella*. The diamondback moth is a major pest of cruciferous vegetables around the world, receives frequent exposure to insecticides, and, not surprisingly, shows extensive resistance to most insecticides in many growing areas. High levels of resistance to Cry1A toxins have been found in populations of the diamondback moth from the Philippines, Hawaii, Florida, and Asia (Tabashnik, 1994a; Tabashnik *et al.*, 1990), with cross-resistance extending to Cry1F at least in Hawaii (Tabashnik *et al.*, 1994a). Resistance appears to be due to one

(Tang *et al.*, 1996b) or at most a few genes (Ferre *et al.*, 1995; Tabashnik *et al.*, 1992), with reduced binding of the toxin to the midgut as the only described mechanism (Ferre *et al.*, 1991; Tabashnik *et al.*, 1994c, Tang *et al.*, 1996a). When tested on broccoli plants transformed with a codon-optimised *cry1Ac* gene similar to those in commercial use, larvae from a resistant strain collected in Florida prospered, even when not exposed to further selection for several generations after collection from the field (Metz *et al.*, 1995; Roush, 1994). Following increased use of the *Bt* subsp. *aizawai*, a strain of diamondback moth with resistance to Cry1C has been found in Hawaii, apparently due to genes other than those conferring resistance to Cry1Ac (Liu *et al.*, 1996).

The results in diamondback moth have closely paralleled those of earlier work on the Indianmeal moth (*Plodia interpunctella*), a pest of stored grain. This species also demonstrated variation in susceptibility to *Bt* in field populations, and was rapidly selected for higher resistance in the laboratory (McGaughey, 1985), also due to a change in the target (van Rie *et al.*, 1990). Subsequent selection in this species showed that resistance could readily be extended to *Bt* strains carrying toxins to which the original resistant strains were susceptible (McGaughey and Johnson, 1992). Furthermore, laboratory selection in other species has also shown that broad cross-resistance is possible (Gould *et al.*, 1992; see Chapter 2).

Principles of Selection for Resistance

Simulation Models

Before discussing specific tactics for managing resistance to transgenic crops, an overview of the factors that influence selection for resistance will be helpful. Where resistance is under the control of a single locus, the number of generations until a control failure occurs will depend on the initial frequency of the resistance allele and the relative fitnesses of the genotypes. When multiple loci contribute to resistance, the interactions between the loci are also important. For brevity, I will focus on a single locus with two alleles, which gives three genotypes: susceptible homozygotes (SS), resistant heterozygotes (RS), and resistant homozygotes (RR). For clarity, I will separate overall fitness into three components: fitness imposed when feeding on transgenic plants, fitness when developing on non-transgenic plants (to incorporate fitness costs that may be associated with the resistance allele), and the percentage of the pest population on each plant type. As a shorthand expression, I will refer to fitness on transgenic plants as equivalent to mortality (which is what would be most easily measured), but the results would be essentially the same if I used fitness in the broader sense (50% mortality of all of the insects with no reduction in reproduction among the survivors has the same effect as a 50% reduction in fecundity when all of the insects survive).

I will explore the general features of selection by presenting results of simple simulation models. Except as noted below, these models assume that resistance is due to a single locus, selection occurs only in the immature stages, the resistant homozygotes are unaffected by the plants, mating is random throughout the populations and, in the absence of selection, the frequencies of the genotypes are based on the Hardy–Weinberg expression (where p represents the frequency of the resistance allele (R) and q the susceptible allele (S), p^2 gives the frequency of RR, $2pq$ for

RS and q^2 for SS). An important implication of the Hardy–Weinberg expression, which is a simple binomial expression based on the assumption of random mating, is: when resistance is rare, the most common carriers of resistance will be heterozygotes. Where some percentage of the population escapes exposure to the toxicant at any effective dose, the model simply sets that fraction of the population aside and assumes that they will be under selection only for their fitnesses on non-transgenic host plants. At the level of resistance allele frequency changes, the models have been checked against similar models (Mallet and Porter, 1992; Tabashnik, 1994b) and give exactly the same results for the same parameter values. One version of the model also tracks population density on both transgenic and non-transgenic host plants.

Experimental biologists have a well-justified skepticism about mathematical models, but these models make so few assumptions that their implications cannot be easily dismissed. Using only algebra (the author will supply the code to anyone who writes for it), the models simply quantify in more detail what seems to be common sense. For example, if 90% of the population is exposed to a selective agent that kills all susceptible individuals but none of the heterozygotes, we would expect that resistance would increase by tenfold each generation. If there were ten resistant heterozygotes in a population of 10 million eggs (i.e., $2pq = 10^{-6}$, with p almost exactly 5×10^{-7}), 100% mortality of the 90% of the larvae exposed to selection would leave only one million larvae, where ten were still resistant heterozygotes, and a tenfold increase in the frequency of the resistance allele. Five more generations of such selection would leave resistance at a frequency of about 50% (each heterozygote has one R and one S allele) with control failures imminent. The simulation model gives essentially the same results, complicated only slightly by the fact that some of the resistant heterozygotes have mated with each other and produced resistant homozygotes. Control failures will occur within a few generations before or after the resistance allele frequency exceeds 50%, depending on assumptions about population growth.

Mortality of Heterozygotes

The most important influence on the rate of resistance evolution is the mortality of heterozygotes. Focusing first on cases where all susceptible homozygotes would be killed by the transgenic plants, the frequency of the resistance allele exceeds 50% quite quickly when transgenic plants cause less than 75% mortality of heterozygotes. On the other hand, the time to resistance increases spectacularly at greater than 95% mortality of heterozygotes (Figure 15.1), provided that there is a significant number of susceptible insects to mate with the survivors of selection. Resistance is delayed because resistant heterozygotes, the most common carriers of resistance, are mostly killed, and the few survivors are most likely to mate with susceptible recruits from the refuges, producing offspring that are also susceptible to the dose used. Although it is probably unrealistic to expect for most transgenic crops that mortality of heterozygotes will exceed 99.5% (the maximum value given in the figures), 100% mortality of heterozygotes could delay resistance problems for more than 200 generations under all of the conditions shown in Figure 15.1, even for only a 2% refuge.

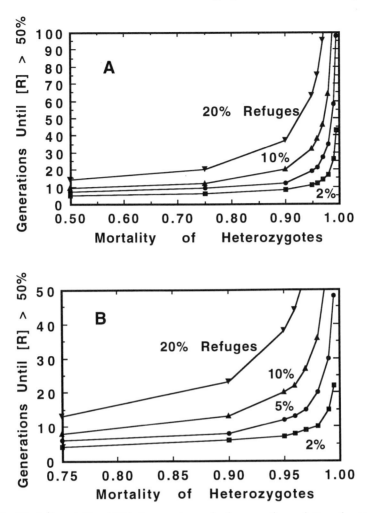

Figure 15.1 Effect of mortality of RS heterozygotes and refuges on the evolution of resistance, as measured when the frequency of the resistance allele [R] exceeds 50%. Results of a simulation model assuming a single locus, random mating, no selective mortality of resistant homozygous larvae, that some fraction of the population escapes exposure (refuges of 2, 5, 10, or 20%) and initial resistance allele frequencies (p_i) of 10^{-6} (A) or 10^{-4} (B). For the sake of legibility, different scales have been used for Figures A and B; data points on both figures include 75, 90, 95, 96, 97, 98, 99 and 99.5% mortality

Initial Allele Frequencies

In contrast, a lower initial frequency of resistance (p_i) delays control failures, but far less than the magnitude of the change (except where mortality of heterozygotes is very near 100%). For example, when 20% of the population develops in refuges and mortality of heterozygotes is 75%, resistance will evolve by generation 20 or 13 for $p_i = 10^{-6}$ or 10^{-4}, respectively (Figure 15.1A versus 15.1B). Alternatively, when only 5% of the population develops in the refuges and mortality of heterozygotes is 99.5%, resistance will evolve in 98 or 48 generations, respectively.

The initial frequency of any allele is determined by the generation of new copies of the allele by mutation and the loss of alleles where there is net selection against them (Roush and McKenzie, 1987). In the case of alleles with high fitness costs, this may mean that the alleles appear sporadically well after the introduction of the toxicant, but it is generally assumed that most resistance alleles are routinely present at low frequencies over long time-scales. So far as I am aware, the frequencies of resistance alleles prior to selection have never been measured for any chemical insecticide, but Gould et al. (1995) present results suggesting that some resistance alleles for Cry toxins may be as common as 10^{-3} in Heliothis virescens, the tobacco budworm (a major pest of cotton bolls). Resistance frequencies this high would threaten the deployment of single Bt toxins under any conditions, but resistance can be delayed for more than 115 generations even when $p_i = 10^{-3}$, provided that more than 10% of the insects escape exposure to the toxicant and mortality of the hetero- zygotes is 100% (Roush, 1994); a 2% refuge gives resistance in just 24 generations. In sum, although there are great uncertainties about the initial frequency of resist- ance, resistance can still be delayed by factors that are under our control, including the expression of toxin in the plant and the percentage of the population that is exposed to the toxin. In this paper, I show the results of simulations for $p_i = 10^{-6}$ and 10^{-4} to emphasize that even with such optimism, resistance management must be aggressively pursued through such manipulations.

Refuge Size

As one might expect, doubling the percentage of the population that escapes expo- sure ('refuge size') tends to delay resistance twofold, but only where mortality of heterozygotes is greater than about 95%, depending on initial allele frequency and refuge size. For example, when $p_i = 10^{-6}$ and only 50% of heterozygotes are killed by the transgenics, resistance evolves by generation 9 with a 10% refuge, but doub- ling the refuge to 20% delays resistance only another five generations (Figure 15.1A). On the other hand, when there is 95% mortality of heterozygotes, with the same doubling of the refuge, the time to resistance is increased from 32 to 63 gener- ations.

Number of Generations Per Year

It is often assumed that species with a greater number of generations per year will be the first to show resistance. This would be true if all other things are equal, but all other things rarely are (Rosenheim and Tabashnik, 1991). Different species could have very different refuge sizes depending on their preferences for alternate hosts, but more importantly, target insects even on the same crop could differ significantly in their sensitivities to the same Cry toxin (e.g., Heliothis virescens versus Helico- verpa zea on cotton). Even for species using the same refuges (e.g., only the non- transformed crop varieties), a species with a heterozygous mortality of 90% could easily evolve resistance faster than a species with 97% mortality but twice as many generations per year (Figure 15.1).

Fitness in the Absence of Toxin

Resistance to insecticides often decreases when selection pressure is removed, but resistance is quickly restored when selection pressure is reintroduced (Metcalf, 1980). Resistance to *Bt* in laboratory-selected populations often shows rapid declines (Tabashnik, 1994a; Tabashnik *et al.*, 1994b,c), but this provides little reason for optimism about the management of resistance to transgenic plants. First, the strains that show the most dramatic declines have all been strongly selected in the laboratory. At least for chemical insecticides, laboratory-selected strains seem to suffer worse fitness costs than strains selected in the field (Roush and Croft, 1986). In contrast to *Bt* resistant strains selected in the laboratory, two field-collected strains have shown relatively slow declines (Liu and Tabashnik, 1996; Tang *et al.*, 1996b). A colony of diamondback moth collected from Florida showed fairly stable resistance to *Bt kurstaki* at about 150–300-fold without additional selection in the laboratory (Tang *et al.*, 1996b). Continued selection could increase the generality of this observation (Tabashnik *et al.*, 1995).

Second, even where resistant strains show significant fitness costs, the more important question is what costs are associated with the heterozygotes. Highly resistant strains may consist almost completely of resistant homozygotes, but it is the heterozygotes that are overwhelmingly the most common carriers of resistance when the resistance frequency is less than 10% (by Hardy–Weinberg, $p^2 = 0.01$ and $2pq = 0.18$). Selection must act against these heterozygotes to slow the evolution of resistance. At least for chemical insecticides, heterozygotes sometimes fail to show any costs of resistance even where the homozygotes suffer significantly (implying that one copy of the susceptible allele is sufficient to provide normal function), but rarely if ever do the heterozygotes show more than half of the disadvantage of the

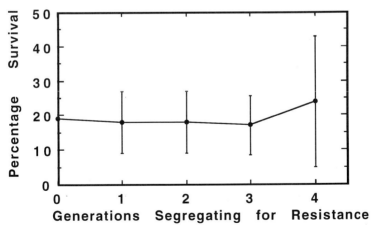

Figure 15.2 Stability of resistance to Cry1Ac in populations of diamondback moth, as measured by survival of second instar larvae on transgenic broccoli. Resistant insects from a collection in Florida were crossed with susceptible insects from New York, and their F_1 offspring allowed to intermate to produce F_2 offspring. As this was the first generation in which resistance could segregate into all three genotypes (RR, RS, and SS), results from the F_2 larvae are given as generation 0. The results are the means of two replicates from each reciprocal F_1 cross (R × S and S × R), where the populations were maintained on non-transgenic broccoli in four cages in the glasshouse. Standard errors are shown by the vertical bars

resistant homozygotes (Roush and McKenzie, 1987). One of the best ways to test for the fitness costs of heterozygotes is to establish a population where resistant and susceptible genotypes are able to mate at random over several generations, and check for a decline in frequencies of resistant genotypes. When the resistant diamondback moths from Florida were outcrossed and reared for four generations on non-transgenic plants and tested for survival on transgenic broccoli, there was no decline in resistance (Tang *et al.*, 1996b; Figure 15.2). The transgenic broccoli plants kill all susceptible homozygotes and heterozygotes (Roush, 1994), so only resistant homozygotes could have survived, and they maintained a 20–25% frequency.

Third, even strong fitness costs cause only a modest delay of resistance. Even where fitness costs are assumed to be high compared to those observed in experiments, say a 50% cost to resistant homozygotes and 25% to heterozygotes (Roush and Daly, 1990; Roush and McKenzie, 1987; Tabashnik *et al.*, 1994b), the effect on selection for resistance is small unless mortality of heterozygotes is greater than about 90% (Figure 15.3). For clarity, I have not assumed that fitness costs automatically reduce the relative fitnesses of the heterozygotes and resistant homozygotes on the transgenic plants, but that the fitnesses on the plants are exactly as given by the mortality values. When the fitnesses on the plants are reduced, it effectively increases the mortality of the resistant heterozygotes. The 'costs' curve of Figure 15.3 increases more sharply, but still only when mortality of heterozygotes is in the 90% range. For example, adding a 25% reduction in fitness to an assumption of 90% mortality of heterozygotes effectively increases the mortality of heterozygotes to 92.5% (10% survival reduced by 25%, where 25% mortality of 10% of the insects has the same effect as a 25% reduction in fecundity for all 10% of the surviving insects), whereas 50% reduction in fitness of the resistant homozygotes gives them a fitness of only 50% on the plants. The result is that resistance evolves in 36 gener-

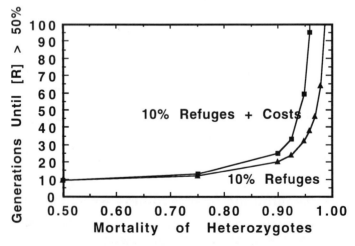

Figure 15.3 Effect of fitness costs on the evolution of resistance. The curve marked '10% refuges' is the same as the 10% curve in Figure 15.1A, with an additional data point shown at 92.5% mortality of heterozygotes. The '10% refuges + costs' curve assumes a 50% cost to resistant homozygotes and 25% to heterozygotes in the untreated refuge. A curve assuming a 50% cost to resistant homozygotes and no cost to the heterozygotes is identical to the '10% refuges' curve for the data points shown

ations compared to the 25 generations shown at 90% mortality for the 'costs' curve in Figure 15.3, but close to the 33 generations shown for the same curve at 92.5% mortality. This also shows that the model is not much affected by fitness of the RR homozygotes (assuming that it is not extremely low), again because selection is driven mostly by fitness of the heterozygotes.

Finally, even strong fitness costs do not readily restore a usable level of suscepti-bility to a population. Again assuming RR and RS to suffer fitness costs of 50% and 25% respectively, resistance drops quickly at first, perhaps even suggesting to the unwary that susceptibility has been fully restored. However, the decline slows when the resistance allele frequency is low and most of the carriers are heterozygotes. In the absence of costs to the heterozygotes, the decline slows even more (Figure 15.4A). As is more clearly seen on a log scale, resistance can remain at frequencies of greater than the worrisome 10^{-3} for up to 35 generations depending on initial allele frequency (Figure 15.4B). Because alleles can linger at frequencies much higher than those before selection ever started (say 10^{-4} or less), it is not surprising that renewed

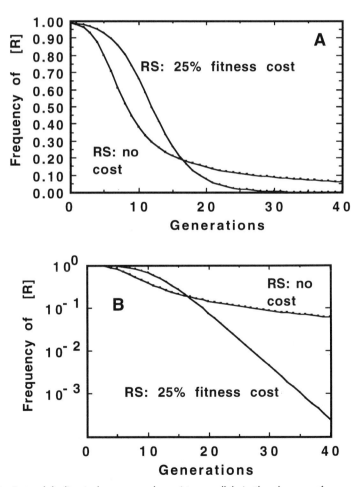

Figure 15.4 Rate of decline in frequency of a resistance allele in the absence of exposure to toxins on a normal (A) and log scale (B). For both curves, a 50% cost to resistant homozygotes was assumed. In one case there was no cost for RS heterozygotes and in the other, a 25% cost

selection often restores resistance much more quickly than in the first use of an insecticide (Metcalf, 1980), including *Bt* (Tabashnik, *et al.*, 1994b).

Conclusions about Selection for Resistance

The initial allele frequency of resistance alleles and costs of resistance are generally beyond our control (barring mass releases of susceptible insects). On the other hand, mortality of heterozygotes can be influenced by the level of expression; higher expression should cause higher mortality of heterozygotes. Thus, excellent molecular biology, and perhaps good crop agronomy (improving the expression of the genes), can help to manage resistance. For example, tenfold increases in expression of *Bt* genes have been achieved by expression in chloroplasts (McBride *et al.*, 1995; see Chapter 2). Expression of *Bt* toxins in cotton seems to be adversely affected by stress for water or light (Fitt and Forrester, personal communication). However, there will likely be technical limits to what can be achieved by the modern equivalent of good chemistry, e.g., overly high expression may limit yields. The size of the refuges is potentially the most easily managed component of this system, either by limiting the proportion of the crop that can be transgenic or by insuring other non-transgenic host plants. The effective and economic application of this approach will require an excellent understanding of insect ecology (e.g., host plant preferences, dispersal, mating patterns). As will be detailed in the next section, resistance management will depend on good ecology *and* good molecular biology.

Resistance Management Tactics for Transgenic Crops

In principle, there are at least five possible ways to slow selection in favor of resistance by transgenic plants. For plants that show constitutive expression:

(1) express toxin genes only moderately strongly, so that not all susceptible individuals are killed;

(2) provide refuges for susceptible insects while expressing the genes as high as possible within acceptable limits for yield effects;

(3) deploy different toxins individually in different varieties; and

(4) deploy plants with a mixture of toxins.

A fifth option may be to modify the expression of the genes in each plant such that they are expressed only when or where needed through tissue-specific, temporal-specific, or inducible promoters. Finally and most importantly, resistance management is most effective when used in the context of an integrated pest management (IPM) program (Metcalf, 1980), which will provide the context for a summary to this chapter.

Moderate Expression

In general, models suggest that a fairly significant proportion of the treated individuals must be allowed to survive if resistance is to be significantly delayed (Curtis,

1985). This can be illustrated with simulations using dose mortality data for *Bt* resistant diamondback moth larvae. For example, a concentration of *Bt* applied to a leaf surface that killed 80% of susceptible larvae killed only 50% of F_1 (heterozygous) larvae; the LC_{95} for the susceptible larvae killed only 70% of the F_1 (Tang *et al.*, 1996b). Using these data to set mortality values for the model shows that there is little benefit to a moderate dose strategy unless the plants kill less than 80% of the larvae, which would probably not be acceptable to growers if better varieties were available (Figure 15.5). *Bt* resistance in the diamondback moth tends to be quite recessive – there is little difference in phenotype between susceptible and heterozygous larvae (Ferre *et al.*, 1995; Tabashnik, 1994a; Tabashnik *et al.*, 1992). If a data set with higher dominance for resistance was used, there would be even less difference between the moderate and high dose curves shown in Figure 15.5.

In the only cases where the low dose approach has been successfully applied to resistance management, it is incorporated with the use of other alternative controls, such as predators, that can further reduce pest populations (Roush, 1989). However, as noted by Gould *et al.* (1991), natural enemies may accelerate the evolution of resistance in a transgenic system if the natural enemies are more effective against susceptible individuals than resistant ones. In experiments, healthy insects often seem less readily attacked by predators than those weakened by feeding on *Bt* transgenic plants (Roush, 1996).

Finally, even though a moderate dose does not appear to provide advantages for managing major resistance genes, it may select more effectively than high doses for minor ones. Laboratory selection programs essentially mimic a moderate dose strategy, since they always allow some survivors (Roush and McKenzie, 1987), and there are abundant examples of successful selection in the laboratory (Tabashnik, 1994a)

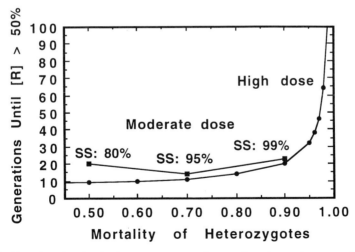

Figure 15.5 Use of moderation in dose to manage resistance, based on resistance to Cry1A toxins in the diamondback moth. The 'high dose' curve is the same as that for a 10% refuge in Figure 15.1A. The percentages over the moderate dose curve show the assumed mortalities of susceptible (SS) insects from data on resistant and susceptible diamondback moth (e.g., at a concentration that killed 95% of susceptible larvae, only about 70% of F_1 larvae were killed; based on Tang *et al.*, 1996b). For both curves, the initial resistance frequency was 10^{-6}, and 10% of the population was assumed to escape exposure each generation

for a range of mechanisms that have not yet been found in the field (Ferre *et al.* , 1995). Especially as transgenic plants probably provide a more uniform dose than sprays, they would likely be more effective in discriminating between minor genotypes, which would enhance their ability to select for polygenic resistance (Roush and McKenzie, 1987). This should be another reason to aim to commercialize only those varieties that effectively kill all susceptible insects tested.

Refuges, Random Mating, and Seed Mixes

Approaches that leave refuges for susceptible insects have been very important for the management of resistance to chemical pesticides, consistent with theory (Tabashnik, 1994a). The refuge approach seems likely to be enhanced by high expression (Figure 15.1). In principle, ultra high expression might kill even all resistant homozygotes, but this would depend on the expression system, sensitivity of the target insect and what resistance genes are available to it. At least with the diamondback moth, plants transformed with the codon-optimized CrylAc gene (with a CaMV 35S promoter) killed only about 10–20% of resistant homozygotes (Metz *et al.*, 1995; Roush, 1994). Thus, I have assumed for my figures that transgenic plants did not kill any resistant homozygotes. The more worrisome problem is not that RR homozygotes are killed, but what happens when 5–20% of SS homozygotes survive (middle to left of Figure 15.5). In such circumstances, even when resistance is assumed to be very recessive, there must be significant survival of heterozygous insects. Only a very large refuge can delay resistance, and it may be prudent to delay deployment of the crop until expression is improved, rather than use up the resource of susceptibility too quickly.

The refuge approach assumes that mating will be random between insects in the refuges and in the transgenic crop. For *Bt* transgenics targeted for Lepidoptera, this assumption would seem to be relatively safe as long as the refuges are sufficiently attractive as oviposition sites (e.g., are not stressed) and deployed within a 0.5–1.0 kilometer of the transgenic plants, given what is known about the dispersal of the key targeted pests (e.g., Schneider *et al.*, 1989; Showers, 1993). For reasons of equity in compliance and to achieve such close proximity of sites, it would probably be best if each grower of transgenic crops was required to provide his or her share of the refuge. The benefits of the refuge could be enhanced by preferential destruction of insects in transgenic patches, such as the use of cultivation and crop residue removal practices to control overwintering stages.

One possible problem is that resistant insects might suffer a disadvantage and develop slower than susceptible insects, putting them out of phase with the susceptible insects (Gould, 1994). At least with diamondback moths in cage experiments, this did not seem to be the case (personal observations), but it does provide another reason to err on the side of larger rather than smaller refuges as a buffer against such effects.

Bt transgenics appear to have no effect on adult moths and would therefore not deter random mating within the transgenic crop. However, this is a more vexing problem for adult Colorado potato beetles, which are severely affected by feeding on transgenic potatoes with a Cry IIIA toxin (Roush, 1996; Chapter 3), and would similarly be an issue for any other transgenic toxin that adversely affected adults (e.g., a toxin active against adults in nectar on which the moths feed). A refuge is clearly needed, but the preferred method of deployment is still under investigation (Chapter 3).

One controversy about refuges for *Bt* transgenic crops is whether the refuge should be outside the crop or inside the crop as a seed mix. Where only the larvae feed and do not move from plant to plant, a seed mix is ideal. It ensures that every crop has a refuge with developmental phenologies and insect control practices consistent across both transgenic and non-transgenic types. It also ensures that the susceptible insects develop in close proximity to the resistant insects, which will help to ensure random mating.

On the other hand, if the larvae move around and feed on different plants, one net effect may be to reduce the proportion of the population in a refuge (a result of 'suicidal dispersal'). Second, heterozygous larvae that could have been killed as neonates may now survive if they move from a non-transgenic to a transgenic plant at a later, less susceptible life stage. This would increase the relative fitness advantage of heterozygotes (i.e., more than 5% of the heterozygotes on the transgenic plants could survive). The result of these factors is that resistance could evolve more quickly to a seed mix than to a similar proportion of susceptible plants outside the crop. However, for this effect to occur, the mortality of heterozygotes must be very high, close to 99%, when they stay on transgenic plants and at least 20% of the larvae must move between plants (Mallet and Porter, 1992; Roush, 1996; Tabashnik, 1994b). Unfortunately, at least for European corn borer (*Ostrinia nubilalis*), target of transgenic corn, and cotton bollworms, more than 20% of larvae appear to move between plants (Davis and Roush, unpublished data). In cotton, not only were more than 70% of *Helicoverpa armigera* larvae observed to move over a mean of three plants, even transgenic plants suffered considerable damage from large (late instar) larvae. Also, there was evidence for 'suicidal dispersal'; even a 25% mix seed produced only about 5% of the pupae found in purely non-transgenic plots (Roush and Fitt, unpublished data) instead of the 25% that would have been expected if the larvae fed only on the plant on which they were laid. This suggests that in moving from plant to plant, many of the larvae fed on neighboring transgenic plants and died.

Multiple Toxins: Mosaics, Rotations, and Sequential Release of Varieties

When multiple toxins with independence of resistance mechanisms are available, there are several options for their deployment. Focusing first on options involving a single toxin, it would probably be impractical to attempt to rotate varieties of the same crop, and in any case, there is little evidence that such a tactic has any advantage over simply releasing the second gene variety only after the first has failed (Curtis, 1987; Roush, 1989). A more likely scenario for the deployment of single toxins, especially for corn, is that two different genes will be released by competing companies, and that these will be used in a mosaic pattern by different growers or even on the same farm. Mosaics are singularly the worst way to deploy two toxin genes, both in theory (Roush, 1989) and in experiments (Figure 15.6). A popular myth is that each of the varieties can serve as a refuge for the other; they cannot since neither is producing many susceptible insects. Even with an untreated refuge, the mosaic system essentially selects for resistance to each of the toxins simultaneously.

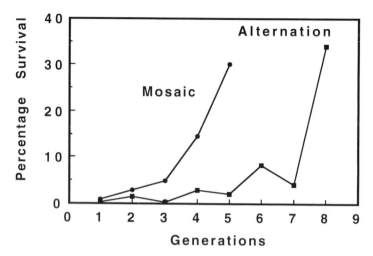

Figure 15.6 A comparison of selection for resistance by a mosaic and alternation (or rotation) of insecticides in a mosquito, *Aedes aegypti*. For each treatment, 10% of each replicate population was held aside from insecticide exposure. For the mosaic, 50% of the population was exposed to DDT and 50% to dieldrin. Using the same concentrations, DDT and dieldrin were applied in alternate generations to the other populations. DDT and dieldrin are due to separate genes and show no cross-resistance (Munsterman and Craig, 1979; Roush, unpublished data)

Mixtures of Toxins: Pyramided Varieties

Contrary to another popular myth, it has long been clear that mixtures of insecticides do not necessarily delay resistance compared to the rotational or sequential use of the same insecticides. Experimental studies have failed consistently to find any advantage to mixtures (Immaraju *et al.*, 1990; Tabashnik, 1989), and theoretical models showed that mixtures will significantly delay resistance only when several conditions are met (Gould, 1986a; Roush, 1989).

Where S_A refers to the allele for susceptibility to toxin A and S_B refers to the allele susceptible to toxin B, the most common two locus genotypes in the population prior to selection will be $S_A S_A S_B S_B$, followed by individuals resistant to one or other of the toxicants, $R_A S_A S_B S_B$ and $S_A S_A R_B S_B$. Mixtures work by the principle of 'redundant killing'; when two toxins are used, each at a level that kills all or nearly all insects susceptible to that toxin, the double susceptible genotype (SSSS) is effectively killed twice, but the carriers of resistance are also killed at least once. When mortality of $S_A S_A$ and $S_B S_B$ is high, the mixtures can work very well, as long as there is a refuge for susceptible insects to 'dilute' the more resistant genotypes (e.g., $R_A S_A R_B S_B$) (Curtis, 1985).

Problems occur when the toxins are used in a way that allows some of $S_A S_A$ and/or $S_B S_B$ individuals to survive, which then allows heterozygotes, $R_A S_A S_B S_B$ and $S_A S_A R_B S_B$, to survive. Similar to the high dose strategy for a single locus, even 10% survival of SS genotypes can dissipate much of the advantages of the mixture (Roush, 1989) even though the pyramiding strategy is more robust to survival of heterozygotes (Roush, 1994). For example, for a 10% refuge with an initial resistance frequency of 10^{-6}, and 30% RS and 5% SS survival for each of the two toxins (as for diamondback moth, see Figure 15.5), resistance evolves in 95 generations. The two toxins used in succession would give 15 generations each for a total from sequential introduction of 30. Using the same assumptions, except that all SS are killed by each of the two toxins used independently, resistance is predicted to take more than 10 000 generations (Roush, 1994). Experiments that achieve high redundant killing, such as laboratory trials with high concentrations of topically applied insecticide, tend to support the mixtures approach; mixtures fail to be very helpful in less intensive selection experiments and field trials (Immaraju et al., 1990) that more realistically allow some survival of partially dosed individuals.

To be most effective, mixtures require a lack of cross-resistance between the toxins, low initial frequencies of the resistance genes (Figure 15.7), refuges (as with the high dose strategy, such that resistant genotypes are rare and can be diluted), and high mortality from each of the toxins when used alone; it helps if at least one heterozygous genotype suffers high mortality and the two resistance loci are not closely linked (Gould, 1986b; Roush, 1989). As will be argued later, toxins that do not provide high levels of mortality can be useful, especially if they are not particularly at risk for resistance evolution – e.g., physical factors such as trichomes (Franca et al. 1994) – but the maximal benefits will be achieved only if each toxin individually causes high mortality.

In contrast to chemical insecticides, at least two features of transgenic plants suggest that mixtures of two or more factors pyramided into the same variety will greatly delay resistance. Transgenic plants offer a way to get consistently high control of SS homozygotes, and Cry toxins provide at least one good candidate gene with consistently fairly recessive inheritance of resistance (Ferre et al., 1995; Tabashnik, 1994a). Pyramids also have several advantages over single gene strategies:

(1) Fitness costs should have a greater impact on delaying resistance to mixtures than single toxins (Gould, 1994; Roush, 1989).

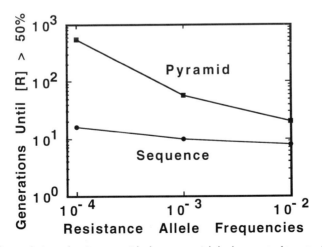

Figure 15.7 The evolution of resistance with the sequential deployment of two toxins compared to the use of the toxins jointly in a pyramided variety for a range of initial allele frequencies. Assumes a 10% refuge, 70% mortality of RS heterozygotes, and 100% mortality of SS homozygotes for each toxin

(2) Pyramids are less sensitive to initial resistance allele frequency; even resistance frequencies of 10^{-3} could allow significant benefits (Figure 15.7).

(3) Pyramids can be greatly improved by manipulating mortality of SS homozygotes, which can be measured.

(4) As noted in the previous paragraph, pyramids are more robust to survival of heterozygotes, which cannot be measured with certainty until resistance has evolved.

(5) Pyramids can be very effective with a smaller refuge. For example, a 10% refuge in Figure 15.7 is much more effective than the 20% refuge in Figure 15.1B (compare the top left point of Figure 15.7 with 75% mortality in Figure 15.1B; both are based on an initial allele frequency of 10^{-4} and 70–75% mortality of heterozygotes). This could relieve some anxieties about the costs of refuges.

Ideally the two toxins to be used in a pyramided variety should be very different to avoid cross-resistance (Gould *et al.*, 1995). Unfortunately, the practical reality is that Cry toxin genes are often going to be the most viable partners in the near term, and the resistance manager is faced with the question of whether to pyramid these toxins or risk a significant benefit for failing to do so until after the frequencies of the resistance alleles have increased (Figure 15.7). As noted above, the diamondback moth probably gives the best insight into the evolution of resistance in the field. That this insect has repeatedly and independently responded to selection in the field with a recessive insensitive target site of limited cross-resistance (Ferre *et al.*, 1995) encourages one to assume that genes for other, broader mechanisms are much rarer or confer a lower mean fitness advantage. Simulations suggest that if there is such a gene for broad cross-resistance, it creates similar problems for both single and multiple toxins (Roush, unpublished). Even then, at what seem to be reasonable values for initial allele frequencies and fitnesses, such a gene seems unlikely to evolve more than about twofold faster to a pyramided variety than if the two toxin genes were deployed one at a time. Compared to the enormous potential pay-off in the absence of a gene that confers broad cross-resistance (Figure 15.7), this seems to be a risk worth taking.

Tissue-Specific, Temporal-Specific, or Inducible Promoters

In principle, specific gene promoters could be used to express genes only in:

(1) the most important tissues ('tissue- or structure-specific' expression), such as the bolls in cotton; or

(2) critical growth periods ('temporal-specific' expression) (Gould, 1988); or

(3) be environmentally induced, perhaps by the spraying of an environmentally benign chemical (Williams *et al.*, 1992).

These tactics are not necessarily exclusive; a temporally specific promoter may be effectively structure-specific if it is turned on only when needed (e.g., late in the season and affecting only the top of a plant). Unfortunately, suitable promoters do not currently seem to be available, and may be difficult to control in a way that manages resistance (Roush, 1996).

Tissue- or structure-specific expression could perhaps leave refuges within each plant. This has been suggested for a maize genotype where the Cry toxin is under control of a PEPC promoter, which provided high levels of expression in green tissue and much lower expression in kernels (Koziel *et al.*, 1993). However, the level of expression in the kernels is still 4–6 times the LC_{50} for European corn borer (*Ostrinia nubilalis*) (calculations from data in Koziel *et al.*, 1993). Damage to corn kernels is tolerable for corn grown for feed, but this level of expression suggests that the ear will not prove to be a very hospitable site for susceptible corn borers.

A slight variation of this idea is to deter insects from feeding on the most economically important plant structures without being killed, thereby lowering selection intensity while protecting the crop (Gould and Anderson, 1991). Cotton is a good example for such work, as considerable damage can be tolerated to leaves but not 'squares' (flower buds) and bolls. In simple experiments designed to mimic plants with toxic squares but non-toxic leaves, *Helicoverpa armigera* larvae showed better survival and weight gain on the 'chimeric' than on transgenic plants. Only 30% of larvae survived on the transgenic plants compared to no mortality on the chimerics (when corrected for controls). However, because of their decreased sensitivity to toxin as they grew in size, the larvae damaged as many squares on the chimeric plants as on the susceptible control (Figure 15.8). Thus, to be fully effective in cotton, expression in the transgenic structures may have to be higher than in

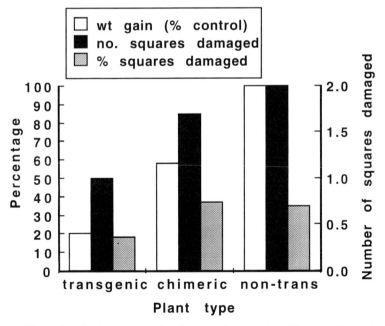

Figure 15.8 Effects of artificial tissue-specific 'chimeric plants' on larval fitness and crop loss in field trials (Narrabri, New South Wales, Australia). Transgenic Cry1A cotton and non-transgenic cotton of the same variety were compared to 'chimeric plants' produced by: stripping the leaves from a transgenic plant; stripping the bolls and squares (developing flower buds) from an adjacent non-transgenic cotton; and tying the stems of the two plants closely together. Third instar *Helicovera armigera* larvae were weighed, caged on the plants, and recovered after one week, when larval weights and plant damage were recorded

current constitutive varieties, and will have to provide sufficiently high expression in all important tissues without affecting the tissues that can be sacrificed. In cotton, for example, both bolls and terminal meristems must be protected (Hopkins *et al.*, 1982; Ramalho *et al.*, 1984; Wilson and Waite, 1982).

Other targeted crops, including potatoes and cotton, can suffer moderate to severe defoliation at some times of the year without yield loss, and might be targeted by inducible or temporal-specific promoters. Unfortunately, for cotton, potato and most other crops, it is the *end* of the season that is least vulnerable. If such a promoter is used, it must 'turn off' in a fashion that does not allow a slowly declining residue in the plant. This would leave the plant for long periods with a moderate dose, and could easily be worse than high consistent expression. Promoters that must be induced with chemical sprays will also likely have a host of potential sociological and registration problems. First, any chemical sprayed in the field will likely have to go through the same extensive and costly registration procedures as a pesticide, which will inhibit their development. Second, due to the perceived risks of crop damage, not to mention increased costs, farmers will likely prefer constitutively expressing over inducible varieties. If only offered the inducible variety, farmers will probably tend to spray the inducer at less than truly economic threshold densities of the pest, as they often do now with synthetic insecticides.

While their use may be complex, it would be premature to dismiss temporal, inducible, or structure-specific expression. On the other hand, because of the technical difficulties inherent in perfecting the expression needed and in the costs for their commercial development, it seems unlikely that plants with appropriate expression will be available in the near future or that the approach will be readily generalized across pests and crops.

Integrated Pest Management

A major cause of extensive resistance to chemical insecticides is that too little emphasis has been placed on non-chemical controls. Because of their efficacy and selectivity, transgenic plants can simplify pest management practices and facilitate the adoption of pest management tools that have been of marginal effectiveness, but they will require a change in thinking. For example, at the high expression levels recommended in this chapter, natural enemies of the pests targeted by the transgenics have an ambiguous role in pest management. They may actually accelerate the evolution of resistance by preferentially feeding on sick susceptible insects on plants with marginal expression, and if too effective, suppress the numbers of susceptible insects in refuges. A more significant role for biological control will probably be in the management of pests not targeted by the transgenic crops. In cotton and potatoes, for example, insecticide applications for key pests like Colorado potato beetle and cotton bollworms typically suppress the natural enemies of other pests such as aphids and mites. Transgenic crops allow increased survival of these beneficial species (see Chapter 3), thereby reducing the need for insecticide treatments against them.

Other tactics seem more compatible with transgenics in an IPM program. Intuitively and in simulation models, the higher the proportion of susceptible plants, the greater the delay of resistance. The problem is in preventing those susceptible plants from suffering yield losses such that there is little disincentive to use them. For species whose population dynamics are driven for all practical purposes by a single host plant – e.g., potatoes for Colorado potato beetle in New York state (Roush and Tingey, 1992) – the transgenic plants themselves could have that effect (Roush, 1996). Rotation with other crops, which remains a widely practiced and very effective technique that helps to manage Colorado potato beetle (Roush and Tingey, 1992), could be used to help protect large refuges from damage and help manage resistance (Gould *et al.*, 1994).

Consider the pink bollworm, *Pectinophora gossypiella*, whose only major host in Arizona and California is cotton. Predators and parasites are minimally effective, and in the absence of overcrowding, population growth seems density independent. In populations uncontrolled by insecticides, various estimates have consistently estimated population growth rates (R_0) of about $1 \times$, $17 \times$, $10 \times$, $5 \times$ and $1 \times$ for the five generations typically observed (Slosser, 1971). The bollworms do very poorly before (and after) small bolls appear on the plant. Without considering overwintering mortality, this means that pink bollworms can increase at about 850-fold per year or the equivalent of just less than fourfold per generation. For simplicity, I will assume that the population would normally increase at fivefold per generation. Under these conditions, assuming random oviposition and that the plants killed nearly all susceptible homozygotes, an 80% transgenic : 20% non-transgenic crop should allow the pink bollworm only to sustain its numbers. Until resistance evolved, four of every five eggs would be laid on an inhospitable host, allowing only one to develop (Figure 15.9). If only 2%, 5% or 10% of the cotton plants are non-transgenic, the population would initially decline, but resistance could evolve before the population collapses.

If on the other hand, an additional more or less permanent control is applied, for example, another host plant resistance (HPR) trait to which resistance is very slow to evolve, the species might go locally extinct before evolving resistance. Even traits that caused only a twofold level of suppression could be valuable in this regard. If a 20% refuge failed to suppress a population sufficiently to avoid damage, the grower could apply another control tactic, such as pheromone disruption, for perhaps the first year (five generations) until the population is suppressed below damaging densities ('20% + Alt'). Pheromone disruption is effective for pink bollworm control, but only when the densities of the adults are sufficiently low. As a consequence, pheromone use is often seen as inconsistent, risky, and only effective in the early season (Stone *et al.*, 1986), i.e., before the population explodes into 17-fold growth. Transgenic plants reduce the risk, and even at only 80% of the crop, could convert a $1 \times$, $17 \times$, $10 \times$, $5 \times$ and $1 \times$ population growth trend to $0.2 \times$, $3.4 \times$, $2 \times$, $1 \times$, and $0.2 \times$, a population that would never appreciably exceed its overwintering density. In principle, pheromone disruption could be used season-long without disrupting biological control of other pests in the cotton ecosystem.

Alstad and Andow (1995) proposed that resistance to European corn borers could be managed by a 50–50 patchwork of early and late planted fields, with Cry toxins restricted to the earlier planted fields, which would then serve as a 'trap' for eggs, protecting the late planted non-transgenic fields from excessive damage even as they provided a refuge. It was assumed that growers would perceive a positive economic incentive to use this strategy (protection of all of their crop while paying a seed premium for only half of it). The success of the Andow–Alstad strategy would depend on the randomness of mating (panmixis) between corn borers from the early and late planted fields, the consistency of attraction of early planted fields (about

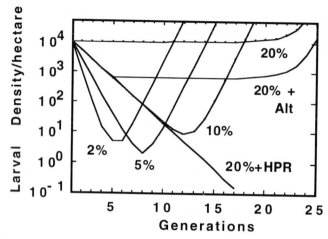

Figure 15.9 Use of transgenic plants to manage population growth. It was assumed that the initial frequency of resistance was 10^{-4} and that 90% of the heterozygotes that fed on transgenic plants were killed. Refuge percentages were 2, 5, 10, and 20%. Alternate controls providing a twofold suppression of population were applied for only five generations in one case (Alt) but more persistently in another case, as for another host plant resistance (HPR) trait. The resistance typically reaches a frequency of about 50% the generation immediately before the populations start to rebound in density

which there is some skepticism), and whether growers perceived an economic advantage (if the seed premium was low, they might still plant a much higher proportion of transgenic seed). A simple 20% non-transgenic crop with a roughly even mix of early and late planted varieties might be more effective, and still regulate population densities along the lines of Figure 15.9.

Growers and government researchers have major concerns about the evolution of resistance, but Figure 15.9 illustrates why the companies producing transgenic plants should also be concerned. The conditions for this simulation are well within the bounds of possibility: an initial allele frequency of 10^{-4} and only 90% mortality of RS heterozygotes. With a 5% refuge, for example, failure could occur in 14 generations (3–7 years for corn borer or pink bollworm) whereas a 20% refuge can provide an extra ten generations. For a 15% loss of market in each of the first 3–7 years, an extra 70% can be gained on the total number of years.

Conclusions

Of the various methods for the management of resistance to transformed plants, the strategy of maintaining completely susceptible plants within the cropping system seems most promising for the near future, especially when the transgenic plants express the insecticidal toxin at a very high level. Even more promising is the use of mixtures of toxins in the resistant plants ('pyramiding' in classical resistance breeding), which also requires that some fraction of susceptible host plants remains in the cropping system.

Even though I have concentrated on transgenics with Cry toxins in this paper, *Bt* genes are probably only the first of many transgenic factors that will be deployed in the field. Cry toxin-based transgenic crops will almost certainly fail in at least some areas with a history of repeated resistance to synthetic insecticides, and due to the same lack of commitment to resistance management. Successes in the management of resistance to Cry transgenics will help to establish sound principles for managing resistance of other transgenic technologies, but failures may only remind us of the importance of implementing what we already know.

Transgenic plants offer a powerful new tool to manage pest populations, but they will fail to live up to their promise if we cannot learn to think of them as more than just another pesticide in an environmentally friendly package. To use these plants wisely requires more research in several areas, not least of which are pest ecology, classical crop breeding, and molecular biology. We need to understand aspects of pest biology that have not concerned us much in the past, such as interplant movement of larvae, as well as others that have always vexed us, such as dispersal and mate finding in adults.

References

Alstad, D. N. and Andow, D. A. (1995) Managing the evolution of insect resistance to transgenic plants, *Science* **268**, 1894–1896.

Curtis, C. F. (1985) Theoretical models of the use of insecticide mixtures for the management of resistance, *Bull. Entomol. Res.* **75**, 259–265.

Curtis, C. F. (1987) Genetic aspects of selection for resistance, In: Ford, M. G., Holloman, D. W., Khambay, B. P. S. and Sawicki, R. M. (Eds), *Combating Resistance to Xenobiotics*, Chichester, England: Ellis Horwood, pp. 151–161.

Ferre, J., Escriche, B., Bel, Y. and van Rie, J. (1995) Biochemistry and genetics of insect resistance to the *Bacillus thuringiensis* insecticidal crystal proteins, *FEMS Microbiol. Lett.* **132**, 1–7.

Ferre, J., Real, M. D., van Rie, J., Jansens, S. and Peferoen, M. (1991) Resistance to the *Bacillus thuringiensis* bioinsecticide in a field population of *Plutella xylostella* is due to a change in a midgut membrane receptor, *Proc. Natl. Acad. Sci. USA* **88**, 5119–5123.

Franca, F. H., Plaisted, R. L., Roush, R. T., Via, S. and Tingey, W. M. (1994) Selection responses of the Colorado potato beetle for adaptation to the resistant potato, *Solanum berthaultii, Entomol. Exper. Appl.* **73**, 101–109.

Georghiou, G. P. (1986) The magnitude of the resistance problem, In: National Academy of Sciences (Ed.), *Pesticide Resistance: Strategies and Tactics for Management*, Washington, DC: National Academy Press, pp. 14–43.

Georghiou, G. P. and Lagunes-Tejeda, A. (1991) *The Occurrence of Resistance to Pesticides in Arthropods*, Rome: Food and Agriculture Organization.

Gould, F. (1986a) Simulation models for predicting durability of insect-resistant germplasm: a deterministic diploid, two locus model, *Environ. Entomol.* **15**, 1–10.

Gould, F. (1986b) Simulation models for predicting durability of insect-resistant germplasm: Hessian fly (Diptera: Cecidomyiidae)-resistant winter wheat, *Environ. Entomol.* **15**, 11–23.

Gould, F. (1988) Evolutionary biology and genetically engineered crops, *Bioscience* **38**, 26–33.

Gould, F. (1994) Potential and problems with high-dose strategies for pesticidal crops, *Biocont. Sci. Technol.* **4**, 451–461.

Gould, F. and Anderson, A. (1991) Effects of *Bacillus thuringiensis* and HD-73 delta-endotoxin on growth, behavior, and fitness of susceptible and toxin-adapted strains of *Heliothis virescens* (Lepidoptera: Noctuidae), *Environ. Entomol.* **20**, 30–38.

Gould, F., Kennedy, G. G. and Johnson, M. T. (1991) Effects of natural enemies on the rate of herbivore adaptation to resistant host plants, *Entomol. Exper. Appl.* **58**, 1–14

Gould, F., Follet, P., Nault, B. and Kennedy, G. G. (1994) Resistance management strategies for transgenic potato plants, In: Zehnder, G. W., Powelson, M. L., Jansson, R. K. and Raman, K. V. (Eds), *Advances in Potato Pest Biology and Management*, St Paul, Minnesota: American Phytopathological Society Press, pp. 237–254.

Gould, F., Anderson, A., Reynolds, A., Bumgarner, L. and Moar, W. (1995) Selection and genetic analysis of a *Heliothis virescens* (Lepidoptera: Noctuidae) strain with high levels of resistance to *Bacillus thuringiensis* toxins, *J. Econ. Entomol.* **88**, 1545–1559.

Gould, F., Martinez-Ramirez, A., Anderson, A., Ferre, J., Silva, F. J. and Moar, W. F. (1992) Broad-spectrum resistance to *Bacillus thuringiensis* toxins in *Heliothis virescens*, *Proc. Natl. Acad. Sci. USA* **89**, 7986–7988.

Hopkins, A. R., Moore, R. F. and James, W. (1982) Economic injury level for *Heliothis* spp. larvae on cotton plants in the four-true-leaf to pinhead-square stage, *J. Econ. Entomol.* **75**, 328–332.

Immaraju, J. A., Morse, J. G. and Hobza, R. F. (1990) Field evaluation of insecticide rotation and mixtures as strategies for citrus thrips (Thysanoptera: Thripidae) resistance management in California, *J. Econ. Entomol.* **83**, 306–314.

Koziel, M. G., Beland, G. L., Bowman, C., Carozzi, N. B., Crenshaw, R., Crossland, L., Dawson, J., Desai, N., Hill, M., Kadwell, S. *et al.* (1993) Field performance of elite transgenic maize plants expressing an insecticidal protein gene derived from *Bacillus thuringiensis*, *Bio/Technology* **11**, 194–200.

Liu, Y-B., Tabashnik, B. E. and Pusztai-Carey, M. (1996) Field-evolved resistance to *Bacillus thuringiensis* toxin CryIC in diamondback moth (Lepidotera: Plutellidae), *J. Econ. Entomol.* **89**, 798–804.

Mallet, J. and Porter, P. (1992) Preventing insect adaptation to insect-resistant crops: are seed mixtures or refugia the best strategy?, *Proc. R. Soc. Lond. Series B* **250**, 165–169.

McBride, K. E., Svab, Z., Schaaf, D. J., Hogan, P. S., Stalker, D. M. and Maliga, P. (1995) Amplification of a chimeric *Bacillus* gene in chloroplasts leads to an extraordinary level of an insecticidal protein in tobacco, *Bio/Technology* **13**, 362–365.

McGaughey, W. H. (1985) Insect resistance to the biological insecticide *Bacillus thuringiensis*, *Science* **229**, 193–195.

McGaughey, W. H. and Johnson, D. E. (1992) Indianmeal moth (Lepidoptera: Pyralidae) resistance to different strains and mixtures of *Bacillus thuringiensis*, *J. Econ. Entomol.* **85**, 1594–1600.

Melander, A. L. (1914) Can insects become resistant to sprays?, *J. Econ. Entomol.* **7**, 167–173.

Metcalf, R. L. (1980) Changing role of insecticides in crop protection, *Annu. Rev. Entomol.* **25**, 219–256.

Metz, T. D., Roush, R. T., Tang, J. D., Shelton, A. M. and Earle, E. D. (1995) Transgenic broccoli expressing a *Bacillus thuringiensis* insecticidal crystal protein: implications for pest resistance management strategies, *Mol. Breed.* **1**, 309–317.

Munsterman, L. E. and Craig, G. B. Jr (1979) Genetics of *Aedes aegypti*: updating the linkage map, *J. Hered.* **70**, 291–296.

Ramalho, F. S., McCarty, J. C., Jenkins, J. N. and Parrott, W. L. (1984) Distribution of tobacco budworm (Lepidoptera: Noctuidae) larvae within cotton plants, *J. Econ. Entomol.* **77**, 591–594.

Rosenheim, J. A. and Tabashnik, B. E. (1991) Influence of generation time on the rate of response to selection, *Am. Natur.* **137**, 527–541.

Roush, R. T. (1989) Designing resistance management programs: how can you choose? *Pestic. Sci.* **26**, 423–441.

Roush, R. T. (1994) Managing pests and their resistance to *Bacillus thuringiensis*: can transgenic crops be better than sprays? *Biocont. Sci. Technol.* **4**, 501–516.

Roush, R. T. (1996) Can we slow adaptation by pests to insect-resistant transgenic crops?, In: Persley, G. (Ed.), *Biotechnology for Integrated Pest Management*, Wallingford, UK: CAB International.

Roush, R. T. and Croft, B. A. (1986) Experimental population genetics and ecological studies of pesticide resistance in insects and mites, In: National Academy of Sciences (Ed.), *Pesticide Resistance: Strategies and Tactics for Management*, Washington, DC: National Academy Press, pp. 257–270.

Roush, R. T. and Daly, J. C. (1990) The role of population genetics in resistance research and management, In: Roush, R. T. and Tabashnik, B. E. (Eds), *Pesticide Resistance in Arthropods*, New York: Chapman and Hall, pp. 97–152.

Roush, R. T. and McKenzie, J. A. (1987) Ecological genetics of insecticide and acaricide resistance, *Annu. Rev. Entomol.* **32**, 361–380.

Roush, R. T. and Tabashnik, B. E. (Eds) (1990) *Pesticide Resistance in Arthropods*, New York: Chapman and Hall.

Roush, R. T. and Tingey, W. M. (1992), Evolution and management of resistance in the Colorado potato beetle, *Leptinotarsa decemlineata*, In: Denholm, I., Devonshire, A. L. and Holloman, D. W. (Eds), *Resistance '91: Achievements and Developments in Combating Pesticide Resistance*, Essex, England: Elsevier Applied Science, pp. 61–74.

Schneider, J. C., Roush, R. T., Kitten, W. F. and Laster, M. L. (1989) Movement of *Heliothis virescens* (F.) (Lepidoptera: Noctuidae) in Mississippi in the spring: implications for area-wide management, *Environ. Entomol.* **18**, 438–446.

Showers, W. B. (1993) Diversity and variation of European corn borer populations, In: Kim, K. C. and McPheron, B. A. (Eds), *Evolution of Insect Pests: Patterns of Variation*, New York: John Wiley and Sons, pp. 287–309.

Siegfried, B. D., Scott, J. G., Roush, R. T. and Zeichner, B. T. (1990) Biochemistry and genetics of chlorpyrifos resistance in German cockroaches, *Blattella germanica* (L), *Pestic. Biochem. Physiol.* **38**, 110–121.

Slosser, J. E. (1971) Population growth of the pink bollworm, *Pectinophora gossypiella* (Saunders) (Lepidoptera: Gelechiidae), PhD Dissertation, University of Arizona, Tucson.

Stone, N. D., Gutierrez, A. P., Getz, W. M. and Norgaard, R. (1986) Pink bollworm control in southwestern desert cotton. III. Strategies for control: an economic simulation study, *Hilgardia* **54**(9), 42–56.

Tabashnik, B. E. (1989) Managing resistance with multiple pesticide tactics: theory, evidence, and recommendations, *J. Econ. Entomol.* **82**, 1263–1269.

Tabashnik, B. E. (1994a) Evolution of resistance to *Bacillus thuringiensis*, *Annu. Rev. Entomol.* **39**, 47–79.

Tabashnik, B. E. (1994b) Delaying insect adaptation to transgenic crops: seed mixtures and refugia reconsidered, *Proc. R. Soc. Lond. Series B* **255**, 7–12.

Tabashnik, B. E., Cushing, N. L., Finson, N. and Johnson, M. W. (1990) Field development of resistance to *Bacillus thuringiensis* in diamondback moth (Lepidoptera: Plutellidae), *J. Econ. Entomol.* **83**, 1671–1676.

Tabashnik, B. E., Schwartz, J. M., Finson, N. and Johnson, M. W. (1992) Inheritance of resistance to *Bacillus thuringiensis* in diamondback moth (Lepidoptera: Plutellidae), *J. Econ. Entomol.* **85**, 1046–1055.

Tabashnik, B. E., Finson, N., Johnson, M. W. and Heckel, D. G. (1994a) Cross-resistance to *Bacillus thuringiensis* toxin CryIF in the diamondback moth, *Appl. Environ. Microbiol.* **60**, 4627–4629.

Tabashnik, B. E., Groeters, F. R., Finson, N. and Johnson, M. W. (1994b) Instability of resistance to *Bacillus thuringiensis*, *Biocontr. Sci. Technol.* **4**, 419–426.

Tabashnik, B. E., Finson, N., Groeters, F. R., Moar, W. J., Johnson, M. W., Luo, K. and Adang, M. J. (1994c) Reversal of resistance to *Bacillus thuringiensis* in *Plutella xylostella*, *Proc. Natl. Acad. Sci. USA* **91**, 4120–4124.

Tabashnik, B. E., Finson, N., Johnson, M. W. and Heckel, D. G. (1995) Prolonged selection affects stability of resistance to *Bacillus thuringiensis* in diamondback moth (Lepidoptera: Plutellidae), *J. Econ. Entomol.* **88**, 219–224.

Tang, J. D., Shelton, A. M., Van Rie, J., De Roeck, S., Moar, W. J., Roush, R. T. and

Peferoen, M. (1996a) Toxicity of *Bacillus thuringiensis* spore and crystal protein to the resistant diamondback moth (*Plutella xylsotella*), *Appl. Environ. Microbiol.* **62**, 564–569.

Tang, J. D., Gilboa, S., Roush, R. T. and Shelton, A. M. (1996b) Inheritance, stability, and lack of fitness costs of field-selected resistance to *Bacillus thuringiensis* in *Plutella xylostella* (L.) (Lepidoptera: Plutellidae) from Florida, *J. Econ. Entomol.* (in press).

Van Rie, J., McGaughey, W. H., Johnson, D. E., Barnett, B. D. and van Malaert, H. (1990) Mechanism of insect resistance to the microbial insecticide *Bacillus thuringiensis*, *Science* **247**, 72–74.

Williams, S., Friedrich, L., Dincher, S., Carozzi, N., Kessman, H., Ward, E. and Ryals, J. (1992) Chemical regulation of *Bacillus thuringiensis* delta-endotoxin expression in transgenic plants, *Bio/Technology* **10**, 540–543.

Wilson, L. T. and Waite, G. K. (1982) Feeding pattern of Australian *Heliothis* on cotton, *Environ. Entomol.* **11**, 297–300.

Acknowledgments

My experimental work on transgenic plants has been supported entirely by New York State and Federal funds, especially by USDA Grant No. 91-37302-6199. I have benefited enormously from the contributions of many collaborators, including G. Fitt, J. Daly, W. McGaughey, W. Tingey, A. Shelton, E. Earle, P. Davis, T. Metz, J. Tang, and M. Whalon, and my assistants P. Beckley, N. Carruthers, C. Richael and M. Burgess. I also thank B. Tabashnik and F. Gould for discussions, correspondence and for sharing unpublished manuscripts. Calgene, Hybritech and Monsanto have provided transgenic potatoes, cotton, and corn, and constructs for transgenic broccoli.

Index